Climate Change in
the Arctic

Maritime Climate Change:
Physical Drivers and Impact

Series Editor:
Neloy Khare

As global climate change continues to unfold, the two-way links between the tropical oceans and the poles will play key determining factors in these sensitive regions' climatic evolution. Now is the time to take a detailed look at how the tropical oceans and the poles are coupled climatically. The signatures of environmental and climatic conditions are well preserved in many natural archives available over land and ocean. Many efforts have been made to unravel such mysteries of climate through many natural geological archives from tropics to the polar region. This series makes an effort to cover in pertinent time various depositional regimes, different proxies – planktic, benthic, pollens and spores, invertebrates, geochemistry, sedimentology, etc., and emerged teleconnections between the poles and tropics at regional and global scales, besides sea-level changes and neo-tectonism. This book series will review theories and methods, analyse case studies, identify and describe the evolving spatial-temporal variations in climate, and provide a better process-level understanding of these patterns. It will discuss significantly, generalisable insights that improve our understanding of climatic evolution across time – including the future. It aims to serve all professionals, students and researchers, scientists alike in academia, industry, government and beyond.

Climate Change in the Arctic
An Indian Perspective
Neloy Khare

Climate Change in the Arctic
An Indian Perspective

Edited by

Neloy Khare

CRC Press
Taylor & Francis Group
Boca Raton London New York

CRC Press is an imprint of the
Taylor & Francis Group, an **Informa** business

First edition published 2022
by CRC Press
6000 Broken Sound Parkway NW, Suite 300, Boca Raton, FL 33487-2742

and by CRC Press
2 Park Square, Milton Park, Abingdon, Oxon, OX14 4RN

© 2022 Taylor & Francis Group, LLC

CRC Press is an imprint of Taylor & Francis Group, LLC

Library of Congress Cataloging-in-Publication Data
Names: Khare, Neloy, editor.
Title: Climate change in the Arctic : an Indian perspective / edited by Neloy Khare.
Description: First edition. | Boca Raton : CRC Press, [2022] | Series: Maritime climate change: physical drivers and impact | Includes bibliographical references and index.
Identifiers: LCCN 2021044897 | ISBN 9780367482695 (hbk) | ISBN 9781032207780 (pbk) | ISBN 9781003265177 (ebk)
Subjects: LCSH: Arctic regions—Environmental conditions. | Climatic changes—Arctic regions. | Pollution—Arctic regions. | Environmental protection—Research—India. | Ecology—Arctic regions.
Classification: LCC GE160.A68 C56 2022 | DDC 551.6911/3—dc23/eng/20211119
LC record available at https://lccn.loc.gov/2021044897

ISBN: 9780367482695(hbk)
ISBN: 9781032207780 (pbk)
ISBN: 9781003265177 (ebk)

DOI: 10.1201/9781003265177

Typeset in Times
by codeMantra

Dedicated to

Late Prof Shashi Bhushan Bhatia

(18.03.1980–01.05.2021)

Prof Shashi Bhushan Bhatia was born on 18 March 1930 at Lucknow. Prof Bhatia post graduated in 1950 from the Lucknow University, obtained Ph.D. degree from the London University and D.I.C. from the Imperial College of London. He joined the Lucknow University as a Lecturer in 1953. In 1958, he was selected in the Geological Survey of India. Subsequently in 1959, Prof Bhatia was appointed as a Reader in the newly created Department of Geology in the Panjab University, Chandigarh, where he superannuated as a Professor and then continued as an Emeritus Professor.

Prof Bhatia's contributions to the Indian geology cover a vast time span. Foraminifera was his original field of specialisation. He is credited with the first discovery of the Permian foraminifera in India. Later he made important contributions to the study of Cretaceous and Paleogene ostracods. His research on the Cretaceous sediments aided in determining the time of extinction of the dinosaurs and initiation of the Deccan Volcanicity. Another important input to the Cretaceous was the report of bryozoa in the Neelkanth Formation – considered at that time a part of the Tal Group. The most important contributions made by Prof S. B. Bhatia, however, are to the Palaeogene sediments of the Himalaya. He studied fauna of the Subathu, Dagshai and Siwalik sequences. His findings precisely fixed the lower age limit of the Kakara sediments at 56.5 Ma, which signifies the time of collision of the Indian and Asian Plates. His research on the Subathu and the Dagshai formations makes significant contributions to the biostratigraphy, event stratigraphy and environment of deposition of that time. He put India on the world map of Chara studies.

Prof Bhatia had around 250 publications in reputed journals. He guided a dozen students for Ph.D. degree. He was a highly decorated scientist. In 1958, he was appointed as Indian Correspondent of the Journal of Micropaleontology. He was a Fellow of the prestigious Indian National Science Academy and recipient of L. Rama Rao award. Oil and Natural Gas Corporation decorated him with a silver plaque. He was an excellent teacher with exceptional skills of communication. He was a perfectionist and demanded the best from his students.

Prof Shashi Bhushan Bhatia breathed his last peacefully on 1 May 2021. His death is a great loss to Indian geology.

Contents

Foreword

Undertaking research in the Polar Regions is complex and challenging. The sustained vitality of polar research is intrinsically linked to the specialised logistics and infrastructural support available, which allows scientists to work in a challenging environment. It necessitates strong international cooperation and sharing of resources and infrastructural facilities. Ny-Ålesund at Svalbard in the Arctic has emerged as a vibrant symbol of international partnership. The Norwegian Government has ably guided many non-Arctic countries to research new areas of Arctic Science.

Having achieved a long and resilient experience in Antarctica, India forayed into Arctic Science over a decade back and recently joined the Arctic Council as an Observer. In collaboration with Norway, India has established its research base 'Himadri' at Ny-Ålesund, the well-known 'Science village' in the Arctic. With more than a decade's effort, many exciting findings and results have emerged, especially in the science of climate change as evident in the Arctic region. With quality scientific output in a short period, India has made its presence strongly felt in Arctic Science and has contributed significantly to the knowledge base of Arctic research. Indian scientists have attempted to unravel the past and understand the present in order to elucidate the future trends on climate change patterns.

The multi-sectoral climate issues associated with Arctic change transcend national boundaries. The stakeholders including governments, indigenous communities, civil society, policymakers and industry need to be equally concerned. For any adaptation and mitigation efforts, informed decisions on the Arctic must be evidence-based and not geopolitically driven. Scientists, therefore, have a cardinal role to play in providing insightful data that can be translated to enable policy paradigm and sound advisory mechanism on climate change and its impact on the globe.

The Indian efforts made in Arctic climate science must be synthesised so that a comprehensive picture could emerge. The present book *Climate Change in the Arctic: An Indian Perspective* is an illustrious effort in this direction, covering a wide range of topics addressing various aspects of climate change in the Arctic. As changes in the Arctic region will have a profound effect on human life on Earth, there is a need for our polity and people including the industry leaders to have up-to-date and robust information on the science of Arctic climate change. This book amply provides a clear picture of the ongoing global climatic changes, their impacts on the Arctic and how the changing Arctic will influence the rest of the world. The message conveyed is clear: we must collectively work to reduce our carbon footprint to save one of the last frontiers on planet Earth, the Arctic.

Arbinda Mitra
Scientific Secretary
Date: June 2021
Place: New Delhi

Preface

The Polar Regions are an essential focus of research as unique systems, but they also play a pivotal role in global Earth systems. Situated at the northernmost part of *Earth*, the Arctic is a spectacular *polar region*. It is a unique area among Earth's ecosystems. The unique characteristics and cultures of the Arctic indigenous peoples are an adaptive mechanism to cope with the cold and extreme conditions. Historically, the Arctic regions' scientific activity is intimately linked with the post-World War II and the subsequent Cold War era's geopolitical circumstances. It has always been a pristine environment owing to least human interventions. For many years, the Arctic region was designated as a North 'dead land' that was not perceived as a place for living or economic activities. However, the situation during the past few decades has changed significantly. Now the Arctic is connected to global natural and anthropogenic chemicals via winds, ice movement and marine currents. Global warming has made the Arctic region more accessible and desirable. It is a complex region that is experiencing unprecedented changes. Based on long-term monitoring, it has been found that the Arctic is warming up much faster than expected, essentially due to Arctic amplification of the worldwide warming scenario. The Arctic holds the world's largest remaining oil and gas reserves, an enormous untapped mineral wealth.

A reduction in Arctic summer ice cover has become more intense in recent years, culminating in an area of a record low of 3.4 million square kilometres in 2012. Ice melting opens a place for new research and economic activities. It allows for exploring the possible amount of natural resources in the Arctic region.

The Arctic's responses to global change are too sensitive. It may be capable of initiating dramatic climatic changes through alterations induced in the oceanic *thermohaline circulation* by its cold, southward-moving currents or through its effects on the global *albedo* resulting from changes in its total ice cover.

Many scientific disciplines and engineering applications have significantly advanced based on the knowledge gained from polar research. Many of the discoveries that took place over Polar Regions have provided a critical understanding of direct benefit to humanity and served the society. It makes sense that access to the polar areas is fundamentally vital from its scientific viewpoints. Evidence continues to accumulate that the scientific value of the Arctic warrants significant investment for polar research. Sustained efforts must be on to continue and indeed flourish over the next several decades.

The Arctic Ocean is primarily surrounded by land and has difficult, terrain and adjacent shorelines. The ice and extreme weather conditions make it further challenging to reach. The logistics constraints make it difficult to conduct research on land and coastal areas; thus, such observations tend to be sparse.

Atmospheric processes in the Arctic are distinctive and scientifically significant. Among many, the transport and disposition of solar radiation, the formation and persistence of clouds, and variability in the atmospheric pressure fields exhibit large-scale patterns that influence the global climate.

A more detailed understanding and representation of these global climate models' processes are essential to improving future climate predictions. It requires intensive observations of the Arctic atmosphere over the oceans with specialised (ice breaker) vessel support.

Similarly, sea ice provides the interface between atmosphere and water, and it is one of the essential components of the system. Inadequate spatiotemporal observations limit our ability to describe the variability, change, and extremes of the polar region environment. Records of past environmental conditions, retrieved from paleo-archives such as ice cores or sediments, provide clues to nature's response to climate change force. Still, these too are incomplete, especially in terms of spatial coverage.

The Arctic Basin plays a pivotal role in the global carbon balance. Nonetheless, the mechanism by which this carbon is transported from the Arctic continental shelf to the deep basin is yet to be adequately understood.

The apparent changes are occurring in the intricate balance between Pacific's and Atlantic's waters. These may bring out a subtle change in the thermohaline profile features because the heat and salt balance prevents melting of surface ice.

The tectonics of ocean gateways, which allow passage of warm and cold currents between oceans, is useful for understanding climate changes in the Arctic region. India takes the honour to have a separate ministry dealing with the Earth System Sciences under which the entire gamut of polar science and logistics falls. Having acute concern on the global climate change issues and the significance both poles carry in a deeper understanding of the climate change system holistically, India opted to adopt a bipolar scientific approach by venturing into Arctic Science. The rich scientific and logistics experience gained through a glorified and meaningful journey of launching more than 40 annual scientific expeditions to the icy continent 'Antarctica' was an ambitious plan to launch unique scientific expeditions to the Arctic region addressing diversified climate-related issues apart from many other scientific objectives to be accomplished through Arctic Science. As a humble beginning, India deputed an exploratory team to visit the Arctic and establish tie-ups with the Norwegian colleagues as the site identified for India was the Norwegian territorial region. Having fruitful, supportive and healthy linkages with the Norway, India was initially allowed to use Norway's International Research Facilities at Ny-Ålesund, Svalbard. And a full-fledged Indian Arctic Research base namely 'Himadri' was finally set up in 2008 at Ny-Ålesund amidst the research bases of other participating countries. During this period, Indian researchers visited the Arctic a number of times in a year and conducted their exceptional dedicated experimentation to obtain observational data. Many significant studies have been undertaken by Indian Climate Scientists focusing on the Arctic Climate changes and their global relevance including Indian tropical climates, specialising in the Indian Summer Monsoon (ISM).

This book titled *Climate Change in the Arctic: An Indian Perspective* highlights some of the most exciting Arctic research conducted by Indian scientists ever since India entered the Arctic Science foray. Indian polar researchers have contributed to a wide range of disciplines, in a bid to get an insight into the essential information about Earth's systems and how they operate. Nevertheless, the research conducted by Indian scientists addressing various facets of the climate change occurring over the

Arctic region has exclusively been collated in 15 chapters to put forth Arctic climate changes through an Indian perspective.

Accordingly, **Khare** introduces the theme in the first chapter highlighting the over-all climate change assessment over the Arctic through the Indian Polar Programme. Under Arctic Atmosphere Weather and Climate Change theme, **Maheskumar and Sunitha Devi** discuss in detail the Arctic weather and climate patterns, **Gwal et al.** investigate GPS-derived total electron content (TEC) and Scintillation Index for the Arctic and Antarctic Stations, and **Sonbawne et al.** monitor the atmospheric aerosols to help assess their sources and effects on Arctic climate.

To get an insight into the signatures of climate change through sediments, **Singh et al.** concentrate on the sedimentary parameters of the Arctic region in space and time to reconstruct past climatic conditions. On the contrary, **Choudhary et al.** have identified biogenic silica as an indicator of paleo productivity in lacustrine sediments of Svalbard, Arctic. They provide a comprehensive review of the studies that dealt with the biogenic silica over the Arctic region. **Tiwari and Megan Da Lima Leitao** touch upon the critical role of persistent organic pollutants and mercury in the Arctic environment. Also, they discuss about the indirect impact of these pollutants on cli-mate change. Besides, **Gopikrishna et al.** studied the fjord sediments for their con-tamination and Implications for the Arctic Climate changes.

The impact of climate change on Arctic life has been assessed by **Hatha and Krishnan** who reported non-polar isolates in the Arctic fjords and Tundra and pointed out a warming trend in the Arctic environment. Simultaneously, **Jasmine Purushothaman et al.** concentrate on the Mesozooplankton community structure in Kongsfjorden, the Arctic, and assess impacts of climate changes through space and time. On the contrary, **Aswathy Shaji and Anu Gopinath** characterise the humic acids isolated from diverse Arctic environments.

The most critical aspect of the great potential Arctic microbes possess for phar-maceutical purposes lies in the biotechnological processes which have various char-acteristics. Such aspects have been detailed by **Dubey and Babel**. Interestingly, bio-optical characteristics in relation to phytoplankton composition and productivity in the Arctic fjord ecosystem have been highlighted by **Tripathy**.

In the Polar Regions, the present-day Circum-Arctic region comprises a variety of tectonic settings due to varied geophysical conditions: from active seafloor spread-ing in the North Atlantic and Eurasian Basin, and subduction in the North Pacific, to long-lived stable continental platforms in North America and Asia. A detailed account on the recent advances in seismo-geophysical studies conducted so far for the Arctic region in different tectonic environs has been provided by **Mishra et al.** in a bid to understand the efficacies of various geophysical and seismological tools.

On the contrary, **Luis and Roy** have addressed the impact of climate change on sea ice. They have studied the decadal Arctic Sea-Ice Variability to observe the pat-tern and its implication to climate change.

As Arctic Science advances, more and more challenging scientific questions will require sustained and continuous observations and measurements in the Arctic regions. It will help us understand the evolution of northern climates. Prediction of future changes can be based only on a full understanding of the Arctic and Antarctic systems.

This book collates recent scientific findings and sights related to the ongoing climate change in and around the Arctic region from the Indian perspective. This book will help satisfy the curiosity; usually, young minds carry. These youngsters and budding researchers will get an insight into the ongoing climate changes taking place over the Arctic region.

Neloy Khare
Date: June 2021
Place: New Delhi

Acknowledgements

Immense supports and help received from all contributing authors are graciously acknowledged; without their valuable inputs on various facets of climate change over the Arctic region, the book would not have been possible. Various learned experts who have reviewed different chapters are sincerely acknowledged for their timely, important, and critical reviews.

I express my sincere thanks to the Ministry of Earth Sciences, Government of India, New Delhi (India) for various inputs, support, and encouragement. Dr M.N. Rajeevan, Secretary, Ministry of Earth Sciences, and Government of India, Dr Arbinda Mitra, Secretary, Office of the Principal Scientific Advisor to Prime Minister, Government of India, New Delhi, and Prof Govardhan Mehta, FRS have always been the source of inspiration and are acknowledged for their kind support. Dr Vipin Chandra, Joint Secretary to Government of India at Ministry of Earth Sciences, New Delhi (India), Dr Prem Chan Pandey, former Director, National Centre for Polar and Ocean Research, Goa (India) and Dr K.J. Ramesh, former Director-General, India Meteorological Department (IMD), New Delhi (India) have always been supportive as true well-wishers. Prof Anil Kumar Gupta, Indian Institute of Technology, Kharagpur (India) and Dr Rajiv Nigam, former Adviser at the National Institute of Oceanography, Goa (India) are deeply acknowledged for providing many valuable suggestions to this book. This book has received significant support from Akshat Khare and Ashmit Khare who have helped me during the book preparation. Dr Rajni Khare has unconditionally supported enormously during various stages of this book. Dr Shabnam Choudhary and Shri Hari Dass Sharma from the Ministry of Earth Sciences, New Delhi (India) have helped immensely in formatting the text and figures of this book, and they have brought the book to its present form. The publisher Taylor and Francis has done a commendable job and is sincerely acknowledged.

Neloy Khare
Date: June 2021
Place: New Delhi

Editor

Dr Neloy Khare, presently Adviser/Scientist G to the Government of India at MoES, has a unique understanding not only of administration but also of quality science and research in his areas of expertise covering a large spectrum of geographically distinct locations like Antarctic, Arctic, Southern Ocean, Bay of Bengal, Arabian Sea, Indian Ocean, etc. Dr Khare has almost 30 years of experience in the field of paleoclimate research using palaeobiology and palaeontology/teaching/science management/administration/coordination for scientific programmes (including Indian Polar Programme). Dr Khare has completed his doctorate (Ph.D.) in tropical marine region and Doctor of Science (D.Sc.) in southern high-latitude marine regions.

Dr Khare has been conferred Honorary Professor and Adjunct Professor by many Indian universities. He has an awe-inspiring list of publications to his credit (125 research articles in national and international scientific journals; three special issues in national scientific journals as Guest Editor; edited a special issue of *Polar Sciences* as its Managing Editor). The Government of India and many professional bodies have bestowed him with many prestigious awards for his humble scientific contributions to past climate changes/oceanography/polar science and southern oceanography. The most coveted award is the Rajiv Gandhi National Award – 2013 conferred by the Honourable President of India. Others include ISCA Young Scientist Award, Boys Cast Fellowship, CIES French Fellowship, Krishnan Gold Medal, Best Scientist Award, Eminent Scientist Award, ISCA Platinum Jubilee Lecture, and IGU Fellowship. Dr Khare has made tremendous efforts to popularise ocean science and polar science across the country by delivering many invited lectures, giving radio talks, and publishing popular science articles. Many books authored/edited on thematic topics and published by reputed international publishers are testimony to his commitment to popularise science among the masses.

Dr Khare has sailed in the Arctic Ocean as a part of 'Science PUB' in 2008 during the International Polar Year campaign for scientific exploration and became the first Indian to sail in the Arctic Ocean.

Contributors

Priyanka Babel
Department of Biotechnology
Mohanlal Sukhadia University
Udaipur, Rajasthan

Purushottam Bhawre
Department of Physics
Govt. Sanjay Gandhi P.G. College
Ganjbasoda, India

Kailash Chandra
Protozoology Section
Zoological Survey of India
Kolkata, India

Shabnam Choudhary
National Centre for Polar and Ocean
    Research
Goa, India

Suryanshu Choudhary
Department of Physics
Rabindranath Tagore University
Raisen, India

Tara Megan Da Lima Leitao
National Centre for Polar and Ocean
    Research
Goa, India

P.C.S. Devara
Centre of Excellence in
    Ocean-Atmospheric Science
    and Technology and
Environmental Science and Health
Amity University Haryana (AUH)
Gurugram, India

Chetan Anand Dubey
Centre of Advanced Study in Geology
University of Lucknow
Lucknow, India

Rajesh Kumar Dubey
UGC-Human Resource Development
    Centre
JNV University Jodhpur
Rajasthan, India

Gopikrishna VG
School of Environmental Sciences
Mahatma Gandhi University
Kerala, India

Anu Gopinath
Department of Aquatic Environment
    Management
Kerala University of Fisheries and
    Ocean Studies
Kochi, India

A.K. Gwal
Department of Physics
Rabindranath Tagore University
Raisen, India

Kannan VM
School of Environmental Sciences
Mahatma Gandhi University
Kerala, India

Neloy Khare
Ministry of Earth Sciences
Govt. of India
New Delhi, India

Krishnan KP
National Centre for Polar and Ocean
    Research
Goa, India

Alvarinho Luis
National Centre for Polar and Ocean
  Research
Goa, India

R.S. Maheskumar
Ministry of Earth Sciences
Govt. of India
New Delhi, India

G. Meena
High Altitude Cloud Physics Laboratory
Indian Institute of Tropical Meteorology
Pune, India

O.P. Mishra
National Centre for Seismology
New Delhi

A.A. Mohamed Hatha
Department of Marine Biology,
  Microbiology and Biochemistry,
  School of Marine Sciences
Cochin University of Science and
  Technology
Kochi, Kerala

Mahesh Mohan
School of Environmental Sciences
International Centre for Polar Studies
Mahatma Gandhi University
Kerala, India

G N Nayak
Marine Sciences, School of Earth,
  Ocean and Atmospheric Sciences
Goa University
Goa, India

G. Pandithurai
High Altitude Cloud Physics Laboratory
Indian Institute of Tropical Meteorology
Pune, India

Haritha Prasad
Protozoology Section
Zoological Survey of India
Kolkata, India

Priya Singh
National Centre for Seismology
New Delhi

Jasmine Purushothaman
Protozoology Section
Zoological Survey of India
Kolkata, India

Rasik Ravindra
National Centre for Polar and Ocean
  Research (Formerly)
Goa, India

Neelakshi Roy
National Centre for Polar and Ocean
  Research
Goa, India

and

Centre of Excellence for Natural
  Resources Data Management
  System, Department of Remote
  Sensing and GIS
Kumaun University
Almora, India

P.D. Safai
Cloud-Aerosol Interaction and
  Precipitation Enhancement
  EXperiment
Indian Institute of Tropical Meteorology
Pune, India

S.K. Saha
High Altitude Cloud Physics Laboratory
Indian Institute of Tropical Meteorology
Pune, India

Aswathy Shaji
School of Ocean Science and
    Technology
Kerala University of Fisheries and
    Ocean Studies
Kochi, India

Anoop Kumar Singh
Centre of Advanced Study in Geology
University of Lucknow
Lucknow, India

Dhruv Sen Singh
Centre of Advanced Study in Geology
University of Lucknow
Lucknow, India

S.M. Sonbawne
High Altitude Cloud Physics Laboratory
Indian Institute of Tropical Meteorology
Pune, India

Sunitha Devi S
India Meteorological Department
New Delhi, India

Anoop Kumar Tiwari
National Centre for Polar and Ocean
    Research
Goa, India

Sarat Chandra Tripathy
National Centre for Polar and Ocean
    Research
Goa, India

Abbreviation List

A.O.	Arctic Ocean
AAS	Atomic absorption spectrometry
ABL	Atmospheric boundary layer
AMDEs	Atmospheric mercury depletion events
AMAP	Arctic Monitoring and Assessment Programme
ANOVA	Analysis of variance
AQUA-GAPS	Global aquatic passive sampling
ASE	Accelerated solvent extraction
ATOS	Antarctic Tourism Opportunity Spectrum
BDL	Below the detection limit
BSTFA	N,O-Bis(trimethylsilyl)trifluoroacetamide
BTBPE	1,2-Bis(2,4,6-tribromophenoxy)ethane
BWP	Brake wear plastics
CA	California
CAA	Canadian Arctic Archipelago
CEC	Capillary electro-chromatography
CFCs	Chlorofluorocarbons
CG	Congeners in the gas phase
CUPs	Current-use pesticides
CVAFS	Cold vapour atomic fluorescence spectrometry
D.M.	Diabetes mellitus
DB_XLB	Type of GC column
DB-5	Type of GC column
DBA	Dibromoanisole
DCM	Dichloromethane
DDC-CO	Dechlorane plus
DDE	Dichlorodiphenyldichloroethylene
DDT	Dichlorodiphenyltrichloroethane
DF	Detection frequency
DOC	Dissolved organic carbon
DPTE	2,3-Dibromopropyl-2,4,6-tribromophenyl ether
ECD	Electron capture detection
EDGARv4.tox2	Emission Database for Global Atmospheric Research v4.tox2
ENV	SPE Cartridge Bond Elut-ENV for extraction of polar residues
ESSO	Earth System Science Organisation
Et-FOSA	N-Ethyl perfluorooctane sulphonamide
Et-FOSEs	Ethyl perfluorooctane sulphonamido ethanols
ETHeBB	2-Ethyl-1-hexyl-2,3,4,5-tetrabromobenzoate
EVA	Ethylene-vinyl acetate copolymer
FLD	Fluorescence detection
FOSAs	Perfluorooctane sulphonamides
FTAs	Fluorotelomer acrylates

FTIR	Fourier transform infra-red spectrometry
FTOHs	Fluorotelomer alcohols
G.C.	Greenland Current
GC-MS or GC/MS	Gas chromatography-mass spectrometer
GEM	Gaseous elemental mercury
GFFs or GF/Fs	Glass fibre filters
GOM	Gaseous oxidized mercury
GRAHM	Global/Regional Atmospheric Heavy Metals
HBB	Hexabromobenzene 2,3-dibromopropyl-2,4,6-tribromophenyl ether
HCBs	Hexachlorobenzenes
HCH	Hexachlorocyclohexanes; Hexachlorocyclohexane
HDPE	High-density polypropylene
HFRs	Halogenated flame retardants
HNPs	Halogenated natural products
HP/H.P.	Hewlett Packard
HPLC	High-performance liquid chromatography
HS-SPME	Headspace solid-phase microextraction
HVS	High-volume samplers
HYSPLIT	Hybrid Single-Particle Lagrangian Integrated Trajectory
ICP/OES	Inductively coupled plasma - optical emission spectrometry
ICP-AES	Inductively coupled plasma atomic emission spectrometry
L/minute	Litre per minute
LCCP	Long-chain chlorinated paraffins
LDPE	Low-density polyethylene
LOD	Loss on drying
LRAT	Long-range atmospheric transport
LVS	Low-volume sampler
MCCP	Medium-chain chlorinated paraffins
Me-FBSA	N-Methyl perfluorobutane sulphonamide
Me-FBSE	N-Methyl perfluorobutane sulphonamidoethanol
Me-FOSA	N-Methyl perfluorooctane sulphonamide
Me-FOSEs	Methyl perfluorooctane sulphonamido ethanols
MeHg	Methylmercury
MEKC	Micellar electrokinetic chromatography
merB, hgcA	Methylation genes and analogues
Mg/a1	Penetration of light/penetration depth unit
MMHG or MMHg	Monomethyl mercury
MoES	Ministry of Earth Sciences
MPs	Microplastics
MSD	Mass spectrometer detector
MSD GC-ECNI-MS	Mass spectrometer detector gas chromatography with electron capture negative ion mass spectrometry
NBFRs	New/novel brominated flame retardants

NCP	Northern Contaminants Programme
NCPOR	National Centre for Polar and Ocean Research
NEEM	North Greenland Eemian Ice Drilling
NIC	Nippon Instruments Corporation
NILU	Norwegian Institute for Air Research
NU	Nunavut
OBPs	Organo-brominated pesticides
OCPs	Organo-chlorinated pesticides
OCS	Oxidation of carbonyl sulphide
ODE	Ozone depletion events
–OH	Hydroxyl functional group
PAD-3	Type of polymer resin cartridge
PAHs	Polycyclic aromatic hydrocarbons
PBBz	Pentabromobenzene
PBDEs	Polybrominated diphenyl ethers
PBM	Particulate bromine mercury
PBT	Pentabromotoluene
PCBs	Polychlorinated biphenyls
PCDD/Fs	Polychlorinated dibenzo-p-dioxins and furans
PCI	Positive chemical ionization
PDM	Pic du Midi
PFAAs	Perfluoroalkyl acids
PFAS	Polyfluoroalkyl substance
PFBA	Perfluorobutyrate; Perfluorobutanoic acid
PFBS	Perfluorobutanesulphonic acid
PFCAs	Perfluorinated carboxylates
PFHxA	Perfluorohexanoic acid
PFHxS	Perfluorohexanesulphonic acid
PFOA	Perfluorooctanoic acid
PFOS	Perfluorooctanesulphonic acid
PFRs	Organophosphorus flame retardants
PFSAs	Perfluoroalkyl sulphonic acids
PFTE	Polytetrafluoroethylene
pH	Potential of hydrogen
PL	Pacific Mode Layer
PML	Polar mixed layer
POPs	Persistent organic pollutants
PPE	Polypropylene
PUFs	Polyurethane foams
QM-A	High-purity grade for particulate matter 10 sampling and analysis
RGM	Reactive gaseous mercury
SBSE	Stir bar sorptive extraction
SCCP	Short chain chlorinated paraffins
SFC	Supercritical fluid chromatography

SIM	Selected ion monitoring
SOAs	Secondary organic aerosols
SOM	Soil organic matter
SPE	Solid-phase extraction
SPSS	Statistical software
SVOCs	Semi-volatile organic compounds
TBA	Tribromoanisole
TBPH	Bis(2-ethyl-1-hexyl) tetrabromophthalate
TCMX	2,4,5,6-Tetrachloro-m-xylene
TGM	Total gaseous mercury
THg	Total mercury
TLC	Thin-layer chromatography
TMS	Trimethylsilyl
TWP	Tyre wear plastics
USEPA	United States Environmental Protection Agency
UV	Ultraviolet
XAD	Adsorbent resin for dioxins and furans

1 Climate Change Assessment over the Arctic Region
Initiatives through Indian Polar Programme

Neloy Khare
Ministry of Earth Sciences

CONTENTS

The Arctic Ocean, surrounding the North Pole, which consists of a large ocean surrounded by land, is like no other ocean on earth because of its unique location and climate. It is the region above the Arctic Circle (at approximately 66° 34′N). The sun does not set on the *summer solstice* and does not rise on the *winter solstice* above the Arctic Circle.

DOI: 10.1201/9781003265177-1

1

The industrial revolution produced an excess of carbon dioxide and other green-house gas emissions. The rising temperatures in the Polar regions result in the rapid melting of the glaciers. The glaciers are diminishing from the land, calving off into the sea (http://www.sciencemag.org/news/2013/08/scienceshot). The impact of changing climate over the Arctic region is reflected in the Arctic amplification and reflected by the Arctic Ocean's shrinking sea ice cover in summer. Decrease in the snow cover over land in the Arctic, especially in spring, and glaciers in Alaska, Greenland and northern Canada is retreating. The permafrost, also known as the frozen ground in the Arctic, is thawing due to warming. Scientists began gathering evidence of changes in Arctic climate since the 1980s, which have become much more pronounced. The Arctic is experiencing unprecedented extremes in sea ice, temperature and precipitation, which remained unreported in the historical records and emerged as an enigma of climate mystery. Indubitably global warming has severely impacted the Arctic's climate, with many strange climatic events such as witnessing a rainy season almost equal to India's and up to 10 months without snow.

Since the late 1970s, the sea ice in the Arctic has decreased dramatically. According to National Snow and Ice Data Center, the Arctic summer sea ice extent in September 2012 was a record low, shown (in white) compared to the median summer sea ice extent for 1979–2000 (shown in orange) (Figure 1.1).

Climate change is a reality and has exhibited dramatic patterns across the Arctic (The National Oceanic and Atmospheric Administration (NOAA) and its partners – Annual Arctic Report Card – 2019). Some salient findings of this report (https://arctic.noaa.gov/Report-Card/Report-Card-2019) are enumerated below:

- The average annual surface air temperature in the Arctic from October 2018 through August 2019 was the second warmest in the observational record. Satellite recorded the second-lowest Arctic sea ice extent in 2019.

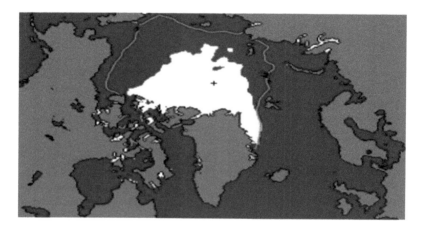

FIGURE 1.1   The Arctic summer sea ice extent measured in 2012 (white outline) compared with the observed changes from 1979 to 2000 (orange outline). In 2013, the Arctic summer sea ice extent rebounded somewhat but was still the sixth smallest extent on record. (Source: National Snow and Ice Data Center)

- The Bering Sea saw record low winter sea ice in 2018 and 2019.
- Birds are being affected, including the breeding population of ivory gull in the Canadian Arctic falling 70% since the 1980s.
- Greenland's ice sheet also experienced rapid melting in 2019, beginning earlier than usual and reaching 95% of the surface.

Arctic's climate changes are significant because the Arctic acts as a barometer of global climate change. Such ongoing changes in the Arctic climate harm the food chain, including phytoplankton and many marine mammals. It includes seals, walrus, whales, and polar bears. Well-known feedback mechanism acting in the Arctic region may lead to further warming. The Arctic *amplified response* to global warming is the repercussion of global temperature rise. Consequently, Greenland's ice sheet is shrinking drastically at an alarming rate (http://news.uga.edu/releases/article/study-2015-melting-greenland-ice-faster-arctic-warming-0616/, Tedesco et al. 2016).

1.1   FEEDBACK MECHANISM AND ARCTIC AMPLIFICATION

Due to sea ice melting in summer, dark open water areas are exposed, absorbing more heat from the sun (Figure 1.2). More ice melts due to excess heat. The sea ice's loss is one of the Arctic amplification drivers (Figure 1.3) (Slivka 2012; Goldenberg 2012). Permafrost may also play a role in positive feedbacks. As the thawing of permafrost starts, plants and animals frozen in the ground begin to decay. Their decomposition releases carbon dioxide and methane back to the atmosphere. It can further induce warming. The shifting Arctic vegetation also affects the surface brightness and adds up to warming. More water vapour is held up due to more warming of the

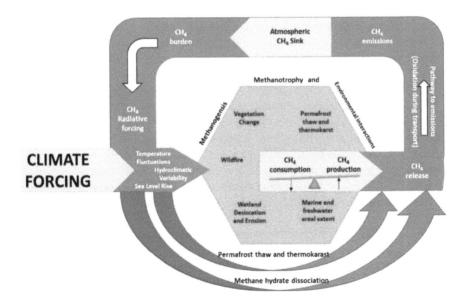

FIGURE 1.2   Feedback mechanism at the Arctic.

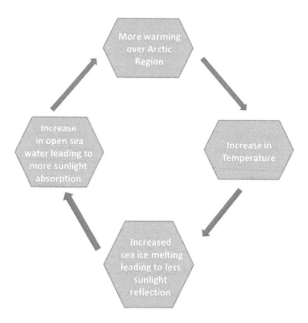

FIGURE 1.3 Arctic amplification.

Arctic atmosphere, which is an important greenhouse gas (Slivka 2012; Goldenberg 2012). In the Arctic, warming is causing further warming in the following manner.

There is no doubt that global warming tends to increase temperatures in the Arctic. It tends to melt ice, decreasing the area covered by sea ice and expanding the scope of darker exposed ocean, which tends to reduce sunlight reflection, as ice is far more reflective than the newly exposed Ocean. It tends to increase the sunlight that is absorbed by the sea. It tends to add to global warming, and the cycle repeats.

Other feedbacks from the loss of Arctic Ocean ice, ranging from a possible slowdown of the so-called "global ocean conveyor belt" to significant shifts in the northern hemisphere's jet stream, could also have severe climatic impacts.

The main consequences of global warming on the Arctic are the increase in temperatures (air and sea), *loss of sea ice* and melting of the *Greenland ice sheet* with a related *cold temperature anomaly*, observed since the 1970s (Foster 2012; Slivka 2012; Goldenberg 2012). Ongoing climate change over the Arctic region is also expected to impact ocean circulation changes, increased input of freshwater (Graeter 2018; Rabe et al. 2011) and ocean acidification (Qi et al. 2017), potential methane releases through the thawing of *permafrost* and methane clathrates (Schuur et al. 2015).

It has also been postulated that due to the potential climate teleconnections to mid-latitudes, these regions are expected to witness a greater frequency of extreme weather events (flooding, fires and drought) (Cohen et al. 2014). It will lead to ecological, biological and phenological changes. Other factors include physical migrations and extinctions (Grebmeier 2012), natural resource stresses, human health, displacement and security issues.

1.2   INDIA AND THE ARCTIC

India's engagement with the Arctic dates back to a century. It was in 1920 when British overseas Dominions signed the Treaty with other signatories like the US, Denmark, France, Italy, Japan, the Netherlands, Norway, Great Britain, Ireland and Sweden concerning Spitsbergen's 'Svalbard Treaty' in February 1920 in Paris.

India has been closely following the Arctic region's developments in the backdrop of the emerging opportunities and challenges due to the global warming-induced melting of the Arctic ice cap. Science, environment, commerce and strategy are the main concerns of India in the Arctic region.

India paved its way in the Arctic by launching a research programme in 2007 to thrust on climate change in the circumpolar north. The primary objectives of the Indian research in Arctic region are as follows:

1. To study the hypothesised teleconnections between the Arctic climate and the Indian monsoon by analysing records archived in the sediment and ice cores from the Arctic glaciers and the Arctic Ocean.
2. To characterise sea ice in the Arctic using satellite data to estimate the effect of global warming on the northern Polar region.
3. To research Arctic glaciers' dynamics and mass budget focusing on the impact of melting glaciers towards sea level changes.
4. To assess the Arctic flora and fauna vis-à-vis their response to anthropogenic activities. Besides, a bi-polar comparison is proposed to be undertaken in life forms.

India launched its first scientific expedition to the Arctic Ocean in 2007. A research base named "Himadri" was established at Ny-Ålesund, Svalbard, Norway, in July 2008 for researching disciplines like glaciology, atmospheric sciences and biological sciences. The area used for the Indian research base is the International Arctic Research Base at Svalbard. A Memorandum of Understanding (MOU) has also been signed with the Norwegian Polar Research Institute of Norway, for cooperation in science, as even with Kings Bay (a company of the Norwegian Government) Ny-Ålesund. It provides logistics and infrastructure facilities for undertaking Arctic research and maintaining the Indian research base 'Himadri' in the Arctic region.

Several scientists from various national institutions have participated in our Arctic programme. India became a member of the Council of the International Arctic Science Committee (IASC) in 2012. India's claim for Observer Status received attention in 2012 with widespread support from all member countries. In recognition of India's commitment and sustained interest in Arctic science, during the Eighth Biennial meeting of the Arctic Council held in Kiruna, Norway on 1 May 2013, under Sweden's Chairmanship, India was provided Observer Status to the Arctic Council.

1.3   BLACK CARBON

In contrast to most atmospheric aerosols, black carbon (BC) aerosols are good candidates for absorbing solar radiation. Due to such absorption, a warming effect on the planet is perceived. On the contrary, other aerosols such as sulphate aerosols provide the cooling effect due to scattering. BC aerosols play a critical role in affecting the climate system by changing and heating the clouds (semi-direct effect) or acting as cloud condensation nuclei (indirect effect). BC aerosols have emerged as third among the most extensive human-generated causes. CO_2 and CH_4, having a present-day radiative forcing of ~0.40 W/m² which is ~25% more of the pre-industrial period (Fifth Assessment Report of the Intergovernmental Panel on Climate Change 2014).

Despite its importance, only a few modelling studies have addressed BC aerosols' effectiveness in warming the planet relative to CO_2 forcing (e.g. Roberts and Jones 2004; Cook and Highwood 2004; Hansen et al. 2005; Stjern et al. 2017; Smith et al. 2018). The results indicate that the BC aerosols are less effective in warming the earth than CO_2 (Yoshimori and Broccoli 2008). The concept of 'efficacy' to measure forcing agents' effectiveness was introduced by Hansen et al. (2005). The efficacy of BC aerosols emitted by burning of fossil fuel and biomass was 0.930 and 0.80, respectively, when effective radiative forcing (ERF) definition is used to estimate the radiative forcing (Hansen et al. 2005). This cloud enhancement in the lower troposphere dramatically reduces the direct warming effect of the BC aerosols. Stjern et al. (2017) found that multi-model median efficacy of the BC aerosols is less than one (0.80).

The decrease in efficacy with BC aerosols' altitude has been found in some recent studies (Ban-Weiss et al. 2012; Samset and Myhre 2015). Ban-Weiss et al. (2012) showed that near-surface BC aerosols cause warming. On the contrary, when BC aerosols are at heights, the varying climate response from BC aerosols at different sizes arises primarily due to various fast climate adjustments. These are defined as the climate response before any change in the average surface temperature globally (Bala et al. 2010; Ban-Weiss et al. 2012). BC warms the surface through diabatic heating. At heights in increased longwave radiation without heating the body, the absorbed solar radiation is lost to space. We note that the efficacy decreases with the size of BC aerosols. BC aerosols' efficiency to exert more significant radiative forcing due to the direct aerosol effect strengthens with altitude (Samset and Myhre 2011, 2015).

Because of the above, it may be postulated that the in-modulating Arctic Climate BC aerosols play an influential role in further strengthening the feedback mechanism, leading to Arctic amplification, and therefore a detailed investigation on this aspect must be made.

The consequences of the impact of rapid changes in the Arctic region go beyond the coastal states. To respond to such challenges warrants the active participation of all those actors who have a stake in global commons' governance. It requires a legitimate and credible mechanism. The interplay between science and policy can significantly contribute to addressing the complex issues facing the Arctic. India, which has significant expertise in polar research matters due to its long experience of launching annual scientific expeditions to the Antarctic and association with the Antarctic Treaty System, can play a constructive role in securing an influential position in Arctic affairs. As a permanent observer in the Arctic Council, India is committed to contributing to evolve and strengthen the effective cooperative partnerships that can contribute to a safe, stable and secure Arctic.

The warming of the Arctic region has recently gained worldwide attention due to its projected impacts on the global climate system. The effect of anthropogenic BC aerosol on snow is of enduring interest due to its role in aerosol radiative forcing (ARF) and further consequences for the Arctic and global climate changes. Having demonstrated its sincere pursuit of Arctic science ever since the Indian research base is set up at Ny-Ålesund (Svalbard), India has been continuously generating data on BC aerosol over the Arctic region. MoES-Indian Institute of Tropical Meteorology (IITM), Pune (India), participated in the Arctic expedition to study BC over the Himadri research base, Ny-Ålesund. Similarly, CSIR-National Physical Laboratory, New Delhi (India), participated in the Arctic expedition to measure the concentration of carbon monoxide (CO) over the Arctic region to compare it with the CO concentration in the Antarctic area. We discuss here briefly some of the Indian contributions to the assessment of BC and CO in a bid to help understand Arctic climate change and associated amplification vis-à-vis its teleconnection with the tropical countries like India.

1.4 ASSESSMENT OF BLACK CARBON AEROSOLS AND SOLAR RADIATION OVER HIMADRI, NY-ÅLESUND

India ventured into assessing the BC and measuring solar spectral at 'Indian Arctic Station, Himadri', Ny-Ålesund, during 2011–2014. The contribution from long-range transport of pollutants from far-away places is found to dominate the local sources such as emissions from shipping and power plants to the annual cycle with maximum BC mass concentration during winter/early-spring season and minimum during the summer season. Moreover, higher BC concentrations were observed during 2012 as compared to other years during the study period. The aerosol optical depth's (AOD) spectral variations observed during the summer months indicate an immense contribution of fine-mode aerosol particles to the BC mass concentration, particularly during 2012. Further, the zenith skylight spectra in the spectral range of 200–1100 nm indicate maximum particle scattered intensity around 500 nm (Dr. S.M. Sonbawne, personal communication). These results play a vital role in the earth-atmosphere radiation balance and hence exhibit profound influence on regional and global climate changes (Stohl et al. 2013).

Raju et al. (2011) attempted to study the BC radiative forcing over the Indian Arctic Station, Himadri, during the Arctic summer of 2012 by using an aethalometer. Measurements of BC aerosols were carried out continuously over the Indian Arctic Station, Himadri, during the Arctic summer (23 July to 19 August) of 2012. The monthly mean BC mass concentration during July and August was 0.093 ± 0.046 and 0.069 ± 0.050 µg/m^3, respectively. BC mass concentration showed maximum loading during 0800–1600 LT. Transport from distant sources (as observed from air mass back trajectories) apart from some local anthropogenic activities (emissions from shipping and power plants) could be the possible sources for the observed BC concentration at Himadri. Using the OPAC and SBDART models, optical properties and ARF in the spectral range of $0.2-4\,\mu m$ for composite aerosol and without-BC aerosol at the top of the atmosphere, surface and atmosphere were computed. The presence of BC resulted in positive radiative forcing in the atmosphere leading to a warming effect ($+2.1\,W/m^2$), whereas cooling was observed at the top of the atmosphere ($-0.4\,W/m^2$) and surface ($-2.5\,W/m^2$). BC formed about 57% of atmospheric ARF.

1.5   PRODUCTION OF CARBON MONOXIDE FROM ICE PACKS

Carbon monoxide is the most critical atmospheric gas which is produced due to the combustion of fossil fuel. It is also produced in large amounts by industries and motor vehicles. Carbon monoxide is a poisonous gas that has a short composition in the atmosphere. The hydroxyl radical (OH) gets combined with it chemically and converts it into non-poisonous material. It helps in monitoring the quantity of the hydroxyl compound. Hydroxyl being an oxidiser controls the composition of many greenhouse gases of the atmosphere. Recent studies show that carbon monoxide is continuously produced and liberated in large amounts in glacier areas. Indian scientists also conducted experiments related to carbon monoxide at Indian Research Centre, Maitri, situated in Antarctic Islands. With the help of various experiments, they know about the regular production cycle of carbon monoxide because of regular consideration of solar actinic rays. Consequently, scientists realised that carbon monoxide production is due to a photochemical reaction in Antarctic glaciers (Dr B.C. Arya, personal communication).

It is considered that some organic materials, like formaldehyde (HCHO), that are entrapped in ice crystals are decomposed through photochemical reactions and produced CO. Carbon monoxide, oxygen, nitrogen oxide analyser, solar photometer, portable climate centre, pyrometer, etc. are the main instruments on which experiments are generally conducted in Polar regions.

Indian scientists performed various experiments to produce carbon monoxide from snowpacks and came to know about the regular alterability in carbon monoxide production in Ny-Ålesund in 2008, especially in March and August. Despite all, Indian scientists also measured the concentration of BC, the size, distribution and composition of aerosols, and the amount of water vapour in the atmospheric air of Ny-Ålesund. In the summer of 2008, ozone analyser was used to measure surface ozone concentration (Dr B.C. Arya, personal communication).

Lucknow University, Lucknow, Birbal Sahni Institute of Paleosciences, Lucknow, Jawahar Lal Nehru University, New Delhi, Indian Institute of Technology, Kharagpur, Wadia Institute of Himalayan Geology, Dehradun, and MoES-National Centre for Polar and Ocean Research, Goa, among many others, have participated in the research and contributed to the understanding of climate change in the Arctic regions and its global impacts. Some of the significant Indian initiatives to address climate change issues impacting Arctic regions are detailed below.

1.6 CLIMATE CHANGE AND ARCTIC GLACIERS

Many changes of glaciers of the Arctic have been noticed in previous years. Indian scientists have also shown much interest to get information related to these changes in glaciers. They have studied West Craig, Borger, and Mindralavan Glaciers in this reference.

1.7 GEOMORPHOLOGY AND SEDIMENTOLOGY OF DIVERSIFIED MORPHOLOGICAL ZONES OF GLACIATED TERRAIN OF THE NY-ÅLESUND

Diversified surface processes of the Ny-Ålesund region carve the landscape and exhibit distinguished Arctic's landforms. Many studies are done on climate change using various proxies. Yet, meagre attention has been paid to geomorphological and sedimentological parameters. Sediment characteristics, AMS^{14}C dates and geomorphic features have been used to reconstruct paleoclimate. Based on the landforms and sediments, this region has been classified into five morphological zones: glacial (moraines GL), proglacial (lacustrine deposits LD), outwash plain (sand deposits OWP), fluvial deposits (FD) and coastal cliff (CC). The glacial moraines (GL) suggest devoid of any sedimentary structures and coarse-grained, matrix-supported boulders, and it is composed of unconsolidated, unstratified, massive.

In contrast, OWP, LD, FD and CC are semi-consolidated, stratified, fine-grained layers of sand, silt, and clay with gravels and faint sedimentary structures. CC's and LD's sediments are very poorly sorted, very positively skewed and very leptokurtic, and is comprised of medium to fine sand, silt and clay. The sediment characteristics of various morphological zones' geomorphic features explain that this region was carved and dominated by glaciers (Prof. D.S. Singh, personal communication).

The poorly sorted sediments of all the geomorphological zones explain the depositional environment's fluctuating energy, especially under warm climate at interglacial stages during 44, 27, 12, 10.5 ka BP. It may be inferred that the prevailing environment was not consistent and persistent for an extended period (Schuur et al. 2015).

1.8 QUARTZ GRAIN MICROTEXTURE AND MAGNETIC SUSCEPTIBILITY ASSESSMENT OF THE NY-ÅLESUND REGION

The quartz grain microtexture reveals predominant glacial activities in the top 40 cm of the section, while the middle 40–55 cm part represents some aeolian activities along with glacial signatures. The bottommost part, in addition to glacial markers, exhibits some aqueous evidence as well. The lithology shows medium-grained sand in the upper leg and coarse-grained sand with occasional shell pieces in the lower leg. Angular gravels (2–12 mm) are present throughout with increasing size from top to bottom. Based on the above observations and ^{14}C AMS dates, it can be summarised that after the Last Glacial Maximum (LGM), the pre-Holocene period shows rapid glacial retreat, followed by a warmer period during the early Holocene. Mid- and late Holocene is marked by a predominantly glacial environment characterised by meltwater streams originating from the glaciers and flowing into the fjord. Magnetic susceptibility studies have also been attempted. Four alternate stages of colder and warmer phases have been established (Dr R. Kar, personal communication). Though some similarities among the different climatic phases are discernible between the quartz grain microtexture and magnetic susceptibility studies, they are not entirely compatible, probably due to their different responses to the climatic variations (http://www.sciencemag.org/news/2013/08/scienceshot-arctic-warming-twice-fast-rest-world).

Indubitably, the sea ice is frozen ocean water that grows and melts in the ocean. On the contrary, icebergs, glaciers and ice shelves float in the sea, having originated on the land. Sea ice is typically covered with snow, and Arctic sea ice keeps the Polar regions cool. It also helps modulate and control the global climate. Having a bright surface, 80% of the sunlight that strikes sea ice gets reflected into space. As sea ice melts in the summer, it exposes the dark ocean surface. Therefore, instead of reflecting 80% of the sunlight, the ocean absorbs 90% of the sunlight. Thus, the oceans heat up, and Arctic temperatures rise further.

A slight rise in temperature at the poles leads to still more significant warming over time, thus making the Polar regions the most sensitive areas to a subtle change in earth's climates. Accordingly, both the thickness and extent of the Arctic's summer sea ice have shown a dramatic decline over the past 30 years, consistent with observations of a warming Arctic. The loss of sea ice can also accelerate global warming trends and change climate patterns.

Sea ice extent is a measurement of ocean area where there is at least some sea ice. Usually, scientists define a minimum concentration threshold to mark the ice edge; the most common cut-off is 15%.

The Arctic sea ice extent is focused more closely than other aspects of sea ice because satellites measure the volume more accurately than other measurements, such as thickness. The Arctic's sea ice minimum is considered when the Arctic's sea ice exhibits the lowest areal extent. It occurs at the end of the summer melting season. The Arctic sea ice maximum is regarded as the day of the year when Arctic sea ice reaches its most considerable areal extent. It occurs at the end of the winter cold season.

1.9   REMOTE SENSING OBSERVATIONS AND MODEL REANALYSIS

Applications of remote sensing techniques and modelling have been applied to assess and quantify the Arctic sea ice loss in July–September, with particular attention to September on a daily, monthly, annual and decadal basis.

Coincidently, the 12 lowest extents in the satellite era occurred in the last 12 years. It is attributed to the impacts of land-ocean warming and the northward heat advection into the Arctic Ocean over the past 40 years (1979–2018); actual warming rates have been identified in the Arctic Ocean in the last 40 years. The study demonstrates the linkages of sea ice dynamics to ice drifting and accelerated melting. It occurs due to persistent low pressure and high air-ocean temperatures, supplemented by the coupled ocean-atmospheric forcing (http://news.uga.edu/releases/article/study-2015-melting-greenland-ice-faster-arctic-warming-0616/). The accelerated decline is recorded in the Arctic sea ice extent and sea ice concentration over the past four decades.

The ocean-atmosphere coupled mechanism plays a vital role in the global climate change. Sea ice variability and trends were computed using satellite and model reanalysis measurements for the whole Arctic and each of its nine regions: (i) Seas of Okhotsk and Japan, (ii) the Bering Sea, (iii) Hudson Bay, (iv) the Baffin Bay/ Labrador Sea, (v) Gulf of St. Lawrence, (vi) Greenland Sea, (vii) Kara and Barents Seas, (viii) the Arctic Ocean and (ix) Canadian Archipelago. Overall, Arctic sea ice declined in all seasons and on a yearly average basis, although the highest and lowest negative trends were recorded in summer and winter/spring, respectively. The study reveals that the Arctic Ocean, Kara and Barents Seas, the Greenland Sea, and the Baffin Bay region are majorly responsible for the total negative sea ice extent trend in the Arctic (Dr. Avinash Kumar, personal communication). The study demonstrated the interannual and seasonal variabilities of Arctic sea ice and interactions among the atmosphere, ice and ocean (Tedesco et al. 2016).

1.10   ASSESSMENT OF SPATIO-TEMPORAL VARIABILITY OF SNOWMELT ACROSS SVALBARD

Indian researchers have monitored snowmelt over the Svalbard region as significant changes in the interannual variation of Arctic snow and sea ice are connected to the global climate changes using active microwave sensors. These sensors are frequently used to detect surface melting because of their sensitivity to the presence of liquid water in snow/ice. Data of QuikScat, OSCAT, ASCAT, and OSCAT-2 are used to map the annual melt duration and summer melt onset for the Svalbard archipelago. It provides one of the most extended and continuous radar backscatter records to esti-mate snowmelt onset and melt duration on Svalbard spanning from 2000 to 2017. A single threshold-based model was used to detect snowmelt timing; the threshold was calculated using meteorological data from the human-crewed weather stations. The results capture the timing and extent of melt events caused by warm air temperature and precipitation because of the influx of moist, mild air from the Norwegian and Barents seas. The highest melt duration and earlier melt onset occurred in southern-most and western Svalbard in response to the influence of the warm west Spitsbergen current. Compared to previous studies, we found considerable interannual variability and regional differences. Though the record is short, there is an indication of an increasing trend in total days of melt duration and earlier summer melt onset date possibly linked to the general warming trend (Dr A.J. Luis, personal communica-tion). Climate indices such as Interdecadal Pacific Oscillation and Pacific Decadal Oscillation are well correlated with onset melt and duration across Svalbard. With the reported year-after-year decrease in sea ice cover over the Arctic Ocean, the trend towards longer snowmelt duration inferred from this study is expected to enhance the Arctic amplification (McCarthy 2011).

1.11   ASSESSMENT OF MASS BALANCE OF THE ARCTIC GLACIERS

Prof. A.L. Ramanathan (personal communication) studied the changes in the area from 1993 to 2018 and mass balance of Vestre Broggerbreen glacier, Ny-Ålesund, Arctic from 2011 to 2017. The glaciated area had decreased from 3.96 km^2 in 1993 to 3.57 km² in 2018. Its range varied between 0.011 and 0.02 km², resulting into a total area loss of 0.39 km² (~10% at 0.016 km² a-1). A comparatively rapid decrease in the glaciated area was found during 1998–2010 (0.02 km² a-1), whereas less retreat rate was found in 1993–1998 (0.011 km² a-1) and 2010–2018 (0.012 km² a-1 13). The Vestre Broggerbreen glacier's mass balance was negative throughout the entire study period (2011–2018). Mass balance ranged between −0.08 (2013–14) and −1.22 m w.e. (2015–2016) with a cumulative mass balance of −4.31 m w.e. (0.016 km² a-1). A strong relationship between mass balance and summer temperature was found with $R^2 = 0.97$ at $P < 0.05$ (Rajmund 2007).

1.12   SCIENTIFIC EXPLORATION OF KONGSFJORDEN

Kongsfjorden, an icy archipelago with a length of about 40 km and a width ranging from 5 to 10 km, is a glacial-fjord in the Arctic (Svalbard) which lies in the N-W coast of Spitsbergen, the main island of Svalbard. It is a site where warmer waters of the Atlantic meet the colder waters of the Arctic. An open fjord without a sill is primarily under the influence of the adjacent shelf processes. The Transformed Atlantic Water (TWA) from the west Spitsbergen current and the glacier-melt freshwater at the inner fjord create intense temperature and salinity gradients along the fjord's length. Southerly winds will result in down-welling at the coast. Such winds also hinder the exchange processes that take place between the shelf and the fjord, while the northerly winds will move the TWA water below the upper layer towards the coast. During summer, the meltwater not only stratifies the upper water column but significantly alters the turbidity.

It also impacts the seasonal changes in the biomass of phytoplankton. Thus, any altered interaction between the Atlantic water and the (turbid) meltwaters from tidal glaciers on a seasonal to interannual timescale is likely to affect the fjord's aquatic ecosystem. The long-term changes in the fjord hydrography and sedimentation will affect the benthic ecosystem.

Against the above backdrop of the fjord system's climate sensitivity, India has evolved a multi-institutional programme of long-term monitoring of the Kongsfjorden. It was initiated by the deployment of an ocean-atmosphere mooring system along regular repeat transects. It was designed to measure seasonal physical, chemical and biological parameters to establish a long-term comprehensive data set on physical, chemical, biological and atmospheric measurements. The influence of interaction between the warm Atlantic water and the cold glacial-melt fresh water and their effects on the biological productivity and phytoplankton species composition and diversity within the fjord are equally essential to be addressed.

1.13   DEPLOYMENT OF UNDERWATER MOORED OBSERVATORY IN THE KONGSFJORDEN FJORD

The Kongsfjorden is a natural laboratory. It is ideal for studying Arctic climate variability. Scientists predict that the melting of the Arctic glaciers will trigger patterns of weather and ocean circulations. Such changes could affect the climate of other parts of the world. One of the significant limitations in the logistics has been to reach the location and collect data, especially during the severe Arctic winter. The IndARC observatory is an attempt to overcome this lacuna. Data collected by IndARC would be used for climate modelling studies to understand the Arctic processes that influence Indian monsoons. The IndARC, the country's first underwater moored observatory deployed in the Kongsfjorden fjord, halfway between Norway and the North Pole, represents a significant milestone in India's scientific endeavours in the Arctic region. The engineers and scientists from the MoES-National Centre for Polar and Ocean Research (NCPOR), MoES-National Institute of Ocean Technology (NIOT) and MoES-Indian National Centre for Ocean Information Services (INCOIS) developed the IndARC. It was deployed from *RV Lance*, a research vessel belonging to the Norwegian Polar Institute. The observatory is moored and anchored at a depth of 192 m. It has an array of 10 state-of-the-art oceanographic sensors strategically positioned at various depths in the water. The sensors were programmed to collect real-time data on seawater temperature, salinity, ocean currents and other vital parameters of the fjord (https://ncpor.res.in/).

The correlation between less and more ice in the Arctic is very close to how the monsoon behaves. Just as we know that the *El Nino effect* (hot ocean temperatures in the Equatorial Pacific) is having a global impact on weather patterns, including the Indian monsoon, we learn that the Arctic ice also has a significant effect. The Arctic precipitation and temperatures from June to October hint at the monsoon likely to occur in the coming year. The oscillation in the air creates the western disturbance as it moves over ice and snow in the Arctic. If there is less oscillation, the air will have less moisture, leading to less rainfall in the monsoons.

1.14 EXPLORING TELECONNECTION BETWEEN ARCTIC CLIMATE AND TROPICAL INDIAN MONSOON

The climate change over the Arctic region and North Atlantic shows a mechanistic link with the Indian Summer Monsoon (ISM) during the Holocene. The marine and continental archives of ISM precipitation suggest significant shifts during the Holocene aligned with the Arctic climate over multi-time scales. The ISM strengthened during the Greenlandian (11.7–8.3 kyr BP), showing variable but overall decreasing precipitation during the Northgrippian (8.3–4.2 kyr BP). Synchronicity exists in palaeoclimatic records. It could be due to possible age errors and resolution and proxy response to the changing climate. During the Meghalayan age (4.2 kyr to recent), the Indian subcontinent witnessed a protracted dry event beginning at ~4.2 kyr BP and ended at ~3.4 kyr BP. Other significant events of the Meghalayan age include the Medieval Climate Anomaly (MCA). The Current Warm Period (CWP) showing a strong ISM, interrupted by the Little Ice Age (LIA) – a cold phase with low precipitation in the Indian subcontinent (Prof. A.K. Gupta, personal communication). The millennial-scale variability in the ISM is associated with the Heinrich and Bond events. The cooling in the Arctic sea, ice expansion in the North Atlantic and weakening of the Atlantic overturning meridional oscillations due to high freshwater flux and ice rafting in the North Atlantic caused weak ISM precipitation over the south and southeast Asia (http://www.rivm.nl/en/Documents_and_publications/Common_and_Present/ Newsmessages/2016/Documentary_Sea_Blind_on_Dutch_Television).

In conclusion, we may submit that India is striving to obtain deep insight into the climate changes occurring over the Arctic region as its impacts are expected to influence Indian climate. Sustained monitoring and observational network shall be an added advantage to strengthen our understanding of Arctic climate and its teleconnection with Indian monsoons.

REFERENCES

Bala G, Caldeira K and Nemani R 2010 Fast versus slow response in climate change: implications for the global hydrological cycle. *Clim. Dyn.* 35: 423–34.

Ban-Weiss GA, Cao L, Bala G and Caldeira K 2012 Dependence of climate forcing and response on the altitude of black carbon aerosols. *Clim. Dyn.* 38: 897–911.

Cohen J et al. 2014. Recent Arctic amplification and extreme mid-latitude weather (PDF). *Nat. Geosci.* 7 (9): 627–637.

Cook J and Highwood EJ 2004 Climate response to tropospheric absorbing aerosols in an intermediate general-circulation model Q. *J. R. Meteorol. Soc.* 130: 175–91.

Foster JM 2012. From 2 satellites, the big picture on ice melt. New York Times.

Goldenberg S 2012. Greenland ice sheet melted at an unprecedented rate during July. In *The Guardian*. London. Retrieved 4 November 2012.

Graeter KA 2018. Ice core records of west greenland melt and climate forcing. *Geophys. Res. Lett.* 45 (C7): 3164–3172.

Grebmeier J 2012. Shifting patterns of life in the pacific arctic and sub-arctic seas. *Ann. Rev. Marine Sci.* 4: 63–78.

Hansen J et al. 2005. Efficacy of climate forcings. *J. Geophys. Res.* 110: 1–45.

McCarthy JJ 2001. *Climate Change 2001: Impacts, Adaptation and Vulnerability. Contribution of Working Group II to the Third Assessment Report of the Intergovernmental Panel on Climate Change.* New York: Cambridge University Press.

Qi D et al. 2017. Increase in acidifying water in the western Arctic Ocean. *Nat. Climate Change* 7 (3): 195–199.

Rabe B et al. 2011. An assessment of Arctic Ocean freshwater content changes from the 1990s to the 2006–2008 period. *Deep-Sea Res. Part I* 56 (2): 173.

Rajmund P 2007. Recent air-temperature changes in the Arctic (PDF). *Ann. Glaciol.* 46: 316–324.

Raju MP, Safai PD, Rao PSP, Devara PCS and Budhavant KB 2011. Seasonal characteristics of black carbon aerosols over a high-altitude station in Southwest India. *Atmos. Res.* 100 (1): 103–110.

Roberts DL and Jones A 2004. Climate sensitivity to black carbon aerosol from fossil fuel combustion. *J. Geophys. Res.* 109: D16202.

Samset B H and Myhre G 2011. Vertical dependence of black carbon, sulphate and biomass burning aerosol radiative forcing. *Geophys. Res. Lett.* 38: L24802.

Samset BH and Myhre G 2015. Climate response to externally mixed black carbon as a function of altitude. *J. Geophys. Res. Atmos.* 120: 2913–2927.

Schuur EAG et al. 2015. Climate change and the permafrost carbon feedback. *Nature* 520 (7546): 171–179.

Slivka K 2012. Rare burst of melting seen in greenland ice sheet. NYTimes.com. Retrieved 4 November 2012.

Smith CJ et al. 2018. Understanding rapid adjustments to diverse forcing agents. *Geophys. Res. Lett.* 45, 12023–12031.

Source: https://arctic.noaa.gov/Report-Card/Report-Card-2019.

Source: https://ncpor.res.in/.

Source: http://news.uga.edu/releases/article/study-2015-melting-greenland-ice-faster-arctic-warming-0616/.

Source: http://www.rivm.nl/en/Documents_and_publications/Common_and_Present/News messages/2016/Documentary_Sea_Blind_on_Dutch_Television.

Source: http://www.sciencemag.org/news/2013/08/scienceshot-arctic-warming-twice-fast-rest-world.

Stjern CW et al. 2017. Rapid adjustments cause weak surface temperature response to increased black carbon concentrations. *J. Geophys. Res. Atmos.* 122: 11462–81.

Stohl A, Klimont Z, Eckhardt S, Kupiainen K, Chevchenko VP, Kopeikin VM, and Novigatsky AN 2013. Black carbon in the Arctic: the underestimated role of gas flaring and residential combustion emissions. *Atmos. Chem. Phys.* 13 (17): 8833–8855.

Tedesco M, Mote T, Fettweis X, Hanna E, Jeyaratnam J, Booth JF, Datta R, and Briggs K, 2016. Arctic cutoff high drives the poleward shift of a new Greenland melting record. *Nat. Commun.* 7: 11723.

Yoshimori M and Broccoli A 2008. Equilibrium response of an atmosphere-mixed layer ocean model to various radiative forcing agents: global and zonal mean response. *J. Clim.* 21: 4399.

2 Arctic Weather and Climate Patterns

R. S. Maheskumar
Ministry of Earth Sciences, Govt. of India

Sunitha Devi S
India Meteorological Department

CONTENTS

DOI: 10.1201/9781003265177-2

2.1   INTRODUCTION

The Arctic is a polar region situated around the North Pole, consisting of the Arctic Ocean and the adjoining land region. The Arctic region is defined as the northernmost part of the globe lying north of the latitude 66.5°N. North of this latitudinal belt, Sun does not set on the June solstice and does not rise on the December solstice. Around the Arctic Circle towards the North Pole, the durations of day and night are shorter. North pole experiences continuous 6 months daylight and 6 months constant night as the Sun rises and sets only once in a year. The Arctic Ocean is a large and unique one partially covered by ice and partly surrounded by the land region belonging to different countries.

The Arctic region near the coastal area experiences maritime-type weather and climate, while the interior land regions experience continental type weather and climate. The winters are mostly stormy and wet with snow, and summers are cool and cloudy, with average temperatures around 10°C over the coastal areas. The interior land regions will be dry and chill with temperatures around −40°C during winter, and summers with long days of sunshine experience an average temperature ranging between 10°C and 30°C.

2.2   GEOGRAPHY

Arctic geography is a varying one due to weather and climatic conditions (Figure 2.1). Many land areas consist of coniferous forest, wooded tundra and glaciers. Permafrost,

FIGURE 2.1   Arctic region.

soil that stays frozen for at least two consecutive years, also increases Arctic area compared to the other delimitations. The ice cover determines the Arctic nature of the marine regions. Sea ice is highest in February–March and lowest in September. The surface of the Arctic ice is monitored almost in real time by satellites. Culturally, the Arctic homes northern indigenous peoples. *As the climate warms*, the Arctic shrinks if defined by temperature, forest line, permafrost or ice cover. Cultural and political boundaries also vary. The Arctic Circle is the most permanent of the delimitations. Also, the polar circle moves very slowly due to the variation of the Earth's axial tilt.

2.3 PATTERNS IN ARCTIC WEATHER AND CLIMATE

The unique geography of the Arctic leads to unusual weather patterns that reappear in the region year after year. Some common weather patterns, such as cyclones or anticyclones, also occur in this region. The Arctic oscillation is an atmospheric circulation pattern that occurs over the mid- to high latitudes of the Northern Hemisphere, including the Arctic.

Weather patterns that recur or persist over multiple seasons are called semipermanent highs and lows and play a significant role in the regional weather cycles. They are:

Cyclones and Anticyclones
Polar Lows
Semipermanent Highs and Lows
Arctic oscillation

2.3.1 CYCLONES AND ANTICYCLONES

In the Arctic, cyclones occur over certain places depending on the time of year. Semipermanent lows in the Arctic include the Aleutian Low, a low-pressure centre that experiences many hurricanes and storms in the winter, and the Icelandic Low, a low-pressure centre located near Iceland (Figure 2.2).

An anticyclone known as the Beaufort High recurs year after year, sitting over the Beaufort Sea and Canadian Archipelago in winter and spring. An anticyclone also frequently appears over Siberia, known as the Siberian High.

2.3.2 POLAR LOWS

Polar lows are small, intense cyclones that form over the open ocean during the cold season (Figure 2.3). From satellite imagery, polar lows can look much like a hurricane, with a large spiral of clouds centred around an eye – for this reason, they are sometimes called Arctic hurricanes. Polar lows range in size from around 100 to 500 km in diameter. Wind speeds average about 80 km/hour, and they can occasionally reach hurricane strength (103 km/hour). Polar lows tend to form when cold Arctic air flows over relatively warm open water. The storms can develop rapidly, reaching their maximum strength within 12–24 hours of formation, but they dissipate just as quickly, lasting on average only for 1 or 2 days.

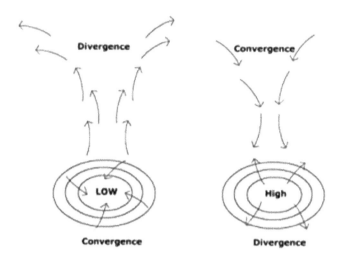

FIGURE 2.2 Left: Cyclone, Right: Anticyclone. (M. Ritter/University of Wisconsin Stevens Point)

FIGURE 2.3 Cyclone over the Arctic Ocean. (NASA/MODIS)

2.3.3 SEMIPERMANENT HIGHS AND LOWS

The circulation and pressure patterns over one area, averaged over months or years, tend to remove the day-to-day variability and bring out long-term ways to affect weather and climate both within and outside the Arctic. Relative strengths of semi-permanent highs and lows can be compared and are reported as indices such as the North Atlantic oscillation and the Arctic oscillation. These indices have been linked to variability in temperatures and sea ice conditions in the Arctic and the South Asian summer monsoon in the tropics.

2.3.4 ARCTIC OSCILLATION

The *Arctic oscillation* refers to a contrasting pattern of pressure between the Arctic and the northern middle latitudes. Overall, if the atmospheric pressure is high in the Arctic, it tends to be low in the northern middle latitudes, such as northern Europe and North America. If atmospheric pressure is lacking in the middle latitudes, it is often high in the Arctic. When pressure is high in the Arctic and low in mid-latitudes, the Arctic oscillation is negative. In the positive phase, the pattern is reversed.

A phase of the Arctic oscillation has an essential effect on weather in northern latitudes. The Arctic oscillation's positive step brings ocean storms farther north, making the weather wetter in Alaska, Scotland, and Scandinavia and drier in the western United States and the Mediterranean. The positive phase also keeps the temperature warmer than usual in the eastern United States but makes Greenland colder.

In the negative phase of the Arctic oscillation, the patterns are reversed. A strongly negative phase of the Arctic oscillation brings warm weather to high latitudes and cold, stormy weather to the more temperate regions. Over most of the past century, the Arctic oscillation alternated between its positive and negative phase. From the 1970s to the mid-1990s, the Arctic oscillation tended to stay in its positive step. However, it has again alternated between positive and negative, with a record negative phase in the winter of 2009–2010 (Figure 2.4).

The semipermanent patterns listed below are centres of action in the Arctic atmosphere, influencing weather patterns in the Arctic and worldwide.

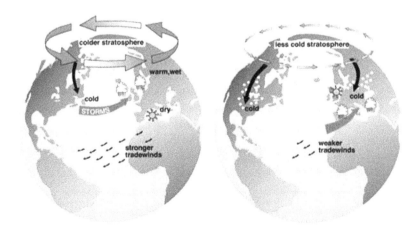

FIGURE 2.4 Left: Effects of the positive phase of the Arctic oscillation. Right: Effects of the Negative Phase of the Arctic Oscillation. (J. Wallace, University of Washington)

Aleutian Low: This semipermanent low-pressure centre is located near the Aleutian Islands. Most intense in winter, the Aleutian Low is characterised by many intense cyclones. Travelling cyclones formed in subpolar latitudes in the North Pacific usually slow down and reach maximum intensity in the area of the Aleutian Low.

Icelandic Low: This low-pressure centre is located near Iceland, usually between Iceland and southern Greenland. Most intense during winter, it weakens and splits into two centres in summer, one near the Davis Strait and the other west of Iceland.

Azores High: The Azores High is a high-pressure pattern that forms in the sub-tropical Atlantic Ocean. Although it occurs outside the Arctic Ocean, it is linked to the Icelandic Low through the North Atlantic oscillation.

Siberian High: The Siberian High is an intense, cold anticyclone that forms over eastern Siberia in winter, associated with frequent cold air outbreaks over East Asia.

Beaufort High: The Beaufort High is a high-pressure centre over the Beaufort Sea, present mainly in winter.

North American High: The North American High is a relatively weak area of high pressure covering most North America during winter. This pressure system tends to be centred over the Yukon but is not as well-defined as its continental counterpart, the Siberian High.

2.3.5 Weather Pattern

2.3.5.1 Temperature

In this schematic drawing of North Atlantic and Arctic Ocean circulation, red arrows represent relatively warm water from lower latitudes entering the Arctic. In contrast, blue arrows show the export of colder water from the Arctic. Shaded white shows the average area covered by sea ice (Figure 2.5).

Temperatures in the Arctic tend to rise during the day, when sunlight warms the ground, and fall at night. Arctic temperatures are warmer in summer when there is more sunlight and colder in winter when the region is dark. The climate of the Arctic is characterised by long, cold winters and short, cool summers. Average January temperatures range from about −34°C to 0°C (−29°F to +32°F), and winter temperatures can drop below −50°C (−58°F) over large parts of the Arctic.

Ocean currents bring heat from warmer regions into the Arctic Ocean. In the Atlantic Ocean, a wind called the Gulf Stream brings warm water up along the coast of North America and across the North Atlantic Ocean towards northern Europe. The Gulf Stream keeps places like Norway and the island of Svalbard much warmer than other areas at similar latitudes in the Arctic.

2.3.5.2 Wind

Winds in the Arctic can vary a lot in strength, but they are typically light. Winds tend to be stronger in the Russian Arctic, where there are more storms than in the Canadian Arctic. Currents can be strong in the Atlantic sector of the Arctic, where there are many storms. Strong temperature inversions form in winter, which slow winds near the ground. Temperature inversions occur where air at the surface is cooler than the air above. These inversions disconnect the surface air from the air above.

FIGURE 2.5 Weather patterns. (G. Holloway, Institute of Ocean Sciences, Sidney, British Columbia)

Although Arctic winds are typically soft, strong gales that can reach hurricane strength can occur and last several days. In the winter, these strong winds scour the snow from exposed areas and form large snowdrifts in sheltered areas. Strong winds increase the wind chill factor. Wind chill refers to the cooling effect of any combination of temperature and wind, expressed as the loss of body heat in watts per square metre of the skin surface.

2.3.5.3 Relative Humidity

Overall, humidity in the Arctic atmosphere is low. Colder air has a lower capacity to hold water vapour than does warm air, and in some places, Arctic air is as dry as air in the Sahara Desert. Humidity tends to be higher over the oceans and in coastal areas in summers when water evaporates from the relatively warm ocean surfaces. Humidity is lower over land areas, such as Canada, where there is less water to evaporate. In winter, humidity is very low because surface temperatures are frigid and very little water evaporates into the atmosphere. At this time of year, sea ice covers much of the Arctic Ocean, preventing ocean water evaporation. However, in areas where there is no sea ice cover, due to high water evaporation, fog formation occurs, making the ocean look as if it is steaming.

2.3.5.4 Precipitation

Throughout the Arctic, the months of summer are cloudier, when the sea ice melts away and exposes open water in the Arctic Ocean. That open water adds more moisture to the air, thus increasing cloud cover. Cloud cover is least extensive in December and January when the ice cover is thickest and temperatures are lowest. However, in the Atlantic sector, clouds are extensive year-round.

Over much of the Arctic, precipitation amounts are low. Some areas are called polar deserts and receive as little precipitation as the Sahara Desert. However, the Atlantic sector of the Arctic, between Greenland and Scandinavia, is an exception. Storms forming in the Atlantic Ocean bring moisture up into this area, especially in winter.

During the winter Arctic Ocean and the land areas experience snow fall. However, rain occurs on rare occasions during winter in the central Arctic Ocean when warm air moves into this region. Snow also falls in summer. More than half of the precipitation events at the North Pole in summer are snowfalls. In the warmer Atlantic sector, snowfall is rare in summer.

2.3.6 CLIMATIC PATTERN

The following description uses the National Centers for Environmental Prediction–National Center for Atmospheric Research (NCEP–NCAR) reanalysis dataset, which assimilates and reprocesses in situ meteorological data and satellite data to produce a comprehensive global dataset of meteorological parameters at 2.5° latitude/longitude resolution (Kalnay et al. 1996). The NCEP–NCAR dataset extends from 1948 to the present, but parameters for the high latitudes are more reliable since the incorporation of satellite-based observations in 1979 (Bromwich and Fogt 2004; Bromwich et al. 2007).

The data for the period 1989 to 2019 are reported here.

2.3.6.1   Surface Temperature (Figure 2.6)

The climatology of surface temperature variation is shown plotted for the period from 1989 to 2019 (Figure 2.6).

2.3.6.2   Vector Wind (Figure 2.7)

The climatology of vector wind variation is plotted and shown for the period from 1989 to 2019 (Figure 2.7).

2.3.6.3   Precipitation Outgoing Long wave Radiation (OLR)

Precipitation through snowfall and rainfall is an important part of the hydrological system, and it modulates the energy and water cycles of the ecosystem and thus has significant socio-economic impacts. Though caused by regional processes mainly, knowledge of precipitation, its underlying procedures, and its high temporal and spatial variations can provide important leads to global energy and water cycle studies. The freshwater input through the rain in the glaciers and ice caps and its impact on their mass balance is an important issue. Unlike the rest of the globe, polar precipitation is shallow and low in intensity, mostly dominated by solid rainfall. Small changes in the atmospheric parameters can significantly influence the precipitation process. Besides, under the immediate and most significant impacts of global warming that the Arctic is exhibiting, quantification of precipitation rate changes and its underlying procedures and precipitation characteristics are some of the particular issues to be addressed for better and more accurate prediction of the future climate (Figure 2.8).

2.3.6.4   Relative Humidity (Figure 2.9)

The climatology of relative humidity variation is plotted and shown for the period from 1989 to 2019 (Figure 2.9).

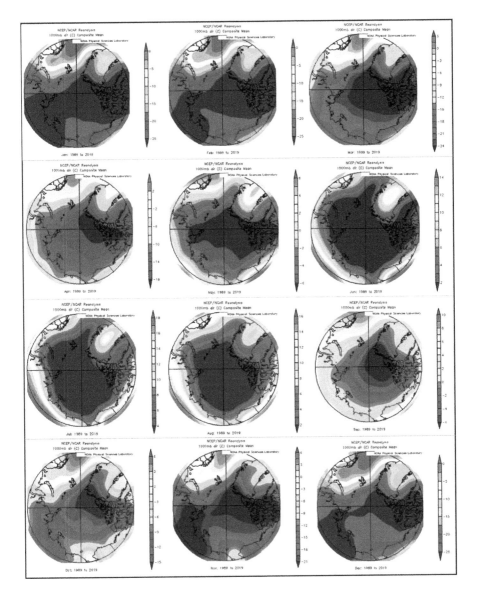

FIGURE 2.6 Surface temperature.

2.3.7 RECENT OBSERVATIONS AND TRENDS

Recent observations of state parameters at AWIPEV, Arctic Research Base Ny-Alusend for the period from 2009 to 2018 are analysed and illustrated below. The AWIPEV Arctic Research Station is located in the far *Arctic* on the *Svalbard* archipelago. This station is part of the International Research Base of Ny-Ålesund, a global scientific base that represents more than ten countries. The AWIPEV station is operated by the *French Polar Institute* and its German counterpart, the *Alfred Wegener Institute*

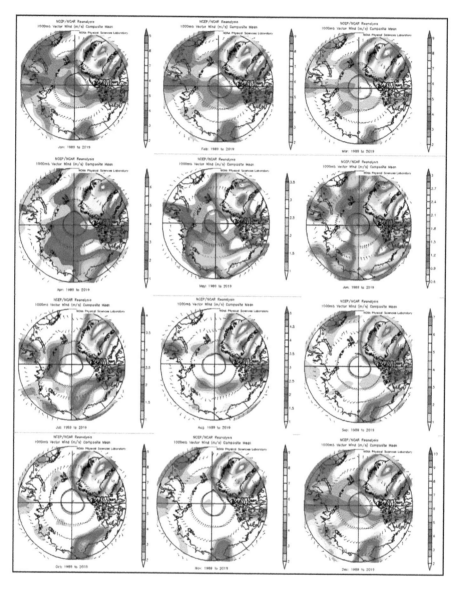

FIGURE 2.7   Vector wind.

for Polar and Marine Research. Ny-Ålesund is a functional base for international research and environmental monitoring settlements situated at 78°55′N, 11°56′ E on the west coast of Spitsbergen in the Svalbard archipelago (Figure 2.1). Twelve institutions from ten different countries have permanent research stations in Ny-Ålesund; three are staffed year-round. A former coal-mining colony, Ny-Ålesund, is a quiet

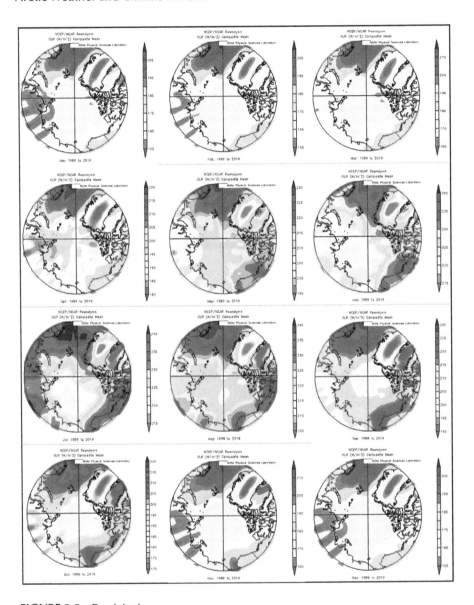

FIGURE 2.8 Precipitation.

research outpost of 30–40 people about 2100 km north of Oslo, growing to over 180 from June, when scientists from around 20 nations will arrive in its short summer. It is owned and run by the Norwegian state firm Kings Bay. The distance to the North Pole is 1231 km. The mean temperature in the coldest month (February) is −14°C, while the warmest month (July) has a mean temperature of +5°C.

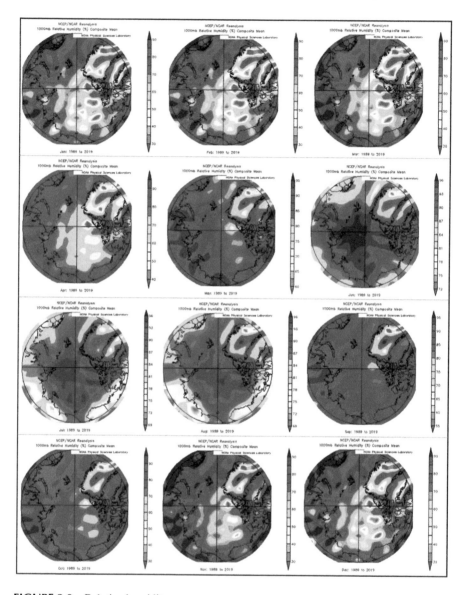

FIGURE 2.9   Relative humidity.

2.3.7.1   Surface Temperature (2009–2018)

The monthly mean variation of maximum and minimum temperatures observed at Ny-Alusend from 2009 to 2018 is shown in the graph plotted in Figure 2.10. Temperatures are lower and more negative during the winter months and higher and positive during the summer months. Maximum temperatures varied between −5°C (during winter) and 7°C (during summer), while the minimum temperatures varied between −13°C (during winter) and 5°C (during summer). From April, the temperature rises and peaks during July, and after that, it starts to fall (Figure 2.10).

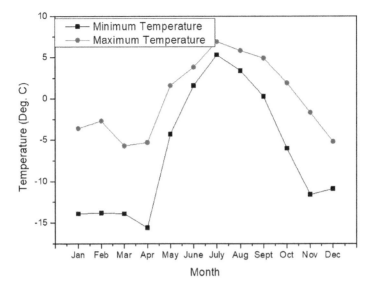

FIGURE 2.10 Monthly mean variation of surface temperature from 2009 to 2018.

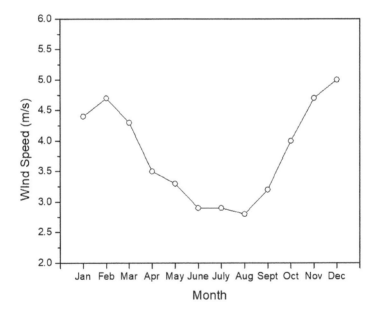

FIGURE 2.11 Monthly mean variation of surface wind speed during the period from 2009 to 2018.

2.3.7.2 Surface Wind Speed

Figure 2.11 depicts the observed wind speed variability during different months in Ny-Ålesund from 2009 to 2018. The wind speeds averaged are maximum during the winter months, while it is minimum during the summer months. These higher wind speeds during the winter months cause blizzards (Figure 2.11).

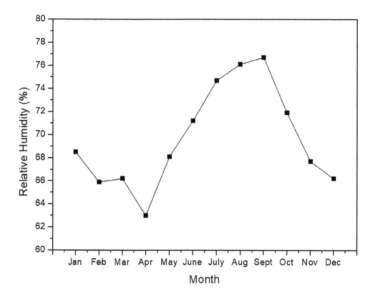

FIGURE 2.12   Monthly mean variation of relative humidity during the period 2009 to 2018.

2.3.7.3   Relative Humidity

The monthly mean variability of relative humidity at Ny-Ålesund station for 2009–2018 is less during winter months and more during summer (Figure 2.12).

2.3.8   Effects of Arctic Weather and Climate

Weather and climate patterns in the Arctic can influence weather and climate world-wide, particularly in the Northern Hemisphere.

2.3.8.1   Climate Effects

The Arctic region acts as a *heat sink* for the Earth: the Arctic loses more heat to space than it absorbs from the Sun's rays. In contrast, lower latitudes get more heat from the Sun than they lose to space. Warm air and water move into the Arctic from tropical and temperate regions. Cold air and water move from the Arctic into lower latitudes; the constant air movement is reflected in day-to-day changes in weather patterns. Over a whole year, and looking at the globe as a whole, the heat gain in lower latitudes gets balanced out, on average, by heat loss in the Polar Regions.

2.3.8.2   Weather Effects

When a winter snowstorm or a cold snap hits temperate regions, people sometimes refer to the frigid temperatures as "Arctic." Cold air moves from the Arctic into other areas, just as warm air from the south moves into Polar Regions. Storms tend to form at the boundaries between cold and warm air. Cold air moving down from the north is experienced as a *cold front*. Arctic blizzards can cause *whiteouts*, making life for people and animals difficult. Some recent studies have argued that long-term changes in Arctic sea ice and climate may impact weather patterns in other parts of the world.

2.3.8.3   Arctic Ice and Indian Monsoon

The glaciers have a significant response to the ongoing atmospheric processes and climate changes in general. These glaciers also impact the local hydrological cycle and fjord ecosystems. Arctic sea ice concentration (SIC) fluctuation has been associated with ocean circulation changes, ecology and the Northern Hemisphere climate. Prediction of sea ice melting patterns is of great societal interest, but such prognosis remains challenging because the factors controlling year-to-year sea ice variability remain unresolved. Distinct monsoon–Arctic teleconnections modulate summer Arctic SIC primarily by changing wind-forced sea ice transport. East Asian monsoon rainfall produces a northward-propagating meridional Rossby wave train extending into the Siberian Arctic. The Indian summer monsoon excites an eastward-propagating circumglobal teleconnection and the subtropical jet, reaching the North Atlantic before bifurcating into the Arctic. The slight Asian monsoon variations induce a dominant dipole sea ice melt pattern in which the North Atlantic–European Arctic contrasts with the Siberian–North American Arctic. The monsoon-related sea ice variations are complementary and comparable in magnitude to locally force Arctic oscillation variability. The monsoon–Arctic link will improve seasonal prediction of summer Arctic sea ice and possibly explain long-term sea ice trends associated with the projected increase in Asian monsoon rainfall over the next century.

2.3.9   Climate Change Impacts on the Arctic

Since the start of the satellite observations, there exists a thorough record of *sea ice extent* in the Arctic since 1979; the decreasing capacity is seen in this record (*NASA, NSIDC*), and its possible link to anthropogenic global warming has helped increase interest in the Arctic in recent years. Today's satellite instruments provide routine views of not only cloud, snow and sea ice conditions in the Arctic but also other, perhaps less-expected, variables, including surface and atmospheric temperatures, atmospheric moisture content, winds and ozone concentration.

Arctic climate change will affect local people and ecosystems, and the rest of the world because the Arctic plays a unique role in the global climate. Permafrost is melting, glaciers are receding and sea ice is disappearing. These changes will affect the rest of the world by increasing global warming further and raising sea levels. Global temperatures are expected to grow also during the 21st century. In the Arctic, this warming is expected to be substantially more significant than the worldwide average. The following changes are expected over the current century: the average annual temperatures are projected to rise by 3°C–7°C (5°F–13°F). The most significant warming occurring in the winter months' precipitation is projected to increase by roughly 20%. Sea ice is expected to decline significantly, reflecting less solar radiation and rising regional and global warming. The area of Arctic land covered by snow is expected to decrease by 10%–20%.

Arctic warming and its consequences have worldwide implications. Changes in the Arctic can influence the global climate through three primary mechanisms:

The Sun's energy reflected to space decreases as snow and ice melt, leading to a more intense surface warming. The melting of Arctic ice and increased regional precipitation can add fresh water to the oceans and potentially affect ocean currents. As warming progresses, more greenhouse gases could be released into the atmosphere by thawing the permafrost. However, warming can increase biological growth and, thus, the absorption of CO_2. By 2100, the melting of Arctic glaciers alone will have contributed to a sea-level rise of roughly 5 cm out of the projected 10–90 cm total rise for this century (IPCC AR4 Projections). Melting of the Greenland ice sheet may increase this number significantly. Access to Arctic resources is likely to be affected by climate change, including wildlife – such as whales, seals, birds and fish sold on world markets – and oil, gas and mineral reserves. Arctic ecosystem changes will impact a global scale, notably by affecting migratory species' summer breeding and feeding grounds.

2.4 INDIAN ARCTIC PROGRAMME

India is the most recent country to commence Arctic research as it established its Arctic research station in 2008. India's Arctic programme aims to contribute to the development, consolidation and dissemination of the current understanding of climate change and its impacts and adaptations in the Norwegian Arctic, Svalbard. India's Arctic research includes atmospheric, biological, marine and earth sciences and glaciological studies. The atmospheric research encompasses investigations into aerosols and precursor gases concerning their radiative, physical-chemical and optical properties and studies of space weather effects on the auroral ionosphere. Biological studies include sea ice microbial communities, and in marine research, phytoplankton pigments, nutrients, pH, DO, seawater salinity and other ecological parameters have been investigated. Earth sciences and glaciological observations include studies of snow-pack production of carbon monoxide and its diurnal variability.

The Indian scientific endeavours in the Arctic realm commenced when a five-member scientific team visited Ny-Ålesund on the Svalbard archipelago of Norway in 2007. India leased a station building at Ny-Ålesund from Kings Bay AS, which owns and manages the International Research Base facilities. The Indian station 'Himadri' was inaugurated on 1 July 2008 (Figure 2.13).

2.5 SUMMARY AND CONCLUSIONS

Once believed to be remote and pristine, the Arctic Ocean is now one of the fastest-warming regions, and the pace/magnitude of environmental change is more significant in the Arctic than at any other location on the Earth. Moreover, the Arctic Ocean and sea ice are crucial parts of the global climate machinery, influencing atmospheric and oceanographic processes and biogeochemical cycles beyond the Arctic region. Over the past several decades, there have been numerous scientific programmes mounted in the Arctic. Nevertheless, critical knowledge gaps still need to be addressed on priority, considering the fast pace of events that happen far north.

FIGURE 2.13   India's Himadri station, Arctic region.

The Arctic fjords are vital systems in the Arctic hydrographical network and serve as pulse points to measure environmental change's cause and effect, may it be fuelled by local disturbances or global processes.

REFERENCES

J. C. Fyfe, K. von Salzen, N. P. Gillett, V. K. Arora, G. M. Flato, and J. R. McConnell, One hundred years of Arctic surface temperature variation due to anthropogenic influence, *Nature Scientific Reports* 3, 2013, 2645.

D.-Y. Gong, S.-W. Wang, and J.-H. Zhu, East Asian winter monsoon and Arctic Oscillation, *Geophysical Research Letters* 28, 2001, 2073–2076, doi: 10.1029/2000GL012311.C.

G. Grunseich and B. Wang, Arctic sea ice patterns driven by the Asian summer monsoon, *Journal of Climate* 29, 2016, 9097–9112.

L. C. Hamilton and M. Lemcke-Stampone, Arctic warming and your weather: public belief in the connection, *International Journal of Climatology* 34, 2014, 1723–1728, doi: 10.1002/joc.3796.

IPCC, *Climate Change 2007: The Physical Science Basis. Contribution of Working Group I to the Fourth Assessment Report of the Intergovernmental Panel on Climate Change*, S. Solomon, D. Qin, M. Manning, Z. Chen, M. Marquis, K. B. Averyt, M. Tignor, and H. L. Miller, eds., Cambridge University Press, Cambridge, UK and New York, NY, 2007, p. 996.

E. Kalnay, M. Kanamitsu, R. Kistler, W. Collins, D. Deaven, L. S. Gandin, M. Iredell, S. Saha, G. White, J. Woollen, Y. Zhu, M. Chelliah, W. Ebisuzaki, W. Higgins, J. Janowiak, C. Kingste, C. Ropelewski, J. Wang, and A. Leetmaa, The NMC/NCAR 40-year reanalysis project. *Bulletin of the American Meteorological Society* 77, 1996.

National Snow and Ice Data Center. All About Arctic Climatology and Meteorology https:// nsidc.org/cryosphere/arctic-meteorology/.

J. E. Overland and M. Wang, Arctic-midlatitude weather linkages in North America, *Polar Science* 16, 2018, 1–9.

J. E. Overland and M. Wang, Large-scale atmospheric circulation changes are associated with the recent loss of Arctic sea ice, *Tellus* 62A, 2010, 1–9, doi: 10.1111/j.1600-0870. 2009.00421.x.

J. Overland, J. A. Francis, R. Hall, E. Hanna, S. J. Kim, and T. Vihma, The melting arctic and midlatitude weather patterns: are they connected? *Journal of Climate* 28(20), 7917–7932. doi: 10.1175/JCLI-D-14-00822.1.

D. W. Thompson and J. M. Wallace, Regional climate impacts of the Northern Hemisphere annular mode, *Science* 293, 2001, 85–89, doi: 10.1126/science.1058958.

T. Vihma, Effects of arctic sea ice decline on weather and climate: a review, *Surveys in Geophysics* 35, 2014, 1175–1214.

3 Investigation of GPS-Derived Total Electron Content (TEC) and Scintillation Index for Indian Arctic and Antarctic Stations

A. K. Gwal and Suryanshu Choudhary
Rabindranath Tagore University

Purushottam Bhawre
Govt. Sanjay Gandhi P.G. College

CONTENTS

3.1 INTRODUCTION

The polar regions area provides a unique perspective and serves as an open natural laboratory for ionospheric study. Users of satellite broadcasts are interested in ionospheric investigations in the Polar regions. Every 30 seconds, data on ionospheric Total Electron Content (TEC) are recorded. In the polar region, the high latitude is directly affected by the energy of charged solar particles and energy; thus, the ionosphere becomes significantly fluctuating. The TEC fluctuations over the polar region depend on various factors, especially the ionisation process in high-latitude regions; the ionisation processes depend on the sun's activities and the sun's zenith angle. TEC was variable with time and season in several particular studies that demonstrated that the TEC strongly depends on Solar Activity (Da Rosa et al. 1973; Soicher 1988;

DOI: 10.1201/9781003265177-3

Van Velthoven 1990; Feitcher and Leitinger 1997; Karnowski et al. 2006). The ionosphere is a nonlinear medium in which RF signals are refracted by the amplitude as well as, depend on the signal frequency, electron density in regions of irregularities on a micro level. Sometimes, the electron density is rapid, and random phase variations can produce phase irregularities in the emerging wave-front, referred to as phase scintillations. Diffraction of signals interference of cross wave-front also leads to variations in signal amplitude referred to as amplitude scintillation (or amplitude fading for degradations in signal strength). These effects are strongest in equator, auroral and polar cap regions. High-latitude auroral irregularities are formed from the precipitation of energetic electrons along the terrestrial magnetic field line into the high-latitude ionosphere. These electrons are energised through a complex interaction between the solar wind and the earth magnetic fields, resulting in optical and UV emissions commonly known as the auroras. This phenomenon characterised the magnetosphere substorm, where associated irregularities in electron density lead to scintillations (Basler et al. 1962; Lansinger et al. 1967; Whitney et al. 1969; Kersley et al. 1995; Aarons 1982). At high latitudes, scintillations are found to be associated with large-scale plasma structures. Experimentally, the two states of the polar ionosphere controlled by the interplanetary magnetic field (IMF) and their association with high-latitude large-scale plasma structures known as patches, blobs and sun-aligned arcs have been discovered since the 1980s (Weber et al. 1984, 1986; Tsunoda 1988; Basu and Valladares 1999). It is well known that ionospheric scintillation is produced by electron density irregularities in the ionosphere, which becomes highly disturbed at times. A radio wave crossing is drifting ionospheric irregularities that suffer a distortion of phase and amplitude. The magnitude of fluctuations varies with the frequency used, magnetic and solar activity conditions, and the day, season and location. These effects are called amplitude and phase scintillation (Biago Forte et al. 2002). Severe amplitude fading and strong phase scintillation affect the reliability of GPS navigational systems and satellite communications. Therefore, it is desirable to obtain a further understanding of ionospheric scintillation and its effects on GPS using a receiver capable of performing such conditions. Ionospheric scintillation is produced by ionospheric irregularity, which affects GPS signals by two factors, namely refraction and diffraction. Both types of effects originate in the group delay and phase advance that a GPS signal experiences as it interacts with free electrons along its transmission path. This chapter describes amplitude scintillation measurements over Indian Antarctica Station Matrix (70°.65 S, 11°.45°E) as part of International Polar Year (IPY) using NovAtel dual-frequency GPS receiver. Stresses on the response of ionospheric TEC during the end of solar activities period over the Arctic and Antarctic GPS stations were used to analyse monthly, seasonal and polar day and polar night TEC behaviours. Nowadays, GPS measurements are commonly used to investigate the structure and dynamics of the ionosphere. The measurement systems are based on the NovAtel GPS, which were stationed at Indian Base Station Maitri, Antarctica. The system was installed in January 2008 during the 27th Indian Antarctic Scientific Expedition. It consists of a 24-channel, high-precision, dual-frequency GPS receiver, GPS antenna modal 702 or 701, a low-noise amplifier (LNA) that boosts the power of the incoming signal to compensate for the line loss between the antenna and the receiver, RF Antenna Cable, 12 V Power Adapter Cable, Null

Modem Data Cable and Data Communications Equipment. The GPS receiver tracks GPS signals at a 1-second sampling rate, and a cut-off elevation angle was set to 400 to eliminate the multipath effect of GPS data. A standard of 30-second data sampling was executed to reduce processing time. The instrumental delay biases observable GPS signals therefore, it is necessary to remove these biases to estimate TEC accurately. The absolute TEC determination can remove both the receiver and the satellite's instrumental biases. Potential errors due to instrument time delay are corrected using the code biases obtained from the Indian base stations Himadri, Arctic and Maitri, Antarctica. GPS-TEC data have been processed for Indian permanent station Maitri, Antarctica, during 2008. The study is divided into three parts:

I. TEC Monthly Behaviour:

The observation for the monthly variation of TEC is based on 12 months of GPS data. In January, TEC fluctuated between 10 and 22 TECU; in February, 7 and 22 TECU; in March, 6 and 20 TECU. In the April month minimum TEC observed 6 TECU and a maximum of 15 TECU observed. At the starting of polar night in May, a minimum of 7 TECU and a maximum of 15 TECU were observed; in the dark month of June, a minimum of 4 TECU and a maximum of 15 TECU were observed; in July, a minimum of 3 TECU and a maximum of 16 TECU were observed; in the start of August when the sun is active, a minimum of 3 TECU and a maximum of 19 TECU were observed; in the spring of Antarctica between September and October, a minimum of 3 and 7 TECU and a maximum of 22 and 22 TECU were observed. Again in the summer months November and December, TEC variation observed between a minimum of 6 and 8 TECU and a maximum of 27 and 26 TECU. This type of behaviour of TEC in the polar region depends on solar zenith angles. Figure 3.1 clearly shows TEC behaviours in all months. TEC pick shift pattern has been noticed in every month from January to May, TEC peak sifted right side, between June and July TEC peak, practically observed overlapping. Still, the sun rising month of August through the peak summer month of December TEC peak is recorded on the left side. The solar zenith angle influences that variety of peak shifting pattern.

II. TEC Seasonal Variation

Based on observation, seasonal variations of ionospheric TEC over Antarctica can be divided into three seasons: first is the summer season (November, December, January and February), second is the winter season (May, June, July and August) and third is the equinox period including two seasons according to solar activities, namely, autumnal equinox (March, April) and vernal equinox (September and October) seasons. In the summer period, TEC monthly median value fluctuates in the range of 11–20 TECU; this type of TEC performance in the summer period is caused by the presence of 24-hours solar activity in the polar region. In winter, monthly TEC median value drops and fluctuates between 8 and 14 TECU. During the winter period, solar activities are negligible as compared to the summer period. The equinox period's study is divided into two parts. First is

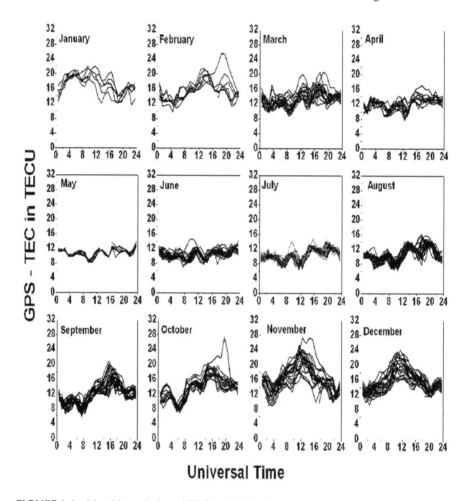

FIGURE 3.1   Monthly variation of TEC at Maitri, Antarctica.

the autumnal equinox from March to April, and the second is the vernal equinox from September to October. In the autumn equinox, monthly TEC median value varies between 8 and 16 TECU because of a partial decrease in solar activities. In vernal equinox period, TEC fluctuates between 8 and 18 TECU due to partial increase in solar activities. Figure 3.2 shows the seasonal variations of TEC over Maitri, Antarctica.

III. Mid-Polar Day and Mid-Polar Night Observation

During the mid-polar night (21 June 2008), it is Antarctic night month, total solar activity absences and TEC variation is significantly less, TEC behaviour is exactly inverse during in the mid-polar day (21 December 2008).This month has sunny days and hence high solar activities are present. Therefore, TEC fluctuation is high compared to polar night. Figure 3.3 shows mid-day and mid-polar night variations.

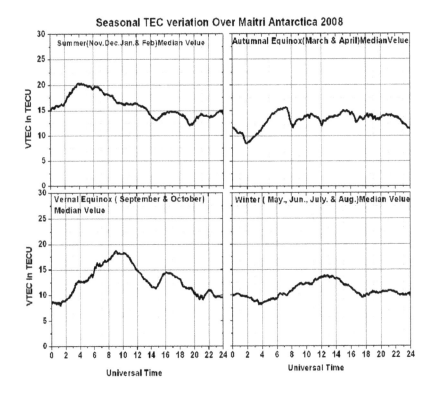

FIGURE 3.2   Seasonal variations in TEC at Maitri, Antarctica.

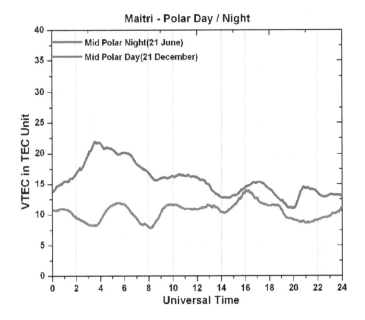

FIGURE 3.3   TEC variations during mid-polar day and night.

3.2 AMPLITUDE SCINTILLATION

Amplitude scintillation is defined by the S4 index derived from detruded intensities of signals received from satellites. The S4 index is computed over 60-second intervals and stored in the Ionospheric Scintillation Monitor Receiver data log along with the phase measurements. The amplitude measurements are filtered using a Low Pass Filter (LPF), and the effects of ambient noise are removed from the S4T. This is achieved by estimating the average signal-to-noise ratio over the 60-second interval. The 60-second estimates are then determined by the expected S4 correction (or S4No) due to ambient noise. This average signal-to-noise ratio (S/No) is feasible because the amplitude scintillation fades did not significantly alter the S/No. Scintillation morphology is described in terms of percentage occurrence in specified threshold level according to intensity and differential phase, respectively, of S4-Index when S4 is more significant than 0.1, 0.3 and 0.5. The study is divided into hourly, monthly and seasonal variations of S4-Index observed in Maitri's GPS signals. Figure 3.4 shows

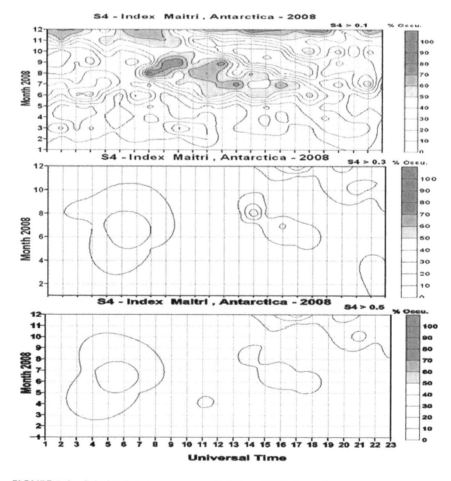

FIGURE 3.4   Scintillation occurrences with different S4 index values.

scintillation occurrences with different S4 index values, i.e. S4 is more significant than 0.1, 0.3 and 0.5. When S4 is more significant than 0.1, this type of slight disturbance is observed most of the time throughout the year during the observation period since the polar ionosphere was disturbed most of the time due to solar and magnetic activity disturbances (Basu et al. 1988). When S4 is more significant than 0.3, the maximum percentage occurrence of 30% is observed in the morning hours (~0400 to 0700UT) of winter months and afternoon time (~1300 to 1800 UT) of summer months. Further, when scintillation index S4 is more significant than 0.5, maximum occurrences up to 28% are observed in the morning hours (0500 to 0700 UT) and the evening hours (1600 to 1800 UT) winter and summer months, respectively.

Figure 3.5 shows the hourly scintillation activity in different seasons at different levels of the S4 index during the summer months (November, December, January and February) when 24-hour sunlight is present in the polar region, and during the day,

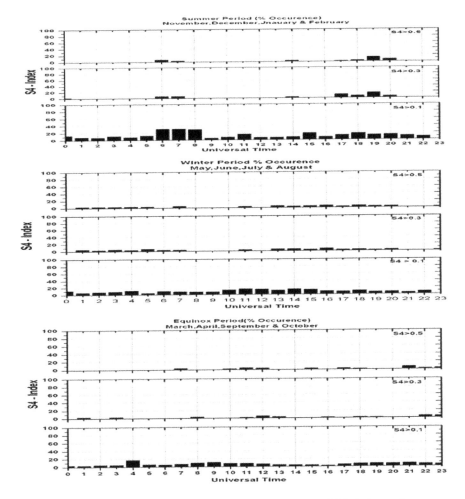

FIGURE 3.5   Hourly scintillation of S4 index during summer at Maitri, Antarctica.

scintillation occurrence is much lower (Aarons et al. 1981). The percentage occurrences of ~30%, 18% and 12% are observed when S4 is more significant than 0.1, 0.3 and 0.5. During the winter months (May, June, July, and August), when there is no daylight (polar nights), the percentage occurrences of ~18% are observed when S4 is more significant than 0.1, whereas at higher S4 is less than 10%. During equinoctial months (March, April, August and September), the occurrence of scintillations is about 19%, 8% and 5%, respectively, with S4 is more significant than 0.1, 0.3 and 0.5 in the morning as well as evening hours.

Overall, from the above observations, it is noted that mostly weak scintillations are observed during the period of low solar activity. GPS data collected from Indian Antarctic Station Maitri during low solar activity conditions show that at high-latitude station, only weak scintillations (S4 > 0.1) are observed both during the day and night hours due to low solar activity conditions. Aarons et al. (1981) reported that maximum scintillation activity is observed at high latitudes in the winter season compared to summer and equinox. In the winter season, when the sun goes down and total solar activity is closed during the night-time in a polar region, maximum scintillation occurs at F-region heights and minimum scintillation occurrence appears in full sunlight months. During the first 2 months of the equinox (March and April), when solar activities decrease, and then the last 2 months (August and September), once the sun rises again solar activities start to increase, as well as scintillation has been visible in the night sector, considerable daytime scintillation has been further observed both during day and night (Rino and Matthews 1980). In general, at high latitude, scintillations occur over night-side auroral oval and the polar cap at all local times. In the winter, polar cap moderate to strong L-band scintillation is observed in association with the so-called contrasting cap patches (Weber et al. 1984). When the IMF is directed southwards, patches with high ionisation density are observed to enter the polar cap from the dayside auroral oval, convected in the anti-sunward direction and eventually exit into the night-side auroral oval. One mechanism by which patch formation is achieved corresponds to the changing plasma convection pattern in response to the IMF component during southward Bz periods (Sojka et al. 1993). One of the formation processes considered the function of giant plasma flows in forming discrete patches has been experimentally substantiated (Rodger et al. 1994; Valladares et al. 1994). The results from the observation of monthly occurrence of L-band scintillation show that more irregularities occurred in the southern summer months from June to December.

3.3   HIMADRI, NY-ÅLESUND, ARCTIC REGION

The TEC is observed during October and November at Himadri, Ny-Ålesund, Arctic Region during the polar night in the North Pole region. Maximum TEC value is observed in October, and minimum TEC value is observed in November. In equinox month, the maximum TEC value fluctuates as compared to polar night month. Figure 3.6 shows the variation in the TEC values at Himadri, Arctic stations. Observed amplitude scintillation is nearly 0.3 in November, but in October, the value of amplitude is 2.5. Figure 3.7 shows the scintillation variations in both months at Himadri, Arctic stations.

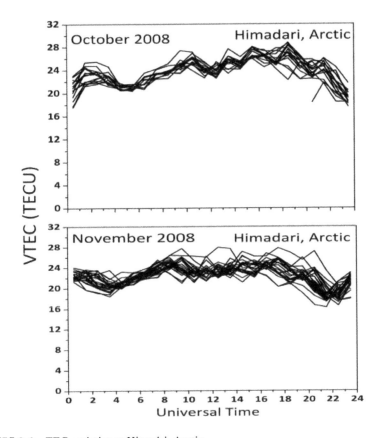

FIGURE 3.6   TEC variation at Himadri, Arctic.

3.4   CONCLUSIONS

One-year GPS data demonstrate a strong correlation between GPS-TEC and its reliance on solar activity. TEC value gets too high during monthly observation in January as compared to June month; TEC value becomes excessively high again in December. Solar activity is highest in January and December compared to June. The seasonal research of TEC reveals that entire solar activities cease and no ionisation processes occur during this season, resulting in a minimal TEC value. The entire solar activities are present during the summer season, which is related to the occurrence of the ionisation process in 24-hour sunshine, therefore the TEC value grows to a maximum. Our previous study of the equinox season is divided into two parts: in the first equinox period (March and April), solar activities are low due to the solar zenith angle, which is the sunset time. In the second equinox period (September and October), the solar zenith angle goes from maximum to minimum, and solar activities start again. The ionisation process progresses, and during this period, TEC variation is high compared to March and April. Ionospheric TEC variation depends on solar activities. Solar zenith angle and TEC variation are high in solar activities, and in the absence of solar activity, minimum solar zenith angles and TEC variation are low,

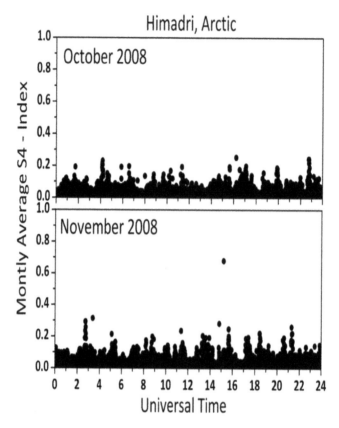

FIGURE 3.7   Monthly average scintillation at Himadri station, Arctic.

whereas maximum solar zenith angles and TEC variation are high. The present study describes scintillation activity as observed during the low solar activity year of 2008 at Maitri, Antarctica, as a part of IPY. The study can be summarised as follows. Weak scintillations (S4 > 0.1) were observed all the 24 hours of the day in almost all the seasons, and during the morning and afternoon hours of the minimum solar period of 2008, slightly higher magnitude scintillations (S4 < 0.5) were observed. Season-wise maximum occurrence was noted during summer months, whereas in the winter and equinox months, scintillations were observed mainly in the early morning and night hours. In terms of month-to-month incidence, the months of June and December are the most common. Scintillation activity is beginning to increase at the beginning of the month this year, commencing in June when the polar night begins. During August and September, the sun rises with minimum solar activity, so the ionisation process starts again, but after October, scintillation activity decreases to a minimum. The present work further confirm the earlier observations at high latitudes reported by various other workers. In Himadri, Arctic region, maximum TEC variation was observed at equinox month compared to polar night months, and maximum amplitude scintillation activities are observed at polar night month compared to equinox months.

ACKNOWLEDGEMENTS

I want to give my sincere thanks to National Center for Antarctic and Ocean Research (NCOAR), Goa, under the Ministry of Earth Sciences, Govt. of India and Maitri's staff for their scientific and logistic supports to research Arctic and Antarctica regions.

REFERENCES

Aarons, J., 1982. Global morphology of ionospheric scintillations. *Proc. IEEE* 70(4), 360–378.

Aarons, J., Mullen, J. P., Whitney, H. E., 1981. UHF scintillation activity over polar latitudes. *Geophys. Res. Lett.* 8(3), 277–280.

Basler, R. P., DeWitt, R. N., 1962. The height of ionospheric irregularities in the auroral zone. *J. Geophys. Res.* 67, 587–593.

Basu, S., Basu, S., Weber, E. J., Coley, W. R., 1988. Case study of polar cap scintillation modelling using DE-2 irregularity measurements at 800 km. *Radio Sci.* 23, 545.

Basu, S., Valladares, C., 1999. Global aspects of plasma structures. *J. Atmos. Sol. Terr. Phys.* 61, 127–139.

Da Rasa, A. V., Waldman, H., Bendito, J., Garriott, O. K., 1973. The response of the ionospheric electron content to fluctuations in solar activity. *J. Atmos. Terr. Phys.* 35, 1429–1442.

Feichter, E., Leitinger, R., 1997. A 22-year cycle in the F-layer ionisation of the ionosphere. *Ann. Geophys.* 15, 1015–1027.

Forte, B., Radicella, S. M., 2002. A different approach to the analysis of GPS scintillation data. *Ann. Geophys.* 45(3–4), 551–561.

Kersley, L., Russell, C. D., Rice, D. L., 1995. Phase scintillations and irregularities in the northern polar ionosphere. *Radio Sci.* 30, 619–629.

Krankowski, A., Shagimuratov, I., 2006. Impact of TEC Fluctuations in the Antarctic ionosphere on GPS positioning oczapowski St. 1,10–957. *Russ Artif. Satell.* 41(1), doi: 10.2478/V10018-007-0005-5.

Lansinger, J. M., Fremeuw, E. J., 1967. The scale size of scintillation-producing irregularities in the auroral ionosphere. *J. Atmos. Terr. Phys.* 29, 1229–1242.

Rino, C. L., Matthews, S. J., 1980. On the morphology of auroral zone radio wave scintillation. *J. Geophys. Res.* 85, 4139.

Rodger, A. S., Pinnock, M., Dudney, J. R., Baker, K. B., Greenwald, R. A., 1994. A new mechanism for polar patch formation. *J. Geophys. Res.* 99, 6425–6436.

Soicher, H., 1988. Travelling ionospheric disturbances (TIDs) at mid-latitude: solar cycle phase dependence. *Radio Sci.* 23, 283–291.

Sojka, J. J., Bowline, M. D., Schunk, R. W., Decker, D. T., Valladares, C. E., Sheehan, R., Anderson, D. N., Heelis, R. A., 1993. Modelling polar cap F-region patches using time-varying convection. *Geophys. Res. Lett.* 20(17), 1783–1786.

Tsunoda, R. T., 1988. High-latitude F-region irregularities: a review and synthesis. *Rev. Geophys.* 26, 719–760.

Valladares, C. E., Basu, S., Buchau, J., Friis-Christiansen, E., 1994. Experimental evidence for the formation and entry of patches into the polar cap. *Radio Sci.* 29, 167–194.

Van Velthoven, P. J., 1990. Medium-scale irregularities in the ionospheric electron content, Ph.D. Thesis, Technische Universiteit Eindhoven.

Weber, E. J., Buchau, J., Moore, J. G., Sharber, J. R., Livingston, R. C., Winningham, J. D., Reinisch, B. W., 1984. F-layer ionisation patches in the polar cap. *J. Geophys. Res.* 89, 1683–1694.

Weber, E. J., Klobuchar, J. A., Buchau, J., Carlson, H. C., Livingston, R. C., de la Beaujardiere, O., McCreday, M., Moore, J. G., Bishop, J. G., 1986. Polar cap F-layer patches: structure and dynamics. *J. Geophys. Res.* 91, 2121–2129.

Whitney, H. E., Aarons, J., Malik, C., 1969. A proposed index for measuring ionospheric scintillations. *Planet. Space Sci.* 17, 1069–1073.

4 Multi-Year Measurements of Black Carbon Aerosols and Solar Radiation over Himadri, Ny-Ålesund

Effects on Arctic Climate

S. M. Sonbawne, G. Meena, S. K. Saha,
G. Pandithurai, and P. D. Safai
Indian Institute of Tropical Meteorology

P. C. S. Devara
Amity University Haryana (AUH)

CONTENTS

DOI: 10.1201/9781003265177-4

4.1 INTRODUCTION

In the present industrialised world, the atmosphere is getting more and more polluted, particularly, the Arctic environment is changing very rapidly. Both natural and anthropogenic activities from low-latitude regions are found to contribute more pollution to the Arctic region through long-range and trans-boundary transport of human-induced activities combined with local feedback mechanisms (e.g., Stohl et al., 2006; Hirdman et al., 2009). When fossil fuels, forest fire, industrial combustion, and other organic material burn, they release soot particles which occur at high latitudes (Quinn et al., 2011; Stohl et al., 2013). Soot particle, which majorly comprises BC aerosol that absorbs solar radiation, increase the atmospheric temperature, leading to global warming. Emission sources of BC in the Arctic have been studied and found to contribute more from long-range transport via the Soviet Union, Europe, North America and East Asia (Sharma et al., 2013). Thus, this topic has become a challenge to the scientific community while assessing BC aerosol impacts on radiation budget due to significant variations in physicochemical and optical properties and loading. BC is mostly known to warm the atmosphere (e.g., Ramanathan and Carmichael, 2008) through its dominant absorption and scattering solar radiation properties. As a result, it attenuates the ground-reaching solar radiation and hence reduces the surface albedo. Lowering of surface albedo could alter the trend towards increasing the glacier's melting rate during the summer season. In the 20th century, it was estimated that about 20% of the warming and snow/ice cover melting is due to change in albedo caused by the BC effect alone in the Arctic (Koch et al., 2011).

The Arctic fresh snow has the highest albedo (as high as 0.9) of the natural substance, but this further reduces the albedo due to BC deposition. This may lead to warming in the Arctic region if it darkens the snow surfaces, while it has a cooling effect if it is located in the upper troposphere. With our present knowledge about Arctic aerosol from ground-based observational data, the aerosol mass concentration is found to be high during late winter and early spring, together with a greater concentration of accumulation mode size particles, referred to as 'Arctic haze', and the lowest concentration during Arctic summer with dominant particles of around 100 nm size (e.g., Croft et al., 2016).

Realising the complex contribution of carbonaceous aerosols to climate change, more research work is being carried out worldwide for better accounting of black carbon (BC) aerosols from the Arctic to the global polar climate. Measurements carried out at Barrow, Alert and Ny-Ålesund stations over a decade have resulted that measured BC concentrations are susceptible to emissions within the Arctic region (Hirdman et al., 2010, Spackman et al., 2010; Granier et al., 2006; Corbett et al., 2010; Lee et al., 2010; De Angelo et al., 2011; Quinn et al., 2011). Koch et al. (2009) reported that BC is located mainly in the Arctic free troposphere with a positive anomaly in the lower troposphere. Studies suggested that in the Ny-Ålesund region, the BC concentration has increased by 45% (Eckhardt et al., 2013). An annual mean BC concentration of 26–49 ng/m^3 has been reported by Hirdman et al. (2010) at Alert and Zeppelin stations, which is higher than the Barrow and Summit values. Recent studies at Ny-Ålesund station by Gogoi et al. (2016) reported the climatologically seasonal BC mean mass loading of $\sim$50.3 $\pm$ 19.5 ng/m^3. The value is approximately three-folds of summer mass loading $\sim 19.5 \pm 6.5$ ng/m^3, which is higher than the lowest BC concentration. Hansen et al. (2004) suggested that BC would produce warming about three times more than an equal forcing of CO_2.

The impact of BC on Arctic air mass has been inferred from back-trajectory analysis (e.g., Raju et al., 2015; Sonbawne et. al., 2021). Studies show that eastern and southern Asia contribute more (Ikeda et al., 2017), while Stohl et al. (2006) and Huang et al. (2010b) found BC as the dominant source of climate change over the Arctic region from the European side. The derivation of aerosol size index (alpha) from a spectral variation of aerosol optical depth (AOD) observations using sun photometers and its relationship with BC air mass concentration has been reported by Rahul et al. (2014) and Safai et al. (2014). BC aerosols are highly variable in space and time in the Arctic environment due to changing circulation and emission patterns that determine the transport pathways. A combination of surface and air-borne observations provides valuable information on the characterisation of aerosols and their climate impacts. Much of our present data on Arctic aerosol originate from ground-based station observations. For example, seasonal cycles of AOD are pronounced and vary from space to space due to changes in emissions, aerosol composition and circulation patterns that affect transport and deposition rates along pathways; Arctic haze and dust are most common in spring, while biomass and forest fire smoke aerosols contribute more to AOD from late spring through autumn (Stone et al., 2014). During the summer months, when the wind blows from ocean to Ny-Ålesund settlement, the particle concentrations increase with wind speed, which is not much observed from other direction (Deshpande et al., 2014). The yearly cycle of aerosols mainly affects the seasonal transport and removal mechanisms, some of which remain poorly constrained (e.g., Arnold et al., 2016).

In the present study, we report and discuss the BC aerosol loading, sources, temporal variations and long-range transport processes observed during 2011–2014 at the Indian Arctic Station, Himadri, Ny-Ålesund (78.9°N, 11.9°E, 8 m a.m.s.l). The results reveal an increase in BC concentration in 2012 as compared to other years under study. In the light of limited studies of the Arctic in India and more so over the globe, the present results constitute an essential piece of work that is essential to a better understanding of the Arctic role in the global climate.

4.2 MEASUREMENTS OF SURFACE BC AEROSOL

The BC aerosol observations have been carried out during 2010–2014 at 'Indian Arctic Station, Himadri', located in the northernmost part of the world at Ny-Ålesund, Norway (78.9°N, 11.9°E, 8 m a.m.s.l), as shown in Figure 4.1, which is surrounded by the biomass of Arctic tundra at the Kongsfjorden and close to the Zeppelin Mountain on the island of Spitsbergen in the Svalbard archipelago. The study over this location allows measuring the pristine, clean Arctic atmosphere with low background values. It experiences various air mass, involving a mixture of atmospheric aerosol types, originating from multiple sources and associated particle transformation and transport.

Continuous observations have been carried out in the Arctic reason using the ground-based optical techniques to investigate the optical-physical characteristics of

FIGURE 4.1   Aerial view of Ny-Ålesund and Indian Station, Himadri. The red open circle in the picture portrays the experimental location.

air-borne atmospheric aerosol particles. Thus, these techniques are commonly used to study the BC mass concentration, which is a crucial element in aerosol–climate interaction research (Dubovik et al., 2002; Markowicz et al., 2008; Mazzola et al., 2012; Zielinski et al., 2004, 2012). These studies have provided valuable datasets and knowledge about the temporal variations, the transformation process of air mass characteristics and information about their Arctic aerosols' trends during the study periods.

4.3   INSTRUMENT, DATA ACQUISITION AND ANALYSIS METHODOLOGY

4.3.1   Aethalometer Measurement of Black Carbon

Aethalometer is a widely used instrument for continuous observation of ambient BC mass concentration in the diverse environment over the globe with a typical uncertainty of <5% (Babu et al., 2011). Two sets of seven-channel Aethalometer (Model AE-42 and AE-30, M/s Magee Scientific Co., USA) were continuously operated during the observations. The Aethalometer AE-42 was used at the Himadri station inside the Ny-Ålesund town. The other Aethalometer AE-30 was employed at Gruvebadet observatory (78.9°N, 11.9°E), situated 1.2 km away at the southern side of the city foothill of Zeppelin Mountain. The AE-42 observations represent more local anthropogenic activity, whereas the AE-30 is away from the local anthropogenic activity. . AE-42 and AE-30 were operated at sample airflow rate of 3 and 5 LPM with time interval of 5 and 30 minutes, respectively. Both Aethalometers were operated at 3 m height above ground. These differences in measuring time interval were put purposely as the AE-42 was operated in town where the local activity is more, as it was restricted to the AE-30 observational site.

Aethalometer provides the fully automatic and unattended operation at seven discrete wavelengths from 370 to 950 nm. The instrument operation has been discussed elsewhere (Sumit et al., 2011). To correct the loading effect, the correction algorithm as given by Virkkula et al. (2007) and the 16.6 m^2/gm mass absorption coefficient were used to obtain the correct BC mass concentration. More details about the measurement method and correction analysis can be found in Safai et al. (2013).

4.3.2 Spectrometer Measurements

The instrument employed in the present study is a portable UV-VIS-NIR spectrometer (Ocean Optics HR2000) with a focal length of 101.6 mm at f/4. The HR2000 spectrometer provides a short power (about 0.5 w). Connected to fibre optic (light guide), light falls on a grating range of 200–1100 nm having an input slit of 50 μm and a spectral resolution of 1 nm (FWHM). The detector is a linear CCD array with 2048 pixels (each 14×200 μm). Spectrometer operating software controls the measurement and stores the spectral data. The instrument's dark current is measured and subtracted from each spectrum according to the corresponding average exposure time. Depending on the intensity of the received scattered sunlight, the exposure time is adjusted automatically to maximise the total signal. The number of accumulations comprising a spectrum also varies to restrict the average time interval between two ranges to about 2 minutes.

Further details covering data acquisition and method of analysis can be found in Meena et al. (2009). Ground-based observations of zenith-sky scattered sunlight have been performed at Ny-Ålesund, Arctic (78.9°N, 11.9°E), in October 2013. The spectra are recorded during the evening hours of 6:10–7:20 pm (local time) in the spectral range of 200–1100 nm.

Figure 4.2 represents zenith-sky scattered light spectra taken at various solar zenith angles; these data were analysed using differential optical absorption spectroscopy (DOAS) technique in the spectral region of 462–498 nm. Trace gas concentration (n) is derived from Lambert–Beer's law: $I = I_o \exp(-\sigma n l)$, where I is the measured spectrum intensity (i.e. zenith-sky scattered light intensity) at the ground,

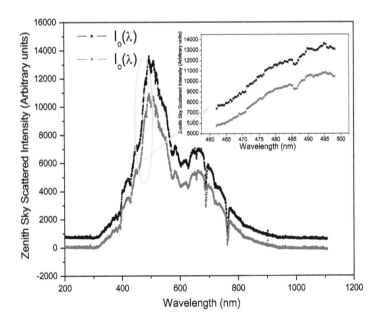

FIGURE 4.2 Zenith-sky scattered light spectra for different solar zenith angles in the 200–1100 nm spectral range.

I_o is the reference spectrum intensity, l is the optical path length (cm), σ is the absorption cross-section (cm^2/molecule) of the absorbing molecule and n is the absorber concentration (molecules/cm^3).

The absorbance concentration is derived from extracted Differential Optical Density (DOD) $= [-ln(I/I_o)] = \sigma'n$, where σ' is the differential absorption cross-section, and 'nl' is the differential slant column density (SCD$_{diff}$) of the absorber (molecules/cm^2), which is derived by SCD$_{diff}$=DOD /σ' (Solomon et al., 1987, Pfeilsticker *et al.*, 1997).

4.4 RESULTS AND DISCUSSION

4.4.1 DIURNAL VARIATION OF BC AEROSOLS

Figure 4.3 depicts the mean diurnal BC mass concentration at Himadri during the period from 2011 to 2014. It can be seen that, on an average, values in each year show a bell-shaped variation with mean BC ranging between 24 and 38 ng/m^3 for the entire period of observations. The diurnal variation in BC might be due to the variation in emissions from local sources, such as ship emission, power plant, vehicular movement, and long-range transport from the neighbouring countries. This variation is also attributed to the changes in surface temperature as 24 hours daylight is present over this location. Similarly, the diurnal pattern is observed between 2011 and 2014, with higher BC values in 2012, but the data that went into averaging are minor. In 2013, the BC concentration was found as low as 24 ng/m^3 compared to previous years.

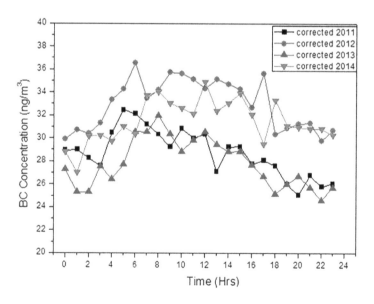

FIGURE 4.3   Mean diurnal variations in BC mass concentration during 2011–2014.

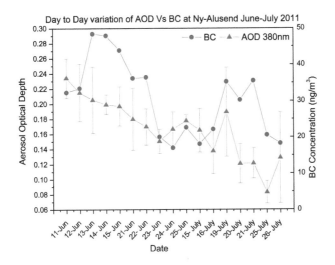

FIGURE 4.4 Daily variation of BC concentration vs $AOD_{380\ nm}$.

4.4.2 ASSOCIATION BETWEEN SYNCHRONOUS BC MASS CONCENTRATION AND $AOD_{380\ NM}$

One of our earlier publications (Safai et al., 2014) revealed the dominant presence of fine-mode aerosol particles in forming carbonaceous aerosols over a tropical urban station. To ascertain this aspect over polar latitudes (Ny-Ålesund in this case), simultaneously measured BC concentration and AOD_{380nm} have been examined in Figure 4.4. The figure shows this comparison for the data recorded over Ny-Ålesund during the summer of 2011. The co-variation is seen as promising in 2011 when the circulation changes during summer over the Arctic are quite different from tropical urban station.

4.4.3 COMPARISON BETWEEN UNCORRECTED AND CORRECTED BC VALUES

The experimentally observed BC values have been corrected for the filter loading effect using the 'K' factor by following the work reported by Seung et al. (2010). The resultant fixed and uncorrected BC concentrations have been compared in Figure 4.5a and b, which depicts the comparison between corrected and uncorrected BC values and the correlation between them, respectively. The results show a good agreement. The diurnal variation of BC mass concentration averaged over the study period of 4 years offers a broad maximum during daylight hours and minimum during night hours. The local boundary layer's evolution during the day and nighttime besides anthropogenic activities such as automobiles, industries and construction/ demolition activities could be a responsible factor for it. Good correspondence with a correlation coefficient of 0.94 confirms the BC data's reliability in inferring different aspects of BC aerosols' influence on climate.

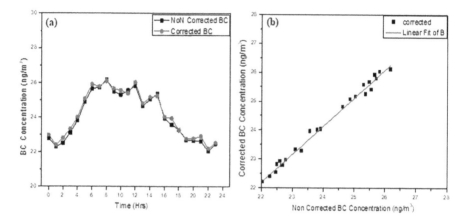

FIGURE 4.5 Mean diurnal variation of 4-year BC: (a) comparison and (b) correlation between corrected and uncorrected BC values.

4.4.4 Cyclic Variation in Long-Term BC Aerosol Mass Concentration

In this section, the monthly mean Arctic BC climatology was examined for the period from 2011 to 2014 to extract the seasonal variability in BC aerosol mass variations (Figure 4.6). The study shows a distinct seasonal variation with higher concentrations during winter/spring and lower concentrations in summer (June–September) each year, mainly due to the Arctic frontal phenomenon. This resultant BC trend is ascribed to BC aerosols' transport from mid-latitude sources (Barrie, 1986; Sirois and Barrie, 1999).

This finding is almost similar to the one reported by Sharma et al. (2006) and Gogoi et al. (2016). It is seen from the figure that during the summer season, BC aerosol concentration decreases from 49.07 to 25.23 ng/m^3 (can be seen from Figure 4.5) but increases in the winter/spring season. The low BC concentration in summer could be mainly due to the frequent wet deposition (Garrett et al., 2011). The summer aerosols in the Arctic air are cleaner than those in winter/spring because of the transport of wet aerosols from lower latitudes. To examine the long-term variability in BC aerosols, trend analysis of the data has been carried out, and the results are explained in the sections to follow.

4.4.5 Trends in Black Carbon Aerosols

Typical ground observed concentrations of BC for 4 years (from 2011 to 2014) have been analysed to extract temporal scales and trend embedded in it as a function of time of year. The results presented in Figure 4.7 show the time series of BC concentration for the 4 years from 2010 to 2014. The long-term fit reveals a cyclic variation with winter maximum (52%) and summer minimum (20.83%). The data are ninth-order polynomial fitted (Chambers et al., 1983). This is consistent with the earlier studies available in the literature (Hirdman et al., 2010; Sharma et al., 2013). Doherty et al. (2010) conclude that estimates of BC concentrations in snow have decreased

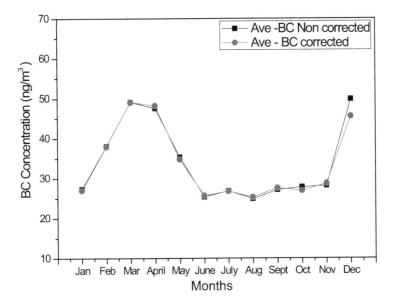

FIGURE 4.6   The monthly average plot of BC for 4 years 2011–2014.

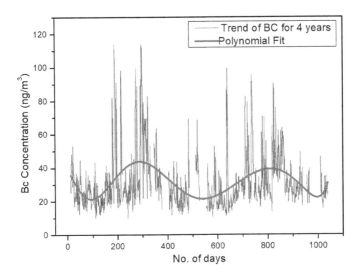

FIGURE 4.7   Long-term variations in BC mass concentration during 2011–2014. The red curve indicates the polynomial fit to the data.

over the Arctic as a whole. The source regions of these particles were viewed as mainly due to long-range transport from combustion activities in Asia, Europe, and North America (Clarke and Noone, 1985). This aspect has been further discussed using air mass back trajectories in the sections to follow.

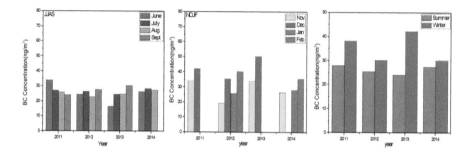

FIGURE 4.8 Bar graph showing seasonal mean BC concentration during summer: (a) winter (b) and seasonal (c) of 2011–2014.

Figure 4.8 displays the long-term monthly (A: Summer, B: Winter and C: Seasonal) BC variations during 2011–2014. Like Figure 4.7, summer minimum and winter maximum BC concentrations can also be noted, but with season-wise concentrations each year. The mean summer and winter BC concentration for the 4 years period is plotted in Figure 4.8c. BC's maximum concentration occurred in summer 2011 and winter 2013, and minimum values were found in summer 2013 and winter 2014.

4.4.6 Frequency Distribution of BC Mass Concentration Bins

The complete BC datasets, archived during the period from 2011 to 2014, have been divided into different concentration bins at an interval of $10\,ng/m^3$. The yearly marching of the frequency distribution of each such bin interval is shown in Figure 4.9. In other words, the frequency distribution of BC mass concentration bins during the summer and winter months is shown in Figure 4.9. It is clear from the figure that 4% of BC is lying below the 4-year mean BC, i.e., ~32 ng/m³. The distribution of BC is seen maximum during 2014 in the range bin of 20 ng/m³. The concentration bins, averaged over 4 years, are shown alongside. The polynomial fit to the variation in the concentration bins shows exponential decay. It also indicates that loading of lower concentration bins is more than higher concentration bins, which indirectly indicates the strength of far-away land/ocean sources as evidenced by the trajectory analysis.

4.4.7 Air Mass Trajectory Analysis

The 7-day backward trajectories used in this analysis were simulated by the Hybrid Single-Particle Lagrangian Integrated Trajectory (HYSPLIT) model, explained in the literature by Draxler and Hess (1998). The model was run for 500, 1000 and 2000 m using 7-day back trajectories, captured at 1300 UTC. The circuits for the mid-summer (August) and mid-winter (December) each year have been examined (Figure 4.10) to understand the probable transport of air mass to the observational site to emphasis the BC pollution from different potential source regions. It can be inferred from the figure that the loading of BC in the Arctic atmosphere is low during

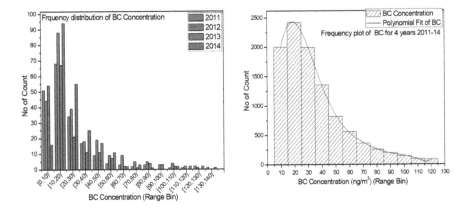

FIGURE 4.9 Frequency distribution of BC concentration bins during different years of observations.

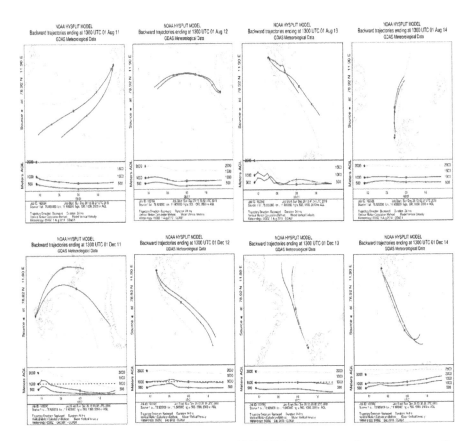

FIGURE 4.10 Back-trajectory analysis for 2011–2014: upper rows show summer months trajectories, and bottom rows show winter months trajectories over Ny-Ålesund, Arctic.

the observation period. The contribution from a local source is less as compared to that of long-range transport. The long-range transport of air mass, contributing more BC aerosols from the Soviet Union and European regions to the Arctic (as the Arctic front extends up to 40°N) during the winter/spring season (Sharma et al., 2004; 2006), could explain the long-term BC mass variations shown in Figures 4.7 and 4.8. Eurasia is also found to contribute to aerosol pollution at an altitude of 0.5–2 km. The air mass, coming from higher altitudes, seems to be responsible for higher BC mass concentration. The Asian emissions contribute to the Arctic only at higher altitudes and minimum at the surface (Shindell et al., 2008; Sharma et al., 2013).

4.5 DIFFERENTIAL AND SLAT OPTICAL DENSITIES OF ATMOSPHERIC CONSTITUENTS

Zenith-sky scattered sunlight spectra recorded at Ny-Ålesund, Arctic (78°9′N, 11°9′E), are analysed to derive Differential Optical Density (DOD) and Slant Column Density (SCD) of NO_2, O_3, H_2O and O_4 on 28 October 2013. Figure 4.11 shows a comparison between the observed and calculated DOD spectra of NO_2, O_3, H_2O and O_4 in the spectral range of 462–498 nm. The DOD is a measure of light passing through the atmosphere and is defined as the logarithm of the fraction of light that is absorbed on a path by the above trace gases. Observed DOD spectrum for

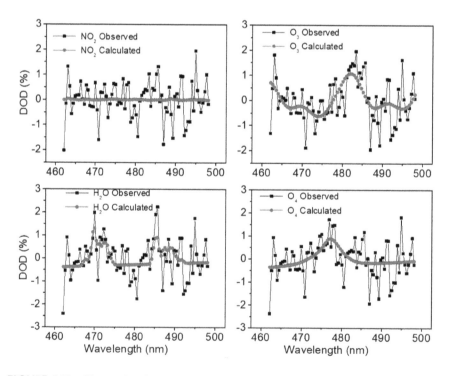

FIGURE 4.11   Observed and calculated DODs of NO_2, O_3, H_2O and O_4 (spectral range 462–498 nm).

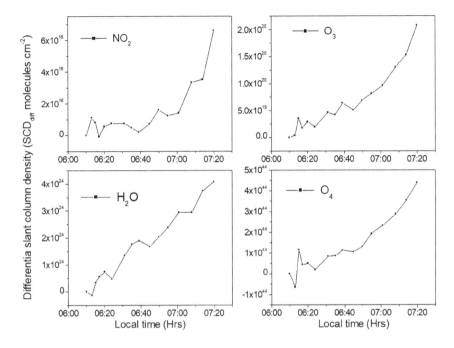

FIGURE 4.12 Differential slant column densities (SCD$_{diff}$) of NO$_2$, O$_3$, H$_2$O and O$_4$ for 28 October 2013 at evening hours.

each species is obtained by removing the theoretically calculated DODs (product of absorption cross-sections and specific concentration of each species) of species from the total observed DODs. The theoretically calculated and observed DODs of O$_3$, H$_2$O and O$_4$ are in good agreement with the calculated DODs, found to be varying up to 1%, 1% and 0.9%, respectively. The calculated DOD of NO$_2$ is 0.2%, which is not significant and also not matching with observed values.

Figure 4.12 shows observed SCD$_{diff}$ of NO$_2$, O$_3$, H$_2$O and O$_4$ for 28 October 2013 at evening hours of 6:10–7:20 pm (local time). The SCD$_{diff}$ are the differences of the column amounts between the column amount of an absorber integrated along the light path through the atmosphere at the time of the measurement and an absorber amount in the reference spectrum, which is taken at evening hours of 6:10 pm of the same day. Therefore, SCD$_{diff}$ of NO$_2$, O$_3$, H$_2$O and O$_4$ seem close to zero at evening hour 6:10 pm. The SCD$_{diff}$ increases with local time because of the growing absorption path length. The NO$_2$, O$_3$, H$_2$O and O$_4$ SCD$_{diff}$ are observed in the order of 6×10^{16} molecules/cm^2, 2×10^{20} molecules/cm^{2}cm^2, 4×10^{24} molecules/cm^{2}cm^2 and 4.5×10^{44} molecules/cm^{2}cm^2, respectively.

4.6 SUMMARY AND CONCLUSIONS

The results of the analysis of the BC mass concentration and skylight spectral data, archived over Himadri, Ny-Ålesund during 2011–2014 and summer 2013 indicate that:

- Daily mean BC mass concentration is maximum during winter and minimum during summer months.
- BC mass concentration is higher in 2012 and lower in 2013 due to changes in contribution of air mass of land and marine origin.
- Higher/lower BC mass concentration bins, observed in the study, indicate contribution of aerosol pollution both from far-away and local sources.
- Long-range transport analysis suggests a mixture of land and marine source contributions to the BC variations noticed during the study period.
- The experimentally measured DOD coincides with that of theoretically calculated.
- The SCD technique developed has been successfully utilised for deriving column density of various constituents that affect the climate.
- The study concludes that ice layers' protection will help reducing the variability in the Arctic climate.

ACKNOWLEDGEMENTS

The authors wish to thank the Director, IITM, for continuous encouragement and suggestions. Thanks,are also due to the authorities of the Amity University Haryana (AUH) , Gurugram, for cooperation and valuable support. We thank Mr. Kundan K. Dani and Dr. M. P. Raju for their help throughout the study. Support from the Director and scientists of NCPOR Goa, is gratefully acknowledged. The authors express their sincere gratitude to the Ministry of Earth Sciences (MoES), Government of India, New Delhi, for continuous support and facilitating the participation of one of the authors (SMS) in a series of Indian Arctic expeditions. The Kings Bay group at Ny-Ålesund is greatly acknowledged for all the logistic support, and NOAA-ARL for the HYSPLIT back-trajectory model analysis results for the study period.

REFERENCES

Arnold, S.R., Law, K.S., Brock, C.A. et al., 2016. Arctic air pollution: Challenges and opportunities for the next decade. *Elementa: Science of the Anthropocene*, 4, 000104, ISSN 2325-1026.

Babu, S.S., Chaubey, J.P., Moorthy, K.K., Gogoi, M.M., Kompalli, S.K., Sreekanth, V., Bagare, S.P., Bhatt, B.C., Gaur, V.K., Prabhu, T.P., and Singh, N.S., 2011. High altitude (~4520 m amsl) measurements of black carbon aerosols over western trans-Himalayas: seasonal heterogeneity and source apportionment. *J. Geophys. Res.* 116, D24201. http://dx.doi.org/10.1029/2011JD016722.

Barrie, L.A., 1986. Arctic air pollution an overview of current knowledge. *Atmospheric Environment.* 20, 643–663.

Chambers, J.M., Cleveland, W.S., Kleiner, B, and Tukey, P.A., 1983. Graphical methods for data Analysis. *Duxbury Press. Wadsworth & Brooks / Cole Statistics/Probability Series.*

Clarke, A.D. and Noone, K.J., 1985. Soot in Arctic snowpack: A cause for perturbation in radiative transfer. *Atmospheric Environment.* 19, 2045–2053.

Corbett, J.J., Lack, J.J., Winebrake, Harder, S., Silberman, J.A., and Gold, M., 2010. Arctic shipping D. A. emissions inventories and future scenarios. *Atmospheric Chemistry and Physics.* 10(19), 9689–9704, doi:10.5194/acp-10-9689-2010.

Croft, B., Martin, R.V., Leaitch, W.R., Tunved, P., Breider, T.J., D'Andrea, S.D., and Pierce, J. R., 2016. Processes controlling the annual cycle of Arctic aerosol number and size distributions. *Atmos. Chem. Phys.* 16, 3665–3682. www.atmos-chem-phys.net/16/3665/2016/; doi:10.5194/acp-16-3665-2016.

DeAngelo, B.E., 2011. An assessment of emissions and mitigation options for black carbon for the Arctic Council. *Technical report of the Arctic Council Task Force on Short-Lived Climate Forcers. Tromso: Arctic Council.*

Deshpande, C.G. and Kamra, A.K., 2014. Physical properties of the arctic summer aerosol particles in relation to sources at Ny-Ålesund, Svalbard. *J. Earth Syst. Sci.* 123 (1), 201–212.

Doherty, S.J., Warren, S.G., Grenfell, T.C., Clarke, A.D., Brandt, and R.E., 2010. Light Absorption from Impurities in Arctic Snow. *Atmos. Chem. Phys.* 10, 1647–11680.

Draxler, R.R. and Hess, G. D., 1998. An overview of the HYSPLIT 4 modelling system for trajectories, dispersion and deposition, *Aust. Met. Mag.* 47, 295–308.

Dubovik, O., Holben, B., Eck, T.F., Smirnov, A., Kaufman, Y.J., King, M.D., Tanré, D., and Slutsker, I., 2002. Variability of absorption and optical properties of key aerosol types observed in world wide locations. *Atmos. Sci.* 59(3), 590–608. http://dx.doi. org/10.1175/1520-0469(2002), 059-590:VOAAOP, 2.0.CO; 2.

Eckhardt, S., Hermansen, O., Grythe, H., Fiebig, M., Stebel, K., Cassiani, M., Baecklund, A., and Stohl, A., 2013. The influence of cruise ship emissions on air pollution in Svalbard — a harbinger of a more polluted Arctic? *Atmos. Chem. Phys.* 13, 8401–8409. http://dx.doi. org/10.5194/acp-13-8401-2013.

Garrett, T.J., Brattstrom, S., Sharma, S., and Worthy D.E.J., 2011. The role of scavenging in the seasonal transport of black carbon and sulfate to the Arctic. *Geophys. Res. Lett.* 38, L16805. doi:10.1029/2011GL048221.

Gogoi, M.M., Suresh Babu, S., Krishna Moorthy, K., Thakur, R.C., Chaubey, J.P., and Nair, V.S. 2016. Aerosol black carbon over Svalbard regions of Arctic. *Polar Science.* 10(2016), 60–70.

Granier, C., Niemeier, U., Jungclaus, J.H., Emmons, L., Hess, P., Lamarque, J.F., Walters, S., and Brasseur, G. P., 2006. Ozone pollution from future ship traffic in the Arctic northern passages, *Geophys. Res. Lett.,* 33, L13807. doi:10.1029/2006GL026180.

Hansen, J., Nazarenko, L., 2004. Soot climate forcing via snow and ice albedos. *Proc. Natl. Acad. Sci.* 101(2), 423–428.

Hirdman, D., Aspmo, K., Burkhart J.F., Eckhardt, S., Sodemann, H., and Stohl, A., 2009. Transport of mercury in the Arctic atmosphere: evidence for a spring-time net sink and summer-time source. *Geophys. Res. Lett.* 36, L12814. doi:10.1029/2009GL038345.

Hirdman, D., et al. 2010. Long-term trends of black carbon and sulphate aerosol in the Arctic: Changes in atmospheric transport and source region emissions. *Atmos. Chem. Phys.* 10, 9351–9368. doi: 10.5194/acp-10-9351-2010.

Huang, L., Gong, S.L., Sharma, S., Lavoué, D., Jia, C.Q., 2010b. A trajectory analysis of atmospheric transport of black carbon aerosols to Canadian High Arctic in winter and spring (1990–2005). *Atmos. Chem. Phys. Discuss.* 10(2), 2221–2244. doi:10.5194/ acpd-10-2221-2010.

Ikeda, K., Tanimoto, H., Sugita, T., Akiyoshi, H., Kanaya, Y., Zhu, C., and Taketani, F., 2017. Tagged tracer simulations of black carbon in the Arctic: transport, source contributions, and budget. *Atmos. Chem. Phys.* 17, 10515–10533. https://doi.org/10.5194/ acp-17-10515-2017.

Koch, D., Schulz, M., Kinne, S., McNaughton, C., Spackman, J.R., Balkanski, Y., Bauer, S., Berntsen, T., Bond, T.C., Boucher, O., Chin, M., Clarke, A., De Luca, N., Dentener, F., Diehl, T., Dubovik, O., Easter, R., Fahey, D.W., Feichter, J., Fillmore, D., Freitag, S., Ghan, S., Ginoux, P., Gong, S., Horowitz, L., Iversen, T., Kirkevåg, A., Klimont, Z., Kondo, Y., Krol, M., Liu, X., Miller, R., Montanaro, V., Moteki, N., Myhre, G., Penner,

J.E., Perlwitz, J., Pitari, G., Reddy, S., Sahu, L., Sakamoto, H., Schuster, G., Schwarz, J.P., Seland, Ø., Stier, P., Takegawa, N., Takemura, T., Textor, C., van Aardenne, J.A., and Zhao, Y., 2009. Evaluation of black carbon estimations in global aerosol models. *Atmos. Chem. Phys.* 9, 9001–9026. doi:10.5194/acp-9-9001-2009.

Koch, D., et al. 2011. Couple aerosol-chemistry-climate twentieth-century transient model investigation: trends in short-lived species and climate responses. *J. Clim.* 24(11), 2693–2714. doi:10.1175/2011JCLI3582.1.

Lee, D.S. and Co-authors, 2010. Transport impacts on atmosphere and climate: *Aviation, Atmospheric Environ.* 44(37), 4678–4734, doi:10.1016/j.atmosenv.2009.06.005.

Markowicz, K.M., Flatau, P.J., Kardas, A.E., Remiszewska, J., Stelmaszczyk, K., Woeste, L., 2008. Ceilometer retrieval of the boundary layer vertical aerosol extinction structure. *J. Atmos. Ocean. Technol.* 25(6), 928–944, http://dx.doi.org/10.1175/2007JTECHA1016.1.

Mazzola, M., Stone, R.S., Herber, A., Tomasi, C., Lupi, A., Vitale, V., Lanconelli, C., Toledano, C., Cachorro, V.E., O'Neill, N.T., Shiobara, M., Aaltonen, V., Stebel, K., Zielinski, T., Petelski, T., Ortiz de Galisteo, J.P., Torres, B., Berjon, A., Goloub, P., Li, Z., Blarel, L., Abboud, I., Cuevas, E., Stock, M., Schulz, K.-H., and Virkkula, A., 2012. Evaluation of sun photometer capabilities for retrievals of aerosol optical depth at high latitudes: the POLAR-AOD inter comparison campaigns. *Atmos. Environ.* 52, 4–17. http://dx.doi.org/10.1016/j.atmosenv.2011.07.042.

Meena, G.S., Londhe, A.L., Bhosale, C.S., and Jadhav, D.B., 2009. Remote sensing 'ground-based automatic UV/Visible spectrometer' for the study of atmospheric trace gases. *Int. Jr. Remote Sensing.* 30(21), 5633–5653.

Pfeilsticker, K., Erle, F., and Platt, U., 1997. Absorption of solar radiation by atmospheric O4. *J. Atmos. Sci.* 54, 933–939.

Quinn, P.K., Stohl, A., Arneth, A., Berntsen, T., Burkhart, J. F., Christensen, J., Flanner, M., Kupiainen, K., Lihavainen, H., Shepherd, M., Shevchenko, V., Skov H., and Vestreng V., 2011. The Impact of Black Carbon on Arctic Climate. *Arctic Monitoring and Assessment Programme (AMAP)*, Oslo, 72 pp., ISBN-978-82-7971-069-1.

Rahul, P.R.C., Sonbawne, S.M., and Devara, P.C.S., 2014. Unusual high values of aerosol optical depth evidenced in the Arctic during summer 2011. *Atmos. Environ.* 94, 606–615. http://doi:10.1016/j.atmosenv.2014.01.052.

Raju, M.P., Safai, P.D., Sonbawne, S. M., and Naidub, C. V., 2015. Black carbon radiative forcing over the Indian Arctic station, Himadri during the Arctic Summer of 2012. *Atmos. Res.* 157(2015), 29–36.

Ramanathan, V., and Carmichael, G., 2008. Global and regional climate changes due to black carbon. *Nat. Geosci.* 1, 221–227. doi:10.1038/ngeo156.

Safai, P.D., Raju, M.P., Budhavant, K.B., Rao, P.S.P., and Devara, P.C.S., 2013. Long term studies on characteristics of black carbon aerosols over a tropical urban station Pune, India. *Atmos. Res.,* 132–133, 173–184. http://doi:10.1016/j.atmosres.2013.05.002

Safai, P.D., Devara, P.C.S., Raju, M.P., Vijaykumar, K., and Rao, P.S.P., 2014. Relationship between black carbon and associated optical, physical and radiative properties of aerosols over two contrasting environments. *Atmos. Res.* 149. http://doi:10.1016/j.atmosres.2014.07.006, 292-299.

Seung, S.P., Anthony, D.A., Hansen, Sung, Y. Cho, 2010. Measurement of real time black carbon for investigating spot loading effects of Aethalometer data. *Atmos. Environ.* 44(11), 1449–1455.

Sharma, S., Ishizawa, M., Chan, D., Lavoué, D., Andrews, E., Eleftheriadis, K., Maksyutov, S., 2013. 16-year simulation of Arctic black carbon: transport, source contribution, and sensitivity analysis on deposition. *J. Geophys. Res.* 118, 1–22. http://doi:10.1029/2012JD017774.

Sharma, S., Lavoué, D., Cachier, H., Barrie, L.A., and Gong S.L., 2004. Long-term trends of the black carbon concentrations in the Canadian Arctic. *J. Geophys. Res.* 109, D15203. http://doi:10.1029/2003JD004331.

Sharma, S., Andrews, E., Barrie, L.A., Ogren, J.A., and Lavoué, D., 2006. Variations and sources of the equivalent black carbon in the high Arctic revealed by long-term observations at Alert and Barrow. *J. Geophys. Res.* 111, D14208, 2989–2003. http://doi:10.1029/2005JD006581.

Shindell, D.T. et al., 2008. A multi-model assessment of pollution transport to the Arctic. *Atmos. Chem. Phys.* 8, 5353–5372. doi:10.5194/acp-8-5353-2008.

Solomon, S., Schmeltekopf, A.Z., and Sanders, R.W., 1987. On the interpretation of zenith sky absorption measurements. *J. Geophys. Res.* 92D, 8311–8319.

Sonbawne, S.M., Devara, P.C.S., Rohini, B., Rahul, P.R.C., Siingh, D., Fadnavis, S., Panicker, A.S. and Pandithurai, G., 2021. Aerosol physico–optical–radiative characterization and classification during summer over Ny-Ålesund, Arctic. *Intl. J. Remote Sens.,* 42(22), 8760–8781, DOI: 10.1080/01431161.2021.1987576.

Spackman, J.R., Gao, R.S., Neff, W.D., Schwarz, J.P., Watts. L.A., Fahey D.W., Holloway, J. S., Ryerson, T.B., Peischl, J., and Brock, C.A., 2010. Aircraft observations of enhancement and depletion of black carbon mass in the springtime Arctic. *Atmos. Chem. Phys.* 10, 9667–9680, www.atmos-chem-phys.net/10/9667/2010/doi:10.5194/acp-10-9667-2010.

Stohl, A., 2006. Characteristics of atmospheric transport into the Arctic troposphere. *J. Geophys. Res.* 111, D11306. doi:10.1029/ 2005JD006888.

Stohl, A., Klimont, Z., Eckhardt, S., and Kupiainen, K., 2013. Black carbon in the Arctic: the underestimated role of gas flaring and residential combustion emissions. *Atmos. Chem. Phys.* 13, 8833–8855. doi:10.5194/acp-13-8833-2013, 2013.

Stone, R.S., Sharma, S., Herber, A., Eleftheriadis, K., and Nelson, D.W., 2014. A characterization of Arctic aerosols on the basis of aerosol optical depth and black carbon measurements. *Elem.: Sci. Anthropocene*2, 000027 (22 pp). http://dx.doi.org/10.12952/journal.elementa.

Sumit, K., Devara, P.C.S., Manoj, M.G., and Safai P.D., 2011. Winter aerosol and trace gas characteristics over a high-altitude station in the Western Ghats, India. *Atmosfera.* 24, 311–328.

Virkkula, A., Makela, T., Hillamo, R., Yli-Yuomi, T., Hirsikko, A., Hemari, K., and Koponen, I.K., 2007. A simple procedure for correcting loading effects of Aethalometer. *J. Air Waste Manag Assoc.* 57, 1214–1222.

Zielinski, T., 2004. Studies of aerosol physical properties in coastal areas. *Aerosol Sci. Technol.* 38(5), 513–524. http://dx.doi.org/10.1080/02786820490466738.

Zielinski, T., Petelski, T., Makuch, P., Strzalkowska, A., Ponczkowska, A., Drozdowska, V., Gutowska, D., Kowalczyk, J., Darecki, M., and Piskozub, J., 2012. Studies of aerosols advected to coastal areas with use of remote techniques. *Acta Geophys.* 60(5), 1359–1385. http://dx.doi.org/10.2478/s11600-011-0075-4.

5 Geomorphology and Landscape Evolution of Ny-Ålesund Region and Its Implication for Tectonics, Svalbard, Arctic

Dhruv Sen Singh, Chetan Anand Dubey, and Anoop Kumar Singh
University of Lucknow

Rasik Ravindra
National Centre for Polar and Ocean Research (Formerly)

CONTENTS

5.1 INTRODUCTION

The Ny-Ålesund region is the northernmost settlement on the planet Earth, inhabited by scientists of different countries. Rugged mountains characterise it with steep flanks, glacial eroded fiord system, small rivers, delta, lakes, coastal low land, raised

DOI: 10.1201/9781003265177-5

beaches, marine terraces and rocks of all the geological ages (Singh and Ravindra 2011a,b; Singh et al. 2018). It exhibits diversified geomorphic features and a complex landscape. The variability in landforms provides the natural laboratory and proxies for the evolution of landscape and reconstruction of tectonics.

The Arctic has experienced several climatic fluctuations during the quaternary period in the form of glacial (stadial) and interglacial (interstadial) cycles during the last few centuries (Overpeck et al. 1997) and drastic environmental changes during the last 30 years (Serreze et al. 2000). There has been a rapid decline in the extent and thickness of sea-ice cover in summer and more recently in winter as well (Rigor and Wallace 2004; Comiso et al. 2008). Glacier expanded considerably during the Little Ice Age (LIA), culminating during the first decade of the 20th century. The LIA terminated by an abrupt rise in temperature around the onset of the 20th century (Svendsen and Mangerud 1997). This region has been analysed for the long-term glacier mass balance (Hagen and Liestol 1990), glacier elevation changes and contribution to sea-level rise (Nuth et al. 2010), geomorphology of the MidreLoven glacier (Singh and Ravindra 2011a), control of glacial and fluvial environments (Singh and Ravindra 2011b), late Pleistocene and Holocene climates (Singh et al. 2011), and climate history of the Ny-Ålesund region (Singh et al. 2018). Many studies have been carried out in this region for climate change. However, no attention has been given to tectonic analysis and the evolution of landforms and landscape.

The interpretation of geomorphic features has a tectonic significance as they preserve and provide information about the area's geological history. However, this area's geomorphic database is limited to a lack of observations, making it challenging to analyse and interpret the geological past. In contrast, the analysis of geomorphic features is a prerequisite for a better understanding of the area's geological history. Geomorphological studies of active tectonics in the late Pleistocene and Holocene are essential to evaluate the tectonic hazards in neotectonically functional areas (Keller and Printer 2002). Therefore, the present study was carried out to understand geomorphology and landscape and its implication for tectonics.

5.2   STUDY AREA

Ny-Ålesund, Spitsbergen Island, Svalbard archipelago is located between Norway and the North Pole in the Arctic Ocean (Figure 5.1). It was uplifted during late Mesozoic and Cenozoic crustal movements with prominent tectonic events of Grenvillian (late Mesoproterozoic), Caledonian (Ordovician-Silurian), Ellesmerian or Svalbardian (late Devonian), Variscan (mid-Carboniferous) and Alpine (early Tertiary) age (Hjelle 1993). The name Spitsbergen (means sharp peaks) is derived from the wild and rugged Heckla Hoek terrain of NW Spitsbergen (Elvevold 2007).

Precipitation usually is low, about 400 mm annually on the western coast of Spitsbergen and half as much in central inland areas and 190 mm/year at Svalbard Airport (Forland et al. 1997). Precipitation is higher on the glaciers due to the orographic effects but seldom exceeds 2–4 m. The mean annual air temperature is about −6°C at sea level and as low as −15°C in the high mountains, and July air-temperatures is 5.9°C. Summer ablation plays a significant role in the total mass balance.

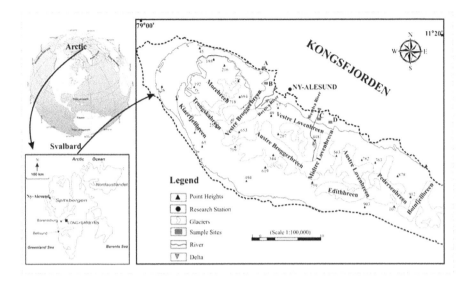

FIGURE 5.1   Location map of the Ny-Ålesund area, Arctic.

5.3   METHODOLOGY

The area's base map was prepared at a 1:100,000 scale using Norsk Polar Institute Tromso maps of 2004 and field observations during the First Indian Expedition to the Arctic in 2007 and subsequently in 2008. Diversified geomorphic features were identified in the field and were documented for their surface processes and sedimentary environments. Geomorphic features were selected, and sediment/shells/peat samples were collected for granulometric analysis by making trenches in the outwash plain, lacustrine deposits, and exposed moraines and coastal cliff.

Lithologs were prepared for the moraines, lacustrine deposits, outwash plain and coastal cliff. The nature of the bedding, sediment grain parameters and matrix percentage was documented in the field and laboratory. The coarser sediments greater than 2 mm in diameter were measured in the area directly using a measuring tape (Wentworth 1922), and finer sediments were analysed in the Sedimentology Laboratory of Centre of Advanced Study in Geology, the University of Lucknow using LPSA, CILAS 1190.

5.4   RESULTS

This region exhibits geomorphic features evolved by the various sedimentary environment and surface processes.

5.4.1   Glacial Environment

The Ny-Ålesund region's glaciers are classified as ice caps, valley glaciers, cirque glaciers, tidal glaciers, rock glaciers and talus cone glaciers (Singh and Ravindra 2011a).

FIGURE 5.2   The permafrost region in the various forms of pattern ground and stone circles.

FIGURE 5.3   (a) Supraglacial streams. (b) Diamicton parallel ridges on the glacier.

Permafrost, indicated by various pattern ground and stone circles, is a widespread feature (Figure 5.2).

The glaciers erode and deposit the sediments to form several geomorphic features characterised by poorly sorted angular coarse grain sediments devoid of any sedimentary structures. The poorly sorted boulders, gravels, sand, silt and clay are deposited to form supraglacial moraines, lateral moraines, recessional moraines, terminal moraines, pushed moraines, hummocky moraines, thrust moraines, outwash plain, etc. (Singh and Ravindra 2011b; Singh et al.2017, 2018, 2019).

The supraglacial streams are present, which may be longitudinal or transverse (Figure 5.3a). These streams may become subglacial and form ice caves that show stalactite and stalagmite like ice features. The supraglacial ephemeral streams enhance the glacier's melting (cf. Singh and Mishra 2001). Supraglacial moraines are preferentially aligned along the length of a glacier in the form of debris strips and diamicton linear parallel ridges (Figure 5.3b).

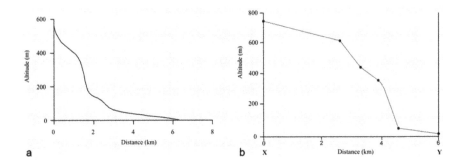

FIGURE 5.4   The longitudinal profiles of the (a) MidreLoven and (b) Vestre Brogger glaciers.

The snout is steep with frontal convexity as observed in the field and indicated in the longitudinal profile. The longitudinal shapes of the MidreLoven (Singh and Ravindra 2011a) and Vestre Brogger glaciers are convex with breaks in slope and abrupt termination (Figure 5.4a and b).

Debris on the base of glacier known as subglacial moraines is raised on the glacier surface by thrusting. Thrusting developed most effectively where there is the transition from sliding melting ice to ice. The mound-shaped moraines on the top surface of a glacier are designated as thrust moraines (Singh and Ravindra 2011b) (Figure 5.5a). Figure 5.5b explains the process of evolution of thrust moraines. Tectonic activity is the only process to form it in the central part of a glacier's top surface, where all other surface processes are absent.

Folded shape moraines having several concentric crests present in front of the glacier, often with a core of ice, are known as frontal moraines. The frontal convexity is persistence in the folded pushed morainic ridges. The folded and pushed/ribbed moraines occur in front of the snout. The continuation of folded made morainic piles and their frontal changes indicate the palaeosurge events.

The mound-shaped complex moraines, formed when the glacier was at its most dynamic stage, are known as hummocky moraines. Mound-shaped moraines, some with ice core, are common in the proximal part of the outwash plain. The hummocky moraines consist of 20% sand, silt and clay and 80% boulders. 65% of the boulders vary in size from 4 to 10 cm in length and 3 to 5 cm in width. 10% boulders are 10–20 cm in length, and only 5% of boulders are more than 20 cm and up to 4 ft in length and 2 ft in width (Figure 5.6). The azimuths of the orientation of the boulders in hummocky moraines were recorded for palaeocurrent analysis. The bimodal rose diagram indicates two directions for the palaeocurrent pattern for the glacier movement in the past (Figure 5.7). The bimodal rose diagram suggests two rules for glacial movement under the direct control of tectonic activity.

The lithology (Figure 5.6) describes that moraines are unstratified deposits. In earlier studies, lateral moraines have been reported as unstratified, devoid of any sedimentary structures (Singh and Ravindra 2011a,b).

The recessional moraines have been reworked by the surging glaciers and the other processes (cf. Singh and Mishra 2002; Singh 2013a,b, 2014; Singh et al. 2017). Recent moraines are present up to a distance of 500 m from the snout. Three stages

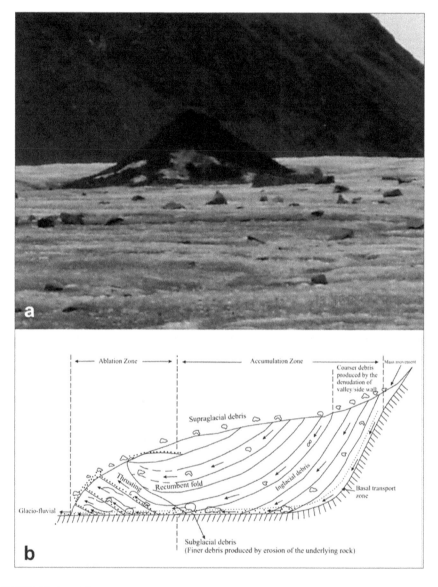

FIGURE 5.5   (a) Thrust moraines on a top surface of the glacier. (b) Process of the evolution of thrust moraines.

of left and two stages of right lateral moraines are present in Vestre brogger breen (Table 5.1).

5.4.2   LACUSTRINE ENVIRONMENT

The terrestrial zone located in front of the glaciers exhibits small-sized circular to semi-circular lakes between snout and terminal moraines. The shape, size and

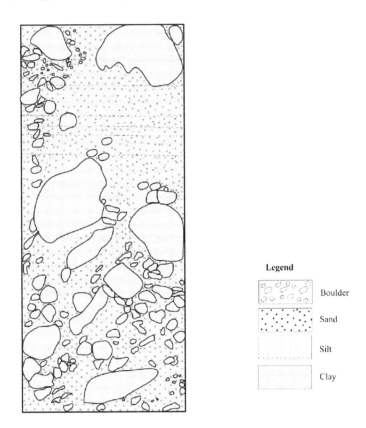

FIGURE 5.6   Schematic lithology of hummocky moraines.

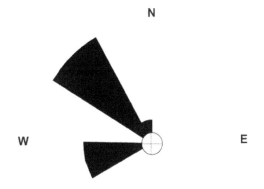

FIGURE 5.7   Rose diagram for hummocky moraines.

TABLE 5.1

Basic Information on Lateral and Recessional Moraine

S. No.	Name of Moraine		Alignment	Length (m.)	Width (m.)	Height (m.)
1	Lateral moraines	1ˢᵗstage	10°	500	30–50	20–25
2		2ⁿᵈstage	20°	600	30–50	10–15
3		3ʳᵈ stage	55°	200	10–15	8
4	Recessional moraines	1ˢᵗstage	70°	200	20–30	3.048–7.62
5		2ⁿᵈstage	70°	300	15–30	1.524–3.048
6		3ʳᵈ stage	60°	250	25–30	4.572–7.62

number of lakes depend on the size of the snout. In MidreLoven glacier, the lakes are two due to large snout. Tiny rivers originate from the glaciers' drain into these lakes, and many small rivers originate from it. These pro-glacial lakes are formed by blocking off the river by recessional moraines. The fine grain sediments are deposited in these lakes.

A trench was made to analyse the lacustrine deposits. The trench, lithology and its sedimentary characteristics are given in Figure 5.8a and b.

5.4.3 GLACIO-FLUVIAL ENVIRONMENT

The outwash plains are reworked by surging glaciers and also by the surface processes which evolve during deglaciation. Ancient outwash plains are present near coastal areas towards the mouth of the river and new ones towards the glacier's snout. The scope of this zone is increasing and shifting towards the hill.

A 1.5-m deep trench was made to analyse the outwash plain sediments and sandur facies. The trench, litholog and its sedimentary characteristics are given in Figure 5.9a and b.

5.4.4 FLUVIAL ENVIRONMENT

The rivers are 3–5 km in length which originate from terrestrial valley glaciers. Rivers are ephemeral and transport the sediments from the mountain to the sandur plain and fjords. Generally, the rivers originating from glaciers are large with high discharge and high sediment load. Still, in this region, snow-fed rivers are tiny with low clearance and low sediment load.

The streams originating from the snout either drain into pro-glacial lakes or join to form the main river, emptying into the Kongsfjorden. These moderately entrenched rivers show varying degrees of sinuosity and flow through outwash plains and coastal plains. The rivers are meandering in the proximal part and braided in the distal portion (Figure 5.10). The abrupt change like the river explains the control of neotectonic activity. The valley width varies from 50 to 100 ft, in which channels are 15–20 ft wide. The depth of the channel varies from 1 to 4 ft. The rivers carry

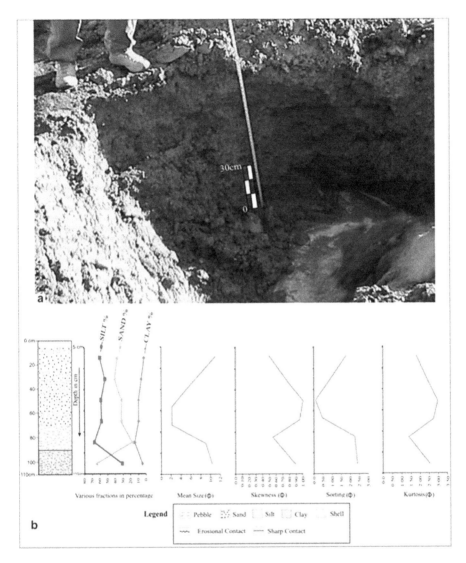

FIGURE 5.8   (a) A trench in the lacustrine area. (b) Lithology and sedimentary characteristics.

bed load, suspended load and dissolved load. The middle and distal part of the river exhibits gravel bar deposits and shows minor scale ripple marks. The rivers drain into Kongsfjorden and form an estuary or delta.

The granulometric analysis of the river indicates that the weight percentage of the finer sediments decreases and coarser fragments increases in the downstream direction, which is contrary to the existing facts (Figure 5.11) (Folk and Ward 1957; Singh and Singh 2005; Singh and Awasthi 2009, 2011). The increase of coarser sediments in the downstream direction explains the neotectonic activity.

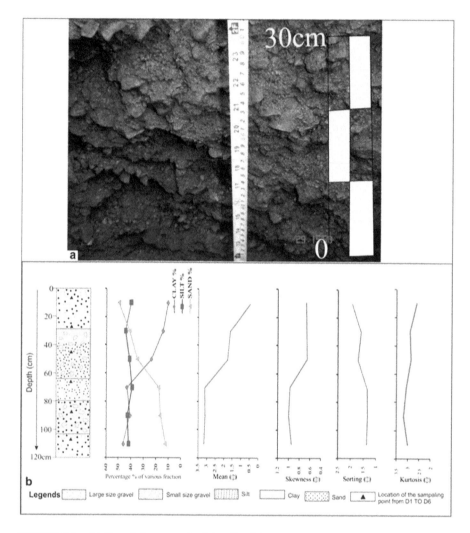

FIGURE 5.9   (a) Trench in outwash plain. (b) Litholog and sedimentary characteristics of the outwash plain.

5.4.5   DELTAIC, COASTAL AND MARINE ENVIRONMENT

The coastline is characterised by raised beaches, wave-cut platforms, wave-cut notches and sea-cliff. The top surfaces of wave-cut media and uplifted marine terraces are sloping away from the Kongsfjorden. Transgression by the sea after deglaciation and upliftment of the area due to tectonic activity evolve marine terraces, raised beaches and cliffs. The beach marks present in the coastal areas are often occupied by raised marine beach terraces (Figure 5.12). About 1 m raised beaches are present in coastal areas.

The Bayalva River originates from the Wester Brogger being glacier and the river originating from the MidreLoven glacier from the delta, whereas many minor

FIGURE 5.10 Abrupt change in sinuosity of the river: (a) meandering in the proximal par-tand (b) braided in the distal portion of the river.

rivers form estuary. Bayalva delta is about 400 m long and 500 m wide with a well-developed distributary mouth bar.

Coastal cliffs are significant and provide information about the palaeoclimate (Figure 5.13a and b). Singh et al. (2018) have explained the palaeoclimate by analysing the 389 cm thick coastal cliff for sediment grain parameters and ^{14}C AMS dates. The analysis illustrates that glaciers and cold climate dominated this region during the stadial stage 47.5, 38, 23, 18, 8.8, and 6.1, 1 ka BP and paraglacial processes and warm weather interstadial stage 44, 27, 12, and 10.5ka BP. The poorly sorted sediments indicate that the energy of the prevailing climate and environments was continuously fluctuating with seven cold climate events and four warm climate events between 47.5 and 1 ka BP.

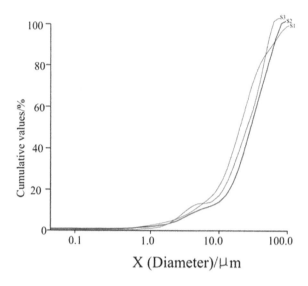

FIGURE 5.11   Granulometric analysis of the fluvial sediments.

FIGURE 5.12   Raised beach marks.

5.5   DISCUSSION

The above results explain the presence of diversified morphological features and various processes and sedimentary environments. The prevailing climate and tectonics control the nature and intensity of each surface process. The geomorphic features present in the area are lateral moraines, hummocky moraines, recessional moraines, supraglacial moraines, thrust moraines, gravel bars, delta, coastal cliff, marine terrace sand raised beaches which owe their origin to glacial as well as other surface processes. Glacial and non-glacial cycles played an essential role in the evolution

FIGURE 5.13   (a) Coastal cliff. (b) Lithology of coastal cliff.

of landforms and this area's landscape. The site is predominantly characterised by glacial landforms; however, the glacio-fluvial, lacustrine, fluvial, coastal, marine and tectonic geomorphic features are also present. They can be identified based on sediment characteristics.

It is difficult to know how many glaciations have occurred since the glacial period on this high latitude because climate change in this region is not well understood. Liestol (1988) has described that there may have been 50 or more glaciations. The ice age is characterised by glaciers, while during interglacial many other surface processes evolve. The modification of geomorphic features by the methods developed during deglaciation has been identified in the Gangotri Glacier region (Kumar et al., 2021a, b; Singh and Mishra 2002; Singh 2004, 2013a; Singh et al. 2017, 2019). Recently in the Kedarnath region, a reworking of sediments and modification of landforms have been identified, affecting the landscape evolution of the area (Singh 2013b, 2014). Geomorphic features are reworked and modified by surging glaciers (Singh et al., 2020) also. The repeated glaciations and inter-glaciations carried out the sculpturing of the area during the quaternary period. The melting of a glacier during the ablation period and thickening of the glacier during the accumulation period washed out the pre-existing supraglacial features every year.

The geomorphic features observed directly in the field have been proved excellent indicators of active tectonics and have been used by various workers in different areas. The fractures on glaciers as well as on the land (Figure 5.14), contours, sinuosity, wave-cut notches, sea cliffs, raised beaches and uplifted marine terraces are well-established pieces of evidence of active tectonics in the field (Burbank and Anderson 2001; Keller and Pinter 2002), whereas thrust moraines, debris strips and longitudinal profile with breaks are identified as new features of active tectonics.

FIGURE 5.14   Fractures on the land surface.

The thrust moraines located on the top surface of the glaciers are evolved by the tectonic activity. There are no surface processes or mass wasting much away from the mountain wall to develop it.

The longitudinal glacier profile with breaks in slope and steep snout with frontal convexity are shreds of evidence of surging glaciers (Singh and Ravindra 2011a). The convex shape and steep slope gradient near the beak of the Midtre Loven breen and Vestre Brogger been glacier with well-marked frontal convexity indicate the surge features (Hagen 1987; Hagen and Liestol 1990; Lovell et al.2018). The surge itself may be induced due to active tectonics, which increases the slope and the glacier's velocity. The breaks in the longitudinal profile have been identified due to tectonic activity in the Ganga Plain (Singh and Singh 2005).

During a warm climate in the Ny-Ålesund region, fluvial processes were dominant, which increased vegetation cover (Singh et al. 2011; Ukkonen et al. 2007). A fluctuation of the glacier front controls the expansion and contraction of the outwash plain area directly to climate change. During the present interglacial period, the glaciers are melting, and so the sea level should rise, and the land area should decrease. The progradation of delta and field experience indicates that the land area is increasing and the sea area falls. It clearly explains that the land area is rising, and the sea area subsides under the direct control of neotectonic activity.

The sinuosity and channel slope parameters in a river system balanced each other, so the rivers are susceptible to change, either climatic or tectonic. Even a slight change may switch a river from one sinuosity to another (Keller and Printer 2002). Tectonic uplift or tilting may change the pattern and sinuosity of a river. The photograph showing channel shape, meandering in the upper part and braided in the distal portion, is due to active tectonics.

The granulometric analysis of the river indicates that the weight percentage of the finer sediments decreases and coarser fragments increase in the downstream direction, which is contrary to the existing facts (Figure 5.11) (Folk and Ward 1957; Singh

TABLE 5.2
Sediment Grain Parameters of Various Morphological Zones

Morphological Zones	Mean Grain Size (Mz)	Sorting (σ)	Skewness (Sk)	Kurtosis (kg)
Lacustrine deposits	8.0–9.0	1.2–1.6	0.53–1.05	2.8–3.6
Sandur deposits	8.0–9.0	1.22–1.47	0.39–0.73	2.4–3.4
Fluvial deposits	9.0–10.0	0.9–1.3	0.5–1.4	2.6–5.7
Coastal deposits	9.0–10.0	1.0–2.0	0.49–1.09	2.2–4.5

and Singh 2005; Singh and Awasthi 2009, 2011). The palaeocurrent pattern should be unidirectionally aligned parallel to the direction of movement of the glacier. The bimodal rose diagram for the moraines indicates two rules for glacial action. The data for sediment grain parameters (Table 5.2) and palaeocurrent pattern are contrary to the existing facts. This abnormality is due to active tectonics. The sedimentary parameters have been used for the first time to identify the tectonic activity of a region.

The coastal cliffs, raised beaches, and uplifted marine terraces are also a manifestation of active tectonics observed in this region's field. It is the place where all the surface processes responsible for the evolution of the landscape of any part of the earth surface are present nearby. A model of the surface process and landform was proposed, which can be used as an analogue to interpret the rock record (Figure 5.15).

5.6 CONCLUSIONS

Based on the above results and discussion, it is concluded that the primary surface process in the area is a glacier, and the secondary surface processes which are active during deglaciations are mass movement, fluvial, aeolian, lacustrine, deltaic, coastal and marine. The dominance of glacial landforms characterises the ice age, whereas the interglacial period by the coexistence of glacial, fluvial, lacustrine, coastal and marine landforms.

The landscape of the Ny-Ålesund, Svalbard, Arctic was evolved by surface processes and sedimentary environments of glacial and the interglacial period. This region has been carved by glacial, glacio-fluvial, fluvial, lacustrine, deltaic, coastal and marine under tectonics' direct control.

The geomorphic features such as an abrupt change in sinuosity, debris strips, steep snout, knickzones along longitudinal profile, raised beaches, and a coastal cliff are identified as field shreds of evidence of active tectonics. Thrust moraines are described as a landform, which is evolved due to active tectonics. The sedimentary parameters such as granulometric analysis and palaeocurrent pattern are used for the first time to measure the tectonic activity in an area also describes that the region is tectonically active. The geomorphic features show contraction and expansion at various stages under the direct control of tectonics. A geomorphic model for this region was proposed, which can be used as an analogue to interpret the other similar areas.

The area's sculpturing was carried out by the sedimentary environments evolved during repeated glaciations and inter-glaciations during the quaternary period under the direct control of tectonic activity.

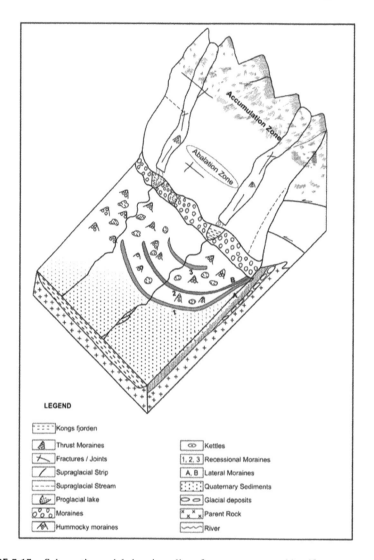

FIGURE 5.15   Schematic model showing all surface processes and landforms.

ACKNOWLEDGEMENTS

Ministry of Earth Science, Government of India, New Delhi and NCAOR, Goa are highly acknowledged for providing an opportunity for the first Arctic expedition in 2007. Centre of Excellence, Government of Uttar Pradesh and Head, Centre of Advanced Study in Geology, University of Lucknow are acknowledged for providing the working facilities. Thanks are due to Dr M. Shiwaji, Dr S.M. Singh, and Dr C.G. Deshpandey, who were part of the First Indian Expedition to the Arctic. Dr Neloy Khare and Dr Rakesh Mishra are thanked for their help during the Arctic expedition in 2008. Dr Anoop Kumar Singh is thankful to SERB N-PDF (PDF/2020/000251) and Mr Chetan Anand Dubey to CSIR (09/107(0384)/2017) for the financial support.

REFERENCES

Burbank, D.W. & Anderson, R.S. (2001). *Tectonic Geomorphology.* Backwell Science.

Comiso, J.C., Parkinson, C.L, Gersten, R., & Stock, L. (2008). Accelerated decline in the Arctic ice cover. *Geophysical Research Letters* 35(L01703), 6.

Elvevold, S. (2007). *Geology of Svalbard.* Norwegian Polar Institute, Norway.

Folk, R.L. & Ward, W.C. (1957). Brazos River bar: a study in the significance of grain size parameters. *Journal of Sedimentary Petrology* 27, 3–26.

Forland, E.J., Hansee-Bauer, I., & Nordli, P.O. (1997). *Climate statistics and long term series of temperature and precipitation at Svalbard and Jan Mayen, 21/97.* Norwegian Meteorological Institute, Oslo, p. 72.

Hagen, J.O. & Liestol, O. (1990). Long-term glacier mass balance investigations in Svalbard 1950–88. *Annals of Glaciology* 14, 102–106.

Hagen, J.O. (1987). Glacier surge in Svalbard with examples from Usherbreen. *Norwegian Journal of Geography* 42(4), 203–213.

Hjelle, A. (1993). *The Geology of Svalbard.* Norwegian Polar Institute, Oslo.

Keller, E.A. & Printer, N. (2002). *Active Tectonics: Earthquakes, Uplift, and Landscape* (2nd Ed.). Prentice-Hall, New Jersey.

Kumar, D., Singh, A.K., Taloor, A.K. & Singh, D.S. (2021a). Recessional Pattern of Thelu and Swetvarn glaciers between 1968 and 2019 Bhagirathi basin, Garhwal Himalaya, India. *Quaternary International* 575–576, 227–235.

Kumar, D., Singh, A.K. & Singh, D.S. (2021b). Spatio-temporal fluctuations over Chorabari glacier, Garhwal Himalaya, India between 1976 and 2017. *Quaternary International* 575–576, 178–189.

Liestol, O. (1988). The glaciers in the Kongsfjorden area, Spitsbergen. *Norsk Geografisk Tidsskrift* 42, 231–238.

Lovell, H., Benn, D.I., Lukas, S., Spagnolo, M., Cook, S.J., Swift, D.A., Clark, C.D., Yde, J.C., & Watts, T.P. (2018). Geomorphological investigation of multiphase glacitectonic composite ridge systems in Svalbard. *Geomorphology* 300, 176–188.

Nuth, C., Moholdt, G., Kohler, J., Hagen, J.O., & Kaab, A. (2010). Svalbard glacier elevation changes and contribution to sea-level rise. *Journal of Geophysical Research* 115(1), F01008.

Overpeck, J., Hughen, K., Hardy, D., Bradley, R., Case, R., Douglas, M., Finney, B., Gajewski, K., Jacoby, G., Jennings, A., Lamoureux, S., Lasca, A., MacDonald, G., Moore, J., Retelle, M., Smith, S., Wolfe, A., & Zielinski, G. (1997). Arctic environmental change of the last four centuries. *Science* 278, 1251–1256.

Rigor, I.G. & Wallace, J.M. (2004). Variation in the age of Arctic sea-ice and summer sea ice extent. *Geophysical Research Letters* 31, L0940.

Serreze, M.C., Walsh, J.E., Chapin, F.S., III, Osterkamp, T., Dyurgerov, M., & Romanovsky, V. (2000). Observational evidence of recent change in the northern high-latitude environment. *Climate Change* 46, 59–207.

Singh, D. S. (2014). Surface processes during flash floods in the glaciated terrain of Kedarnath, Garhwal Himalaya and their role in modifying landforms. *Current Science* 106(4), 594–597.

Singh, D.S. & Awasthi, A. (2009). Impact of land use and landscape change on the environment. *Urban Panorama* 8(2), 72–78.

Singh, D.S. & Awasthi, A. (2011). Natural hazards in the Ghaghara River area, Ganga Plain, India. *Natural Hazards* 57, 213–225.

Singh, A.K., Kumar, D., Kumar, V. & Singh, D.S., (2020). Study of temporal response (1976–2019) and associated mass movement event (during 2017) of Meru glacier, Bhagirathi valley, Garhwal Himalaya, India. *Quaternary International* 565, 12–21.

Singh, D.S. & Mishra, A. (2001). Gangotri Glacier characteristics, retreat and processes of sedimentation in the Bhagirathi valley. *Geological Survey of India Special Publication* 65(III), 17–20.

82   Climate Change in the Arctic

Singh, D.S. & Mishra, A. (2002). Role of tributary glaciers on landscape modification in the Gangotri Glacier area, Garhwal Himalaya, India. *Current Science* 82(5), 101–105.

Singh, D.S. & Ravindra R. (2011a). Geomorphology of the MidreLoven Glacier, Ny-Alesund, Svalbard, Arctic. In Singh, D.S. & Chhabra N.L. (Eds.), *Geological Processes and Climate Change* (pp. 269–281). Macmillan Publishers India Limited.

Singh, D.S. & Ravindra R. (2011b). Control of glacial and fluvial environments in the Ny-Alesund region, Arctic. *Mausam* 62(4), 641–646.

Singh, D.S. & Singh, I.B. (2005). Facies architecture of the GandakMegafan, Ganga Plain, India. *Journal of Paleontological Society of India* 12, 125–140.

Singh, D.S. (2004). Late quaternary morpho-sedimentary processes in the Gangotri Glacier area Garhwal Himalaya, India. *Geological Survey of India, Special Publication* 80, 97–103.

Singh, D.S. (2013a). Causes of Kedarnath tragedy and human responsibilities. *Journal Geological Society of India* 82(3), 303–304.

Singh, D.S. (2013b). Snow Melt Ephemeral Streams in the Gangotri Glacier Area, Garhwal Himalaya, India. In Kotlia, B.S. (Ed.), *Holocene: Perspectives, Environmental Dynamics and Impact Events* (pp. 157–164). NOVA Publisher, USA.

Singh, D.S., Dubey, C.A., Kumar, D., Kumar, P., & Ravindra, R. (2018). Climate events between 47.5 and 1 ka BP in the Ny-Alesund region's glaciated terrain, Arctic, using geomorphology and sedimentology of diversified morphological zones. *Polar Science* 18(2018), 123–134.

Singh, D.S., Dubey, C.A., Kumar, D., Vishawakarma, B., Singh, A.K., Tripathi, A., Gautam, P.K., Bali, R., Agarwal, K.K., & Sharma, R. (2019). Monsoon variability and significant climatic events between 25 and 0.05 ka BP using sedimentary parameters in the Gangotri Glacier region, Garhwal Himalaya, India. *Quaternary International* 507, 148–155.

Singh, D.S., Tangri, A.K., Kumar, D., Dubey, C.A., & Bali, R. (2017). Pattern of retreat and related morphological zones of Gangotri Glacier, Garhwal Himalaya, India. *Quaternary International* 444, 172–181.

Singh, V., Farooqui, A., Mehrotra, N.C., Singh, D.S., Tewari, R., Jha, N. & Kar, R. (2011). Late Pleistocene and early Holocene climate of Ny-Alesund, Svalbard (Norway): a study based on biological proxies. *Journal of Geological Society of India* 78(2), 109–116.

Svendsen, J.I. & Mangerud, J. (1997) Holocene glacial and climatic variations on Spitsbergen, Svalbard. *Holocene* 7, 45–57.

Ukkonen, P., Arppe, L., Houmark-Nielsen, M., Kjaer, K.H. & Karhu, J.A. (2007). MIS 3mammoth remains from Sweden-implications for faunal history, palaeoclimate and glaciations history. *Quaternary Science Review* 26, 3081–3098.

Wentworth, C.K. (1922). A scale of grade and class terms for clastic sediments. *Journal of Geology* 30(5), 377–392.

6 Biogenic Silica Indicator of Paleoproductivity in Lacustrine Sediments of Svalbard, Arctic

Shabnam Choudhary
National Centre for Polar and Ocean Research

G N Nayak
Goa University

Neloy Khare
Ministry of Earth Sciences, Govt. of India

CONTENTS

6.1   INTRODUCTION

Primary productivity in the past can be reconstructed by multiple biogeochemical proxies such as nutrients, amino acids, pigments and isotopes. Besides many other proxies, biogenic silica (bSi) has emerged as a potential paleoproductivity proxy and is widely used to reconstruct palaeoceanography and paleoclimate changes.

DOI: 10.1201/9781003265177-6

83

Biogenic silica in lake sediments has received considerably more attention in recent years as it indicates climate and environmental changes over a wide geographical area on a different time scale (Carter and Colman 1994; Colman et al. 1995).

Biogenic silica is a type of amorphous silica derived from diatoms and other siliceous microorganisms (Liu et al. 2014). Biogenic silica documents direct measure of biological products from the siliceous algae and diatoms (Conley 1998; Kaplan et al. 2002). It is produced in the euphotic zone by the diatoms, predominantly in the high-latitude areas. Although the biogeochemical cycle of silicon is slow as compared to other elements like oxygen, carbon and nitrogen, its sedimentation rate is rapid (Shan et al. 2011); therefore, silicon can be preserved in the form of biogenic silica in sediments (Birks et al. 2004; Shan et al. 2011). In lakes, nutrient concentration in water and surface temperature affects productivity. Therefore, variation in biogenic silica content can reflect changes in past climate conditions. Earlier studies from lakes of different parts of the globe such as Lake Baikal, Lake Pipa, Lake Huguangyan Maar and lakes from the Arctic region reported that a high concentration of bSi represents warm and humid climate conditions. Simultaneously, lower values indicate cold and dry climatic conditions (Shan et al. 2011). These studies suggested that changes in diatom production are related to changes in climate. Therefore, biogenic silica can be used to reconstruct past climate changes in lake sediments.

6.2 BIOGEOCHEMICAL CYCLE OF SILICA

The silicon cycle is a complex interaction of biological, chemical and geological processes occurring on a wide variety of spatial and temporal scales. The transport of Si from the land to the oceans through terrestrial ecosystems and river catchments to the estuarine and coastal zone is controlled by a complex set of terrestrial and aquatic processes. The weathering of silicates is an important sink for atmospheric CO_2. Silicon is also an essential element in terrestrial ecosystems, in soil formation processes and in regulating species composition of vegetation. The import of Si into coastal zones from the terrestrial environment is essential to sustain diatom growth. Diatoms play a vital role in the oceanic carbon sink and eutrophication of coastal zones. Because of its global environmental impact, the silicon cycle is receiving considerable attention in recent decades.

Atmospheric CO_2 plays an essential role in silicate weathering. During silicate weathering, dissolved CO_2 is consumed and dissolved silicate (DSi), i.e., ortho-silicic acid (H_4SiO_4), is released from the crystalline structure of silicate minerals. For example, in the weathering of anorthite (over kaolinite) to gibbsite, DSi is produced, and CO_2 is consumed (Stumm and Morgan 1974).

$$CaAl_2Si_2O_8 + 2CO_2 + 8H_2O \rightarrow Ca2^+ + 2Al(OH)_3 + 2H_4SiO_4 + 2HCO_3$$

$$(Anorthite) \qquad\qquad\qquad (Gibbsite) \ \left(Ortho\text{-}silicic\ acid\right)$$

Further, DSi (H_4SiO_4) is transported through rivers and released eventually to the lakes and oceans. The significant silica sources include rivers, groundwater flux, seafloor weathering inputs, *hydrothermal vents*, and atmospheric deposition (*aeolian flux*)

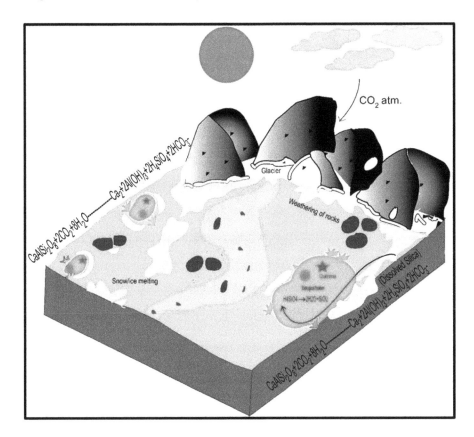

FIGURE 6.1 Cycling of silica in the lake environment.

and glacial weathering of rocks. Rivers are the largest silica source to the marine environment, contributing up to 90% of the total silica delivered to the aquatic systems (Gaillarde et al. 1999). In contrast, in the closed systems like lakes, silica is mainly derived through runoff and glacial weathering of rocks in the high-latitude regions (Figure 6.1). The DSi (H_4SiO_4) released to the lakes and oceans is consumed by primary producers like diatoms (40% of the total phytoplankton) as they extract silica from the water and use for the construction of their hard part known as diatom frustule in the form of biogenic silica or hydrated silica ($SiO_2.nH_2O$). They are responsible for extracting a large amount of silica from the water column (Wang et al. 2014), and the diatoms are critical players in the global biogeochemical cycle.

6.3 MATERIALS AND METHODOLOGY

6.3.1 Study Area and Sample Collection

The samples were collected from lakes around Ny-Ålesund, the International Arctic Research Base, located in the Spitsbergen Island of Svalbard archipelago (Figure 6.2) as part of the Indian Arctic Programme (summer phase) during August 2016.

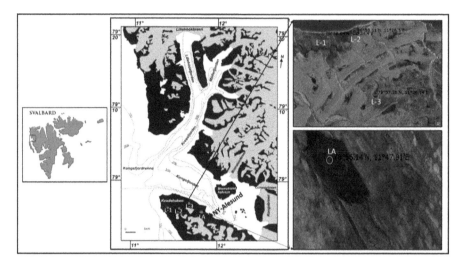

FIGURE 6.2   Map showing the study area. (Modified after Svendsen et al. 2002.)

Sediment core samples were collected from four shallow freshwater lakes. The lakes were fed by meltwater from glaciers, snowdrift and local precipitation. For the sediment core collection, a PVC hand-held corer was inserted by hammering manually into the lake sediment bed and then retrieved, and the cores were sub-sampled at 2 cm. Samples were labelled and transported to the laboratory in a frozen condition. In the laboratory, each sub-sample was dried at 60°C in the oven and utilised for further analysis.

6.3.2 SAMPLE ANALYSIS

Biogenic silica concentration in the sediment was measured by the wet alkaline extraction method, modified by Mortlock and Froelich (1989) and Muller and Schneider (1993), where the blue silicon-molybdenum complex intensity was measured using a visible spectrophotometer (UV-1800 Shimadzu) at 810 nm. The reagents used for the determination of biogenic silica were prepared using the following procedures.

The sulphuric acid solution was prepared by adding 250 ml of concentrated sulphuric acid slowly and continuously to 750 ml of Milli-Q water while stirring. 38 g of ammonium heptamolybdate tetrahydrate, $(NH4)6Mo7O_{24}$, was dissolved in 300 ml of Milli-Q water, and this solution was added to 300 ml of sulphuric acid solution to prepare ammonium molybdate solution. 10 g of oxalic acid was dissolved in 100 ml of Milli-Q water. 2.8 g of ascorbic acid powder was added to 100 ml of Milli-Q water to prepare an ascorbic acid solution.

After the preparation of all the reagents required, 0.2 g of the sediment sample was weighed accurately and transferred to a 50 ml polypropylene centrifuge tube with 25 ml of 1% Sodium Carbonate (Na_2CO_3) solution as the alkaline treatment (Na_2CO_3) has a high bSi extraction efficiency (Liu et al. 2014). The sample was then placed in a preheated water bath at 85°C for 5 hours along with the blank to extract the biogenic silica from the sediment. Then, the sample was centrifuged to remove the suspended matter, and 1 ml of supernatant was released in a 100 ml volumetric flask containing 35 ml of Milli-Q water. To this solution, four drops of 1:100 HCl was added to bring the pH 7–8, i.e., close to seawater's pH. Subsequently, 1 ml of mixed reagent was added and kept for half an hour to form the yellow silicomolybdate complex. 1 ml of oxalic acid was added to the solution immediately (to decompose silicomolybdate complex) followed by 1 ml of ascorbic acid, which reduces the acid complex to a blue-coloured complex. Then, the solution was allowed to cool at room temperature for the reaction to complete, and the entire content was made up to 100 ml using Milli-Q water. After a wait of 30 minutes, the sample solution was taken in the cuvette of 1 cm. Against a reagent blank at a wavelength of 810 nm, the blue colour intensity was measured using the spectrophotometer. Using a standard Sodium Hexafluorosilicate, the stock solution was prepared, known as the working solution, which is utilised to prepare a calibration curve.

6.3.3 QUALITY CONTROL (QC) AND QUALITY ASSURANCE (QA)

The biogenic silica concentration was estimated from the calibration curve. Duplicate measurements were conducted on each sample, and relative error was noted to be less than 3%.

6.4 SOURCE AND DISTRIBUTION OF BIOGENIC SILICA (bSi) IN ARCTIC LAKES

Biogenic silica documents direct measure of biological products from the siliceous algae and diatoms (Conley 1998; Kaplan et al. 2002; Choudhary et al. 2018a–d), and biological productivity in lake sediment is characterised mainly by the diatom production (Birnie 1990; Roberts et al. 2001; Choudhary et al. 2018a). bSi concentration was found low in lake LA, L-1, L-2 and L-3 (Figure 6.3a–d), suggesting the presence of a low concentration of siliceous microfossils in these lakes, maybe due to the carbonate-rich sediments resulting in poor preservation of diatoms as also reported by Guilizzoni et al. (2006). However, increased concentration of biogenic silica from 6 cm to the surface and clay in core LA and L-2 (Figure 6.3a and c) indicated high productivity suggesting warming conditions in the region due to the lakes' exposure to the ice meltwater influx. In core L-1, bSi was relatively high in the middle portion of the core from 22 to 6 cm along with TOC and clay suggesting deposition of finer particles from suspension facilitating high productivity. The average C/N ratio in all the cores was found to be >15, indicating both autochthonous and allochthonous terrestrial sources for organic matter. Relatively large input of terrestrial organic matter derived from mosses growing in the catchment region of the lake is responsible for variation in bSi in the Arctic lakes. During the warmer conditions, lake water becomes warmer suitable for the growth of phytoplankton and zooplanktons resulting in overall increased productivity. Increased diatoms production can lead to an increased accumulation of biogenic silica in sediments, ultimately resulting in a decline in the water column reservoir of DSi (Conley et al. 1993). However, during the colder conditions, catchment area of the lakes would be covered with snow resulting in less solar radiation and nutrients, making it difficult for plants to survive and ultimately low diatom production. Therefore, the change in concentration of biogenic silica in the sediment column is related to the climatic conditions, nutrient concentration and variations in solar radiation. Paleolimnological evidence based on an accumulation of biogenic silica in the sediments has been used to study

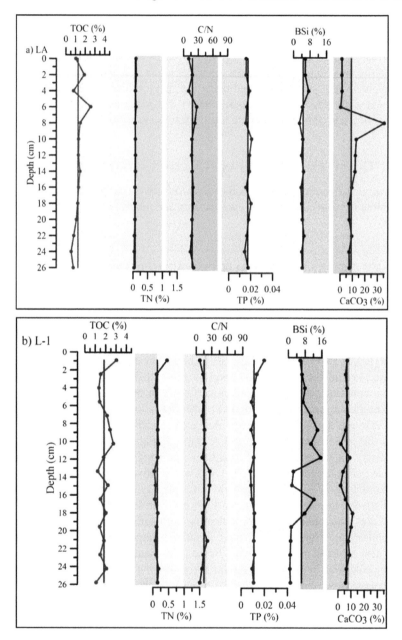

FIGURE 6.3 Depth-wise distribution of organic components of cores: (a) LA, (b) L-1, (c) L-2 and (d) L-3.

(*Continued*)

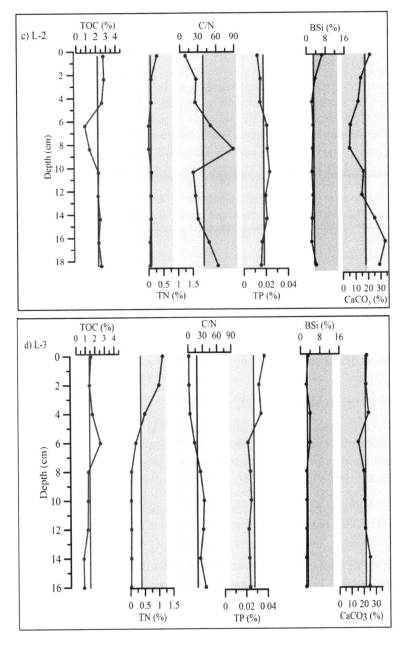

FIGURE 6.3 (CONTINUED) Depth-wise distribution of organic components of cores: (a) LA, (b) L-1, (c) L-2 and (d) L-3.

the change in climatic conditions and time course of depletion of DSi in the water column (Schelske et al. 1983, 1986a). Therefore, biogenic silica is an ideal proxy to understand the past environmental conditions archived in lacustrine sediments of the Arctic (Jiang et al. 2011).

6.5 FACTORS CONTROLLING THE DISTRIBUTION
OF BIOGENIC SILICA IN LAKES

6.5.1 Grain Size

In a closed environment like lakes where the exchange of water is relatively less and the supply of nutrients is limited, only small-sized diatoms can survive and constitute most phytoplankton. Hydrodynamic condition and resulting grain size affect the growth of diatoms and, ultimately, bSi concentration (Wang et al. 2014). Availability of relatively coarser sediments may reduce the concentration of bSi due to the dilution effects. However, in some regions, where the sediment is relatively coarse to medium, bSi concentration may also be high due to high primary productivity and low sedimentation rate. The high abundance of diatoms reduces the dilution effect of terrigenous sediments. Therefore, the sediment's grain size should be taken into account while using bSi as a proxy for paleo-productivity.

6.5.2 Biogeochemical Proxies (TOC, TN and TP)

Biogenic Silica and TOC often show similar variations indicating that biogenic silica has originated from the organic source. C/N ratio <10 suggests the source of organic matter exclusively derived from algae. A close association between bSi and other geochemical proxies indicates that the lake's organic matter is predominantly autochthonous and derived from algae (Choudhary et al. 2020).

6.5.3 Detrital Al and Al/SiO$_2$ Ratio

A similar distribution of biogenic silica, sand and Al in sediment indicated that dissolved silica either due to the incorporation of aluminium in diatom frustules (Van Bennekom et al. 1988; Choudhary et al. 2018a) or low silica solubility in a freezing environment might have caused precipitation of dissolved biogenic silica with Al in sediment (Rickert et al. 2002; Choudhary et al. 2018b). Low values of Al/SiO$_2$ ratio in the sediment support precipitation of biogenic silica with detrital Al, while higher values of Al/SiO$_2$ indicate replacement of Si by Al in diatom frustules.

6.5.4 Availability of Carbonate-Rich Sediments

Generally, in the Polar regions, CO$_2$ dissolution is high, leading to an increase in carbonic acid concentration, which explains the low CaCO$_3$ content. However, the availability of carbonate-rich sediments in the catchment area may dilute biogenic silica concentration, resulting in low siliceous microfossils (Choudhary 2019).

6.6 CONCLUSIONS

In Arctic lakes, bSi concentration was low, suggesting the presence of low siliceous microfossils in these lakes, which may be due to the carbonate-rich sediments resulting in poor preservation of diatoms. However, a relatively high concentration of bSi in the upper section indicated high primary productivity due to the warmer conditions and glacier retreat. Therefore, bSi is a robust proxy to understand the past environmental changes archived in lacustrine sediments of the Arctic. bSi content in Arctic lacustrine sediments is influenced by the input of terrestrial organic matter and many other factors like grain size, different biogeochemical proxies and detrital Al in the water column.

ACKNOWLEDGEMENTS

The authors place on record to thank the Director, National Centre for Polar and Ocean Research (NCPOR), Goa, for providing the opportunity to one of the authors (SC) to participate in the Indian Scientific Expedition to Antarctica and Ministry of Earth Sciences (MoES) for providing the logistic support required for the collection of samples. One of the authors (SC) thanks the University Grant Commission (UGC) for awarding a national fellowship (F1–17.1/2015–16/MANF-2015–17-UTT-51739). Another author (GNN) thanks the Council of Scientific and Industrial Research (CSIR) for granting the Emeritus Scientist position.

REFERENCES

Birks HJB, Jones VJ and Rose NL (2004). Recent environmental change and atmospheric contamination on Svalbard as recorded in lake sediments–synthesis and general conclusions. *Journal of Paleolimnology* 31(4), 531–546.

Birnie J (1990). Holocene environmental change in South Georgia: evidence from lake sediments. *Journal of Quaterny Science* 5(3), 171–187.

Carter SJ and Colman SM (1994). Biogenic silica in Lake Baikal sediments: results from 1990–1992 American cores. *Journal of Great Lakes Research* 20(4), 751–760.

Choudhary S (2019). Multiproxy paleoclimate reconstruction from the sediments of high latitude regions. Unpublished PhD. Thesis. Goa University, India.

Choudhary S, Nayak GN and Khare N (2018d). Provenance, processes and productivity through the spatial distribution of the surface sediments from Kongsfjord to Krossfjord system, Svalbard. *Journal of Indian Association of Sedimentologists* 35(1), 47–56.

Choudhary S, Nayak GN and Khare N (2020). Source, mobility and bioavailability of metals in fjord sediments of Krossfjord and Kongsfjord system, Arctic, Svalbard. *Environmental Science and Pollution Research* 27, 15130–15148.

Choudhary S, Nayak GN, Tiwari AK and Khare N (2018b). Sediment composition and its effect on productivity in Larsemann Hills, East Antarctica. *Arabian Journal of Geosciences* 11(15), 416.

Choudhary S, Nayak GN, Tiwari AK and Khare N (2018c). Source, processes and productivity from distribution of surface sediments, Prydz Bay, East Antarctica. *Polar Science* 18, 63–71.

Choudhary S, Tiwari AK, Nayak GN and Bejugam P (2018a). Sedimentological and geochemical investigations to understand the source of sediments and recent past processes in Schirmacher Oasis, East Antarctica. *Polar Science* 15, 87–98.

Colman SM, Peck JA, Karabanov EB, Carter SJ, Bradbury JP, King JW and Williams DF (1995). Continental climate response to orbital forcing from biogenic silica records in Lake Baikal. *Nature* 378, 769–771.

Conley DJ (1998). An inter-laboratory comparison for the measurement of biogenic silica in sediments. *Marine Chemistry* 63(1), 39–48.

Conley DJ, Schelske CL and Stoermer EF (1993). Modification of the biogeochemical cycle of silica with eutrophication. *Marine Ecology Progress Series* 101, 179–192.

Gaillardet J, Dupré B, Louvat P and Allègre CJ (1999). Global silicate weathering and CO2 consumption rates deduced from the chemistry of large rivers. *Chemical Geology* 159 (1–4), 3–30.

Guilizzoni P, Marchetto A, Lami A, Brauer A, Vigliotti L, Musazzi S, Langone L, Manca M, Lucchini F, Calanchi N and Dinelli E (2006). Records of environmental and climatic changes during the late Holocene from Svalbard: palaeolimnology of Kongressvatnet. *Journal of Paleolimnology* 36(4), 325–351.

Jiang S, Liu X, Sun J, Yuan L, Sun L and Wang Y (2011). A multi-proxy sediment record of late Holocene and recent climate change from a lake near Ny-Ålesund, Svalbard. *Boreas* 40(3), 468–480.

Kaplan MR, Wolfe AP and Miller GH (2002). Holocene environmental variability in southern Greenland inferred from lake sediments. *QuaternaryResearch* 58(2), 149–159.

Liu B, Xu H, Lan J, Sheng E, Che S and Zhou X (2014). Biogenic silica contents of Lake Qinghai sediments and their environmental significance. *Frontiers of Earth Science* 8(4), 573–581.

Mortlock RA and Froelich PN (1989). A simple method for the rapid determination of biogenic opal in pelagic marine sediments. *Deep-Sea Research, Part A Oceanography Research Papers* 36(9), 1415–1426.

Muller PJ and Schneider R (1993). An automated leaching method for the determination of opal in sediments and particulate matter. *Deep-Sea Research Part I: Oceanography Research Papers* 40(3), 425–444.

Rickert D, Schluter M and Wallmann K (2002). Dissolution kinetics of biogenic silica from the water column to the sediments. *Geochemica et Cosmochimica Acta* 66 (3), 439–455.

Roberts D, Van Ommen TD, McMinn A, Morgan V and Roberts JL (2001). Late-Holocene East Antarctic climate trends from ice-core and lake-sediment proxies. *The Holocene* 11(1), 117–120.

Schelske CL, Conley DJ, Stoermer EF, Newberry TL and Campbell CD (1986a). Biogenic silica and phosphorus accumulation in sediments as indices of eutrophication in the Laurentian Great Lakes. *Hydrobiologia* 143, 79–86.

Schelske CL, Stoermer EF, Conley DJ, Robbins JA and Glover R (1983). Early eutrophication in the lower Great Lakes: new evidence from biogenic silica in sediments. *Science* 222, 320–322.

Shan J, XiaoDong L, Liqiang XU and LiGuang S (2011). The potential application of biogenic silica as an indicator of paleo-primary productivity in East Antarctic lakes. *Advances in Polar Science* 22(3), 131–142.

Stumm W and Morgan JJ (1970). *Aquatic Chemistry: An Introduction Emphasising Chemical Equilibria in Natural Waters*. Wiley- Interscience, New York.

Svendsen H, Beszczynska-Moller A, Hagen JO, Lefauconnier B, Tverberg V, Gerland S, BorreOrbaek J, Bischof K, Papucci C, Zajaczkowski M and Azzolini R (2002). The physical environment of Kongsfjorden–Krossfjorden, an Arctic fjord system in Svalbard. *Pollution Research* 21(1):133–166.

Van Bennekom AJ, Berger GW, Van der Gaast SJ and De Vries RTP (1988). Primary productivity and the silica cycle in the Southern Ocean (Atlantic sector). *Palaeogeography, Palaeoclimatology, Palaeoecology* 67(1–2), 19–30.

Wang L, Fan D, Li W, Liao Y, Zhang X, Liu M and Yang Z (2014). Grain-size effect of biogenic silica in the surface sediments of the East China Sea. *Continental Shelf Research* 81, 29–37.

7 Role of Persistent Organic Pollutants and Mercury in the Arctic Environment and Indirect Impact on Climate Change

Anoop Kumar Tiwari and
Tara Megan Da Lima Leitao
National Centre for Polar and Ocean Research

CONTENTS

DOI: 10.1201/9781003265177-7

7.1 INTRODUCTION

The north pole comprises landmasses above the 60°N latitude contributed by countries such as Russia, United States of America, Alaska, Norwich, Sweden, Finland, Greenland, Iceland, Canada and Norway. The region from the Arctic towards the circumpolar equator includes polar and boreal ecosystems. They have their distinct climatic signatures. The Arctic Ocean is one of the five major oceans. It consists of the Barents Sea, Beaufort Sea, Baffin Bay, Chukchi Sea, East Siberian Sea, Greenland Sea, Hudson Bay, Hudson Strait, Laptev Sea, Kara Sea, and the White Sea, which are connected to the Pacific Ocean by Bering Strait and to Atlantic Ocean through the Greenland Sea and Labrador Sea (Pidwirny 2006).

Resident indigenous peoples are enduring extreme conditions daily and are dependent on local produce for survival. Unlike Antarctica, the Arctic is also home to many animals such as the Arctic foxes, polar bears, Arctic caribou, Siberian cranes, etc., which thrive off the food chains' webs in this peaceful environment.

7.1.1 POLLUTANTS IN THE ARCTIC

The world is responsible for generating an immense quantity of diverse persistent pollutants that travel and condense in low-temperature regions, specifically the poles. The introduction of pollutants in this region has increased mortality rates of the ecosystem's apex members. Arctic Monitoring and Assessment Programme (AMAP) focuses on key issues and create a change to be presented to policymakers. AMAP reports enclose hazard profiles for persistent and bio-accumulative Persistent Organic Pollutants (POPs) upon which planetary boundary assessments have been conducted (Reppas-Chrysovitsinos et al. 2017).

7.1.2 ATMOSPHERE

The sunlight in the Arctic is governed by the Earth's tilt towards or away from the Sun. It leads to longer days or nights. The lower Arctic atmosphere is ~two km above sea level, where most photochemical reactions result in the troposphere's cycling and lead to ozone depletion (Barrie et al. 1988). The latent heat stored in the Arctic Ocean as well as ocean inevitability released during autumn into the troposphere, affected melting of sea ice cover to a large extent in the past decade, and therefore late summer water opening continues (Overland and Wang 2010). Air pollutants travel to the Arctic by three pathways: low-level transport followed by an ascent to the Arctic, low-level transport alone and uplift outside the Arctic followed by a descent into the Arctic (Stohl 2006). Winds are deflected to the Coriolis force's right in high-pressure regions, current circulates clockwise, and in low-pressure areas, current circulates counterclockwise. These patterns affect the deposition of POP and mercury (Hg) in the Arctic.

7.1.3 EXTERNAL EFFECTS OF HALOGENS ON MERCURY AND POP

Pollutants may be produced at the poles or travel over considerable distances through the atmosphere and oceans and get deposited in the Polar Regions. According to

Shiraiwa et al. (2017a, b), atmospheric secondary organic aerosols (SOAs) travel in semi-solid or solid state through interaction with ice nucleation centres and settle in liquid phase in the Polar areas of high relative humidity. In the recent past, experiments were conducted for five consecutive nights at Station Nord, Greenland, to reveal elemental mercury (Hg^0) production due to rapid oxidation of bromine species (Br^- and BrO^-) to form $Hg(II)$ in the air present in the snow up to a depth of 40 cm (Ferrari et al. 2004).

Researchers found that the depletion of atmospheric mercury occurs in the lower troposphere of the Arctic during springtime (Schroeder et al. 1998, 1999) due to bromine radical photochemical oxidation (Holmes et al. 2006) by which mercury enters terrestrial and marine environments through wet (Sanei et al. 2010) or dry deposition (Lindberg et al. 2001).

7.2   DISCUSSION

7.2.1   PERSISTENT ORGANIC POLLUTANTS

These persistent pollutants can be classified as organic and inorganic. POPs are created from incomplete combustion of fossil fuels (Ma et al. 2017) and forest or biomass burning (Luo et al. 2020), oil spills and unbridled use of pesticides, insecticides and weedicides in the agricultural industry, industrial and manufacturing activities (Balmer et al. 2019a; Bartlett et al. 2019). Poly-aromatic hydrocarbons (PAHs) in marine Arctic water and sediment originates from natural underwater seeps, and those measured in air, freshwater and terrestrial environments are predominantly from atmospheric and combustion-derived sources (Balmer et al. 2019a).

POP levels are well defined in the Stockholm convention signed on the 22nd day of May 2001 and was enforced on the 17th day of May 2004. POPs are a diverse group of long-range pollutants. These include congeners of polychlorinated biphenyls (PCBs), PAHs, organo-chlorinated pesticides (OCPs), chlorofluorocarbons (CFCs) and organo-brominated pesticides (OBPs). Chlorinated kinds of paraffin exist as a mixture of short-chain (C_{10}–C_{13}), medium-chain (C_{14}–C_{17}) and long-chain ($>C_{18}$) congeners and can be measured by mass spectroscopy (Vorkamp et al. 2019b).

Halogenated natural products (HNPs) are a subset of POPs. It contains bromine, chlorine and iodine. It rarely contains fluorine, along with oxygen and nitrogen. It consists of the following classes: bromophenols, bromoanisoles, hydroxylated and methoxylated bromodiphenyl ethers and polybrominated dibenzo-p-dioxins (Bidleman et al. 2019).

Shrestha et al. (2009) discovered hexachlorobenzene and phenanthrene in contaminated soils (Shrestha et al. 2009). Polychlorinated biphenyls (PCBs) are expressed as 209 congeners (Mullins et al. 1984; Frame 1997). The study also includes dichlorodiphenyltrichloroethane (DDT) to dichlorodiphenyldichloroethylene (DDE) (Zepp et al. 1977), toxaphene and congeners (Casida et al. 1974), mirex (Holdrinet et al. 1978), endrin (Katsoyiannis and Samara 2004), methoxychlor (Monitoring 2014), chlorinated benzenes (Savinov et al. 2011), chlordane compounds (Rigét et al. 2010), polychlorinated dibenzo-p-dioxins and furans (PCDD/Fs) (Wittsiepe et al. 2008;

Hoferkamp et al. 2010; Savinov et al. 2011). Current-use of pesticides including 2-methyl-4-chlorophenoxyacteic acid, metribuzin and pendimethalin. Phosalone, quizalofop-ethyl, tefluthrin and triallate have also been reported recently (Balmer et al. 2019a, b).

7.2.2 LONG-RANGE ATMOSPHERIC TRANSPORTATION (LRAT)

POPs' presence in high concentration in the Arctic proves its ability to disperse long distances in gaseous or particulate form, either free or adsorbed on a particle's surface and sediment in the rain, snow or ice. Polluted air parcels from long distances originated from Asia and Africa from desert areas and, mixed with natural and anthropogenic emissions, make their way to the Arctic (Pacyna and Keeler 1995). In the Arctic, Western Russia was a dominant source for PCB-28 and PCB-101 and in Central Europe for PCB-180, the U.S.A. contributed 15% of PCB-180 to Alert Station (Ubl et al. 2012).

Northern Contaminants Programme (NCP) created a baseline monitoring platform specifically for POPs (Hung et al. 2005), which depicted atmospheric flow patterns in spring, LRAT of POPs to the Arctic (Kallenborn et al. 2015). Lindane (γ-HCH) was used as a pesticide for canola seed in Canadian prairies with dangerously high atmospheric loading levels (Waite et al. 2005). Samples from the poles revealed a decreasing trend in PCDD/F concentrations at Zhangmu-Nyalama > Arctic > Antarctica suggesting air parcels with significant emissions introduced into the Arctic and Tibetan plateau environments (Jia et al. 2014).

Long-range transportation of atmospheric PFRs in the Arctic was higher than that of HFRs. The dry deposition of Halogenated Flame Retardants (HFRs) in seawater and ice and high volatilisation rates of Organophosphorus Flame Retardants (PFRs) in seawater were observed by Vorkamp et al. (2019b). Atmospheric concentrations of chlorinated kinds of paraffin are as follows: short chain chlorinated paraffins (SCCP) > medium chain chlorinated paraffins (MCCP) > long chain chlorinated paraffins (LCCP), SCCP and MCCPS, also seen in the Antarctic indicated long-range transportation (Vorkamp et al. 2019a). Long Range Atmospheric Transport (LRAT) for HBB, DPTE and PBT was studied by Möller et al. (2010, 2011). Barents Sea sediment cores revealed the LRAT hypothesis of PAH and PCB congeners (Zaborska et al. 2011). LRAT of new brominated flame retardants (NBFRs), polybrominated diphenyl ethers (PBDEs) and dechlorane plus (DDC-CO) in Arctic communities have been suggested (Carlsson et al. 2018). Long-range transportation in the Arctic of atmospheric PFRs was higher than HFRs.

The dry deposition of HFRs in seawater and ice and high volatilisation rates of PRFs in seawater were observed by Vorkamp et al. (2019b). K_{ow}/K_{aw} with acidity/basicity of all POP compounds at relevant pH values (in microenvironments) is an excellent screening assessment tool of contaminants in the Arctic (Rayne and Forest 2010). Due to constant evaporation and deposition events in the poles, POPs demonstrate a 'grasshopper' effect and get deposited into deep benthic sediments through oceanic processes (Gouin and Wania 2007; Muir and de Wit 2010). Level III fugacity model of Bering Sea simulated processes of transfer and fate of PAHs in air and water, highlighting a summer air-to-sea flux determining the ocean as a summer sink for PAHs (Ke et al. 2017).

7.2.3 POPs in the Environment

The prevalence of POPs in environmental matrices depends on their octanol: water partition coefficients (Kelly and Gobas 2003), fugacity ratios (Sweetman et al. 2005), enantiomer fraction (Eljarrat et al. 2008) and photochemical oxidation, amongst other *in situ* factors. Cold condensation by latitudinal fractionation of long-range transport volatile pollutants determines the distribution of specific POPs (Malmquist et al. 2003).

Atmospheric temperatures control specific organo-halogen pesticides concentrations, wherein DDE showed a high concentration during winter in the Canadian and Russian Arctic (Halsall et al. 1998).

Brominated congeners of some POPs associated with particles were studied for their volatilisation fates at Alert by Su et al. (2008). A long-term monitoring of air concentration module exploring loading and transport in the Arctic was conducted by Li et al. (2004). The flux of γ-HCHs in the High Arctic atmosphere is dependent on snowfall, and its re-emission into the atmosphere is amplified by wind conditions (Halsall 2004). Fu et al. (2009) documented the decrease in aerosol pollution after polar sunrise (Fu et al. 2009a).

A comparison on atmospheric monitoring data from Alert, High Arctic and Nam Co, Tibetan plateau showed three new flame retardants 1,2-bis(2,4,6-tribromophenoxy) ethane (BTBPE), 2-ethyl-1-hexyl-2,3,4,5-tetrabromobenzoate (ETHeBB) and bis(2-ethyl-1-hexyl) tetrabromophthalate (TBPH) at high concentrations (Xiao et al. 2012). Temporal trends were studied at four Arctic stations for POPs using high-volume air samplers and environmental persistence; long-range transport was determined by Hung et al. (2016). Atmospheric measurements from 2006 to 2014 revealed perfluorinated carboxylates (PFCAs) and perfluoroalkyl sulfonic acids (PFSAs) are a growing concern, along with fluorotelomer alcohols (FTOHs); perfluoro butyrate (PFBA) and perfluoro octanoic acid (PFOA) are the most prominent over the Arctic environment (Muir et al. 2019).

7.2.4 POPs in Ice

Wet deposition in the form of snow for PAHs accounts for significant PAH atmospheric losses in winter (Halsall et al. 2001). A study of POP interaction within the ice in different microenvironments in the Barents Sea suggested that ice is not the right candidate for long-range transportation of PCBs (Gustafsson et al. 2005). Semi-volatile organic compounds (SVOCs) in snow revolute during snow melting but possess high interfacial air partitioning coefficients or lower Henry's law constants, allowing them to penetrate deeper snow layers (Herbert et al. 2006).

Polyfluoroalkyl substances (PFASs) in seawater revealed north Atlantic Ocean > Greenland Sea > Southern Ocean with elevated levels of perfluorooctanoic acid (PFOA), perfluorohexanesulphonic acid (PFHxS), perfluorohexanoic acid (PFHxA), perfluorooctanesulphonic acid (PFOS) and perfluorobutanesulphonic acid (PFBS);

melting snow and sea ice in the Greenland Sea is a source for PFASs in the marine ecosystem (Zhao et al. 2012). α-HCH was measured in the Beaufort Sea in the polar mixed layer (PML), and the Pacific Mode layer (PL) suggested that ice formation led to an increase in concentration of this pollutant in the water beneath the ice and that majority of it is degraded longer than the atmosphere by a 20-year lag (Pućko et al. 2013). It is also noticed that meltwater ponds were considered loading sites (legacy OCPs, CUPs) from the atmosphere by sea ice deposition and subsequent delivery to the Beaufort Sea (Pućko et al. 2015).

Organic tracers from ice cores have been deflated to the atmosphere from regional sources, mixed by aeolian processes and scavenged by snow (wet deposition) and dry deposition and incorporated into firm glacial archives (Seki et al. 2015). Novel brominated flame retardants NBFRs such as hexabromobenzene (HBB), pentabromotoluene (PBT) and pentabromobenzene (PBBz) in Arctic air and soil from Ny-Ålesund during 2014–2015 were studied for their air-soil fugacity and to determine the rates of deposition (Hao et al. 2020). Net air-snow fluxes of PFASs (FTOHs and FTAs) in Arctic snow showed a positive value for net volatilisation fluxes into the atmosphere just as soon as wet deposition occurs (Xie et al. 2015). PCBs distribution on leeward and windward faces of Mt. Sygera, in the Southeast of Tibetan Plateau, depends on wind factor and precipitation driven by the Indian monsoon, and is associated with total organic carbon (Meng et al. 2018).

7.2.5 POPs in Lakes

High Arctic lakes are in-efficient sinks for POPs, and the mobility of these contaminants depends on the hydrological regime (partial flow through melting), minimal scavenging and retention in sediments with low carbon settling and sediment particles. It resides in the water column (Diamond et al. 2005). Transport of PFAS in Svalbard, Norway, through ice core analysis (1997–2000) indicated the distribution of contaminants into Arctic locations through the surface, melted glacier and river water (Kwok et al. 2013). Melt pond concentrations in sea ice in the Arctic during summer varied due to precipitation, gaseous exchange and sea ice melting over time (Pućko et al. 2017).

7.3 METHODOLOGY

7.3.1 Sampling Points

Land-based Arctic stations: Alert on Ellesmere Island, Canada (Hung et al. 2016), Storhofði on Iceland (Hung et al. 2010), Zeppelin at Ny-Ålesund in Svalbard (Ma et al. 2011), Pallas in Finland (Hansson et al. 2006), Villum station in Station Nord (Skov et al. 2020), Greenland, Dunai Island (Stern et al. 1997), Valkarkai (Monitoring 2014) Tiksi in the Arctic (Konoplev et al. 2012), Tagish/Little fox lake in Canada (Helm et al. 2004) and AndØya in Norway (Bidleman et al. 2017) are used for atmospheric air modelling of POPs.

Datasets of atmospheric concentrations of α- and γ-HCHs at Zeppelin Mountain Air monitoring station, Ny-Ålesund, Svalbard, Norway and Alert, the Canadian

Arctic station (Becker et al. 2008) are also used for mercury and POP data collection and research. Offshore (cruise expeditions) atmospheric and hydrosphere data collection in the Arctic and northern oceans was carried out by Harner et al. (1998) and Möller et al. (2010, 2011).

Kongsfjorden in Svalbard, Norway, is an area featuring compounded Arctic microenvironments: benthic sediments (Szczybelski et al. 2016), water masses (marine) (Pouch et al. 2017), marine apex predators (Lydersen et al. 2016), seabirds (Murvoll et al. 2006; Haugerud 2011; Ortiz 2016) and fjord ecosystems (Hallanger et al. 2011). Liefdefjorden in Svalbard, Norway (Evenset et al. 2009; Vieweg et al. 2010; Warner et al. 2010; Hallanger et al. 2011; Carlsson 2013; Carlsson et al. 2014) and Adventfjorden (Warner et al. 2010; Pouch et al. 2017; Carlsson et al. 2018; Johansen 2019) are similar sampling areas.

Surface waters are collected from East-West Lake, Cape Bounty and Allen Bay on Barrow Strait and deep water mooring stations in Eastern and Western Fram straits (Ma et al. 2017). Surface waters from lakes and streams in the vicinity of Polish Polar Station in Hornsund (Svalbard) situated in the southern part of Spitsbergen on Horsund Fjord's shores were studied by Pawlak et al. (2019). In colder regions other than the Arctic, below 60°N, Kragujevac in Serbia was analysed for POPs post-war (Turk et al. 2007).

7.3.2 Air Sampling

Active sampling is a convenient means by which air can be sampled, requiring a power source and constant maintenance. In contrast, the passive selection is inexpensive and easy to operate with no pre-requisite power source. Data from active sources are quantitative, whereas data from passive sources are semi-quantitative (Hung et al. 2010, 2016). Conventional high-volume air sampling (HVS) assembly includes a glass-fibre filter (GF/F) situated upstream to trap particle-bound contaminants followed by a polyurethane foam (Eriksson et al. 2013) or XAD resin to collect gas-phase organic pollutants (Galarneau et al. 2006). Choi et al. (2008) deployed XAD resin, passive samplers for 1 year (2005–2006) at the Korean Polar Research Station at Ny-Ålesund, Norway (Choi et al., 2008).

7.3.2.1 Active Air Sampling

Air samples are collected using high-volume samplers with glass-fibre filters and polyurethane foams (PUF's). In order to trap particle and gas-phase POPs over a period of time, in three locations in Kragujevac, Turk et al. (2007) used high-volume air samplers with an assembly consisting of GV2360 Thermoandersen TSP which is made of stainless steel boxes for GF/F and polyurethane foam filter holders using a 1200 W motor on the inverter to draw in air with its flow measured by Sierra 620 fast flow insertion mass flow meter with totaliser.

The average air sampled by the background sampler was $1600\,m^3/24$ hours and by the other two samplers, glass fibre, was $900\,m^3/24$ hours. Becker et al. (2008) used two HVS at two stations with 1 GFF and 2 PUF plugs at a rate of $900–12,000\,m^3$ over a 48-hour weekly basis (Zeppelin) and $13,000\,m^3$ over seven days weekly basis.

Fu et al. (2009a) collected total suspended particles using a high-volume sampler at the Arctic from February to June 2008. Xie et al. (2015) conducted air sampling in Ny-Ålesund using a high-volume air sampler at ~15 m³/hour for seven days using GF/F (gas phase) and PUF/XAD-2 cartridge (particle phase).

Onboard a stationary Amundsen cruise ship, Wong et al. (2011) collected air and water samples using stainless steel pipes above sea level connected to a low-volume sampler passed through GF/Fs plugged with PUF plugs. In contrast, the lower deck had high-volume air samplers at the ship's bow and filled in the same manner. Wu et al. (2014) used an HVS, guided by a wind sensor, to collect a wind-blowing sample over the bow. High-volume air sampler on the upper deck functioned only to sample masses of air coming straight from the ocean and eluded contamination from the ship's smokestack using a wind vane attachment (Galbán-Malagón et al. 2012).

7.3.2.2 Passive Air Sampling

Passive samplers are ideal for remote locations and inaccessible routes. Passive air sampling over 10–60 days (2003–2004) indicated that PCB accumulation is linear over time (Sakin et al. 2017). Turk et al. (2007) used passive air samplers with PUFs in chambers prewashed and rinsed with solvent acetone before installation. The global aquatic passive sampling (AQUA-GAPS) used passive samplers to monitor POPs in the Arctic waters (Lohmann and Muir 2010).

7.3.3 Water Sampling

Due to global distillation and condensation at higher latitudes, POPs arrive at the poles through the atmosphere and worm their way into the hydrosphere (Connell et al. 1999). In 2013, water was considered for global assessment of hydrophilic instead of the previous hydrophobic POPs (Muir and Lohman 2013). Evenset et al. (2007b) collected terrestrial lake water at EllasjØen, BjØrnØya, Norway, in 1999 and filtered it through GF/F filters and PUF for suspended particulates and dissolved fractions. Wong et al. (2011) collected marine water samples on international polar year expeditions (2007–2008) to the Canadian Arctic (Labrador Sea, Hudson Bay and the southern Beaufort Sea), using low-volume sampler (LVS) with GF/F followed by SPE containing ENV cartridges.

In contrast, solid adsorbent (XAD-2) resin was used in HVS for dissolved phase analytes. For a composite study, Galbán-Malagón et al. (2012) collected atmospheric (aerosol and gases), seawater (dissolved and particulate) and plankton samples between Iceland and Arctic ice-margin Zone and Svalbard archipelago in 2007. For seawater samples, the HVS was equipped with XAD-2 resin (dissolved) and GF/F (particulate). Pawlak et al. (2019) collected surface lake water and stream samples close to Polish Polar Station in Hornsund (Svalbard) in 2010, at multiple points, to re-create the watercourse of the lake.

7.3.4 Snow and Ice Sampling

Snow can be transient (seasonal snowpack) or permanent (permafrost and ice caps), precipitates as snowfall, hailstones, graupel and rime, and interferes with diffusive exchange (Wania et al. 1998). Contaminant fates are strongly dependent on snow/ice formation in the Arctic as most land and seas are covered (Daly and Wania 2004). Some factors that aid in these fates are snow ageing, wind conditions, volatilisation rates and snowfall rate (Halsall 2004). Precipitation of snow leads to accumulation/ entrapment of contaminants in the seasonal snowpack, which on melting reverts back into freshwater and marine reservoirs (Wania et al. 1998). Snow samples were collected in Ny-Ålesund (approximate. 4 hours after snowfall) by Xie et al. (2015). Ice cores from Site-J, Greenland and Ushkovsly were collected and processed by lipid extraction (Seki et al. 2015).

7.3.5 Pretreatments

Oehme and Stray (1982) described the usage of pretreated PUF giving excellent recoveries between 80% and 115% with efficiency affected by temperature, humidity and sampling volume. Turk et al. (2007) pretreated GF/Fs at 400°C for 5 hours and PUFs by Soxhlet extraction 1:1 acetone:hexane (Merck suprasolv) by Foss Tecator Soxtec 1045 HT-2 apparatus for 4 hours at 120°C with GFFs and PUFs wrapped in aluminium foil and placed in a zipped bag. Evenset et al. (2004) pre-burned glass jars and transferred pooled guano samples and froze them at −20°C; all filters, PUFs and wrapping materials were precombusted at 450°C for 8 hours before use. Choi et al. (2008) rinsed resin meshes with 10 mL of toluene/acetone (80:20 ratio), and the rinsate was added to the Soxhlet apparatus. Turk et al. (2007) prewashed filters.

They cleaned the filters (8-hour extraction in acetone and 8-hour extraction in dichloromethane) wrapped in two layers of aluminium foil, labelled, placed into zip-lock polythene bags transported in an icebox at 5°C and placed in lab freezer at −18°C. Gas chromatography coupled mass spectroscopy was conducted after that. Fu et al. (2009b) precombusted quartz fibre filters at 450°C for 3 hours. They placed them in precombusted glass jars with Teflon-lined screw cap at −20°C in darkness and were stored from 1991 till 2007.

Wong et al. (2011) used PUFs precleaned by rinsing with water, acetone wash and Soxhlet extraction in petroleum ether for 22 hours each; GF/Fs were pre-baked at 400°C overnight; XAD-2 resin was cleaned by Soxhlet extraction in acetone, hexane and dichloromethane (DCM) each for 2 days; and PUF plugs and SPE cartridges were stored at −22°C and XAD-2 resin was stored in a refrigerator at 4°C until extraction.

Net trawl (mesh size 200 μm) was cleaned with pre-filtered seawater and un-salted water and placed into a zip-sealed bag and glass-fibre filter. Whatman filter paper of 47 mm, 2.7 μm pore size was precleaned as per standard protocol. After washing these filters were folded and wrapped in aluminium foil and placed in zip-locked bags and kept in the freezer (−20°C) (Galbán-Malagón et al. 2012). Besides this, he used precombusted (4 hours at 450°C) and pre-weighed quartz microfibre filter (QMA 203×254 mm, 1 μm nominal pore size, Whatman) and a PUF plug (100 mm diameter×120 mm, Klaus Ziemer GmbH) precleaned using Soxhlet with hexane/DCM (1:1) for 24 hours, and QMAs were precombusted for 4 hours at 450°C in foil sleeves and placed into zip-locked bags and stored in the refrigerator on

board. Galbán-Malagón (2012) pre-cleaned XAD-2 resin columns with methanol and dichloromethane, and kept at 4°C until sampling These cartridges were
re-sealed and placed in 4°C refrigerator lab on boards. Aerosol filters were kept at
−20°C until analysis. Wu et al. (2014) used precleaned PUFs (Soxhlet extraction
for 24 hours using acetone and hexane (1:1 $^v/_v$) and then exposed to ambient air over
the sampling period. Xie et al. (2015) extracted 5–8 L meltwater through a polymer
resin cartridge (40 g PAD-3) packed into a glass column at ~300 mL/minutes for
neutral PFASs analysis, and columns were stored at 4°C and GF/Fs at −20°C.

7.3.6 SPIKING

Turk et al. (2007) spiked each filter with surrogate recovery standards before extraction and internal measures such as terphenyl for PAHs and PCB-121 were added for
analyses. Choi et al. (2008) spiked samples with a mixture of $^{13}C_{12}$-labelled PCBs
and a combination of labelled OCPs as internal standards. Fu et al.'s (2009b) analyses
gave recovery for PAHs > 80%.

Wong et al. (2011) fortified the samples with d6-α-HCH before extraction, and
internal standard mirex was added post-extraction. Galbán-Malagón et al. (2012)
spiked 10 ng of PCBs to samples. Internal standards were spiked into GFF and PUF/
XAD-2 cartridges before extraction. Xie et al. (2015) and Pawlak et al. (2019) used
four internal standards injected into the samples PCB 28 C^{13}, PCB 180 C^{13}, naphtha-
lene-d_8 and benzo[a]anthracene-d$_{12}$ (Supelco).

7.3.7 EXTRACTION

Turk et al. (2007) extracted passive samplers with DCM in Buchi system B-811
automatic extractor. Choi et al. (2008) removed the spiked samples by Soxhlet
extractor for 16 hours with 350 mL toluene/acetone (80:20). Extracts were split quan-
titatively at 3:1, a more significant fraction for PCB analysis and smaller for OCP.
Extraction of filter aliquots was done thrice with dichloromethane/methanol (2:1, $^v/_v$)
by ultrasonication for 10 minutes and then filtered through quartz wool-packed
Pasteur pipettes (Fu et al. 2009b). Wong et al. (2011) proceeded with extraction with
Soxhlet and with 400 mL of petroleum ether. Galbán-Malagón et al. (2012) extracted
samples recovered from gas, aerosol, particulate and phytoplankton with Soxhlet
for 24 hours wherein PUF extraction used solvents acetone:hexane in the ratio of
3:1, and for aerosol, particulate and phytoplankton filters were used with solvents
hexane:dichloromethane in the ratio of 1:1. Xie et al. (2015) transported filters in
aluminium bags sealed tightly and stored at 4°C and −20°C before extraction with
modified Soxhlet apparatus (M.S. extractor) for 16 hours with DCM.

Lipid extraction was done with methylene chloride/ethyl acetate (2:1) mix, and
saponification was done with 0.5 M KOH/methanol (Seki et al. 2015). Two-stage
liquid-liquid extraction was done with DCM by adding 15 dm^3 of DCM followed by
30 minutes of shaking (Pawlak et al. 2019). Wu et al. (2014) used accelerated solvent
extraction (ASE 200, Dionex) with hexane and DCM (1:4 v/v) at 110°C and 1500
psi for two static cycles with heating time (T_h) of 6 minutes and fixed time (T_s) of 3
minutes to give 1 mL extract.

7.3.8 Concentration

Extracted samples were concentrated with Heipemdolf VV micro-concentrator and agitated for 1 minute with concentrated sulphuric acid (1:1), and their volume was reduced under a gentle nitrogen stream at ambient temperature (Turk et al. 2007). Choi et al. (2008) reduced extract volumes by rotary evaporation and re-collected the extracts in 5 mL of DCM/hexane. The rotary evaporator was used by Galbán-Malagón et al. (2012) for concentrating samples till 0.5 mL, and dissolved seawater samples were extracted using elution of XAD-2 columns using 200 mL of methanol followed by 300 mL of DCM followed by liquid-liquid extraction of methanol fraction thrice with hexane and hexane and DCM fractions were combined and subjected to rotary evaporation.

Fu et al. (2009b) used rotary evaporator to concentrate filtered extract, and dried it under pure nitrogen gas followed by the addition of 50 μL of N, O-bis-(trimethylsilyl) trifluoroacetamide (BSTFA) with 1% trimethylsilyl chloride and 10 μL of pyridine to concentrates placed at 70°C for 3 hours to derive –COOH and –OH groups to TMS esters. Xie et al. (2015) evaporated extracts to 5 mL with hexane and added 3 g of Na_2SO_4 to remove water and further reduce to 200 μL under a gentle stream of nitrogen. Pawlak et al. (2019) transferred extracting solvent to a glass vial and evaporated to a volume of $1 \, cm^3$ in a gentle stream of N_2.

7.3.9 Column Clean-Up

In the study by Turk et al. (2007), hexane was passed through an activated florisil column (6 g) using 200 mL of petroleum ether with 1.5 g of Na_2SO_4 at the top. The solvent flow was 2 mL/minute regulated by nitrogen pressure followed by sample concentration to 1 mL, and fractionation was conducted on silica gel column, and sulphuric acid-modified silica gel column was used for PCB/HCH/DDT samples.

Choi et al. (2008) used silica gel chromatography (with layers of essential, neutral, acidic, neutral silica) for column clean-up, followed by activated alumina chromatography. Alumina (3 g of 3% deactivated) and anhydrous sodium sulphate (1 g) were used for gas, aerosol and seawater, and each sampled was divided into three fractions. The first fraction contained PCBs in hexane (5 mL) this was evaporated to 1–2 mL and concentrated under a gentle stream of N_2 to 150 μL and sealed. Phytoplankton samples were fractioned using a column with 0.5–1 g of anhydrous sodium sulphate over 3 g of 3% deactivated neutral alumina, and 5 g of silica gel (silica 60, 200 mesh) activated at 250°C for 24 hours. The elution of the first fraction using 40 mL of Hexane/DCM (1:2) was used for PCB quantification by Galbán-Malagón et al. (2012). Wu et al. (2014) eluted the extracted 1 mL sample through an activated silica gel (at 450°C for 4 hours) column and deactivated alumina with 80 mL hexane and 80 mL DCM mixture (1:1 v/v).. Lipid extraction and saponification were followed by silica gel column chromatography for column clean-up (Seki et al. 2015).

7.3.9.1 Post-Clean-Up Spiking

Choi et al. (2008) concentrated cleaned extracts to 10 μL and spiked with ^{13}C-labelled method performance standard ($^{13}C_{12}$-PCB-111) before analysis. Derivatives were mixed with 140 µL of n-hexane containing internal standard before Gas Chromatography-Mass Spectrometer (GC-MS) analysis (Fu et al. 2009a). Galbán-Malagón et al. (2012) added 5 ng of PCBs (30, 142) before injection. Wu et al. (2014) spiked elutes with 2,4,5,6-tetrachloro-m-xylene (TCMX) as internal standard and concentrated it to 100 µL under nitrogen stream. Xie et al. (2015) spiked concentrates with 1 ng of 9:1 FTOH injection standard before quantification.

7.3.10   STORAGE OF SAMPLES

Evenset et al. (2007b) stored filters and PUFs wrapped in aluminium foil and packed in separate plastic bags and stored at −20°C directly after sampling. Galbán-Malagón et al. (2012), selected filters were wrapped in foil and placed in zip sealed bags at −20°C. Wu et al. (2014) kept all extracts in sealed vials at −20°C before analysis.

7.4   QUANTIFICATION OF POPs

Gas Chromatography (GC) with electron capture detection (ECD) was used for quantitative analysis of air samples for HCHs by Hewlett Packard (HP) 5890 Series II chromatography instrument coupled with a Trio 1000 mass spectrometer as a detector with SIM mode and an external standard for quantification of POPs (Turk, M. et al., 2007). ECD supplied with Quadrex-fused PCB 180 and OCPs associated with Gas chromatography/ mass spectrometer (GC/MS- H.P. 6890-HP5972) was used to determine 16 polycyclic aromatic hydrocarbons (PAHs) stipulated by the United States of America Environmental Protection Agency. Turk et al. (2007) analysed sample intervals using one laboratory blank and one reference material every ten samples for efficient runs.

Fu et al. (2009a) effectively analysed samples using a HP 6890 GC interfaced with HP 5973 Mass Spectrometer Detector (MSD) and a fused silica capillary column (Dimensions: 30 m×0.25 mm internal diameter, 0.25 µm film thickness). Wong et al. (2011) analysed various congeners of Hexachlorohexanes (HCH), Hexachlorobenzenes (HCB), Dibromoanisole (DBA), Tribromoanisole (TBA) and Mirex in air samples using gas chromatography coupled with electron capture negative ion mass spectrometry (Agilent 6890 GC-5973 MSD GC-ECNI-MS), and the selected column for quantitative analysis was a DB-5 Agilent. In order to analyse certain POPs like DBA and TBA, GC oven programme was kept at $T_i = 90°C$, $T_R = 170°C$ at 20°C/minute followed by 240°C at 3°C/minute with a splitless mode of injection of the sample with 250°C inlet temperature, and 2 µL sample injection volume and Quadrupole and ion source temperatures were 150°C with transfer line temperature as 250°C, to provide the best possible recovery for the samples.

Galbán-Malagón et al. (2012) used an Agilent 7890 Gas Chromatograph coupled with an ECD detector and a capillary column (Type-Wcot Sil 8CB 60 m, Dimensions-inner diameter 0.25 mm, film thickness, 0.25 µm). Francis et al. (2017) selected as T_H (holding temperature) 90°C for 1 minute, and raised to 190°C at 20°C/minute and then raised to 310°C at 3°C/minute holding 310°C for 18 minutes. Using an alternative approach, Wu et al. (2014) quantified HCBs with Gas Chromatograph (Agilent 7890A) with mass-selective detector (Agilent 5975c) in negative chemical ionisation mode with methane as a reactant gas along with a fused silica capillary column (Type-DB_XLB Make-J&W Scientific, Inc., Folsom, CA; Dimensions-0.25 mm internal diameter×60 m×0.25 μ film). On the other hand, Xie et al. (2015) used Agilent 6890 gas chromatograph–5973 mass spectrometry with positive chemical ionisation (PCI) using the same reagent gas, methane.

The GC/MS analyses for ice core trapped POPs were modified to trimethylsilyl (TMS) esters and esters with bis (trimethylsilyl) trifluoroacetamide to retain 75% recovery in a lipid fraction prior to analyses (Seki et al. 2015). Pawlak et al. (2019) determined PCBs and PAHs using Agilent Technologies 5975C gas chromatography coupled with an Agilent Technologies 7890A mass spectrometer on selected.

7.4.1 Results

Active atmospheric sampling in the firstarea Kragujevac 1 (KG-1), near two transformers contained elevated DDT and corresponding metabolites due to evident historical pollution and POPs generated during intensive combustions, explosions and burning (Turk et al. 2007). In contrast, passive air sampling showed ~7 μg of the sum of several indicator congeries at the KG-1 site, which was higher than those in other areas.

Due to the variation of seabird visitations to Lake Oyangen, BjørnØya and and Lake EllasjØen, a population approximate was taken and PCB levels in the guano of little auk were determined to be ~25 ng/g wet weight (feeds on marine zooplankton), kittiwake ~124 ng/g ww (feeds on zooplankton and small fish), glaucous gull ~1224 ng/g ww (feeds on prey species higher in the food chain, opportunistic feeder) owing to their dietary compulsions; calculations based on specified specific guano excretion rates and seabird population numbers estimate 4.4 g/year of total PCBs in guano excretion at Lake EllasjØen contributing 20% to the lake; LRTP generates 28 g/year of PCB's to EllasjØen and 3 g/year to Lake Oyangen; contamination patterns reveal that lake EllasjØen is contaminated by seabird guano dominated by persistent congeners 153 > 138 > 118 > 180 > 99 (Evenset at al., 2007b).

Choi et al. (2008), estimated average congener PCB levels as 7.3 pg/m³ at polar stations (Alert, BjørnØya, Dunai, Eureka, Ny-Ålesund and Tagish), as 65.7 pg/m³, Devon Island as 34.2 pg/m³ owed to local and regional pollution and were compared to Antarctic regions such as King George Island as 2.5 pg/m³ due to South American depositions.

Fu et al. (2009a) detected 17 PAHs (three to seven rings) from phenanthrene to coronene with concentrations ranging from 1 to 1031 pg/m³ with a wintertime maxima of 190–470 pg/m³ with dominant species of fluoranthene phenanthrene and pyrene.

Wong et al. (2011) studied the concentrations of fluoro-telomer alcohols (FTOHs) at Alert air samples which accounted for the total Per and Polyfluororalkyl Substances (PFASs); FTOH were found below the detection limit (BDL); Methyl and ethyl perfluorooctane sulfonamido ethanols (Me- and Et-FOSEs) were <0.10–4.8 pg/m³, and perfluorooctane sulfonamides (FOSAs) were <0.014–0.82 pg/m³, respectively; FTA's was 16 times lower than FTOHs. Perfluorobutanoic acid (PFBA) was the most dominant Perfluoroalkyl acids (PFAAs) with 1.7 pg/m³ with a 99% detection frequency (DF) and Perfluoro octanoic acid (PFOA) and Perfluorooctane sulfonic acids

(PFOSs) were seen in 89% and 96% of the samples with ranges of <0.0063–1.3 pg/m^3 and <0.0063–2.8 pg/m^3, respectively, whereas at Zeppelin and AndØya in the Norwegian Arctic – PFOSA had a DF of 56% with a range of <0.086–4.9 pg/m^3 at Zeppelin and 9.6% DF at AndØya whereas PFOA had DF of 59% with a range of <0.12–4.0 pg/m^3 at Zeppelin and 48% DF with a range of <0.043–0.43 pg/m^3 at AndØya; PFASs had low DF at Zeppelin and AndØya.

Galbán-Malagón et al. (2012) found high concentrations of poly chlorinated biphenyls (PCBs) near Iceland and lower near Svalbard islands, but with some variation, dissolved PCBs were >1 pg/m^3 first reported in the Greenland Current (G.C.) (between Greenland and Iceland, flowing northeast); particulate phase concentrations in ice-free seawater from G.C. and Arctic Ocean (A.O.) ranged from 0.01 to 12.29 and 0.004 to 9.73 pg/L, respectively; and plankton showed an increasing trend with latitude but variability nearer to the margin of ice. Atmospheric concentrations of a total of 27 congeners in the gas phase (C_G) were 15.7–86.9 and 3.5–81.7 pg/m^3 in the air over G.C. and A.O., respectively, whereas aerosol-bound PCBs account for 23% of total atmospheric concentrations due to POP sorption at low temperatures.

Wu et al. (2014) detected Hexachlorobenzenes (HCBs) between a range of 24–180 pg/m^3 and sites with the minimum concentrations whereas maximum concentrations were found in the Central Arctic Ocean indicative of long residence time, Central Arctic Ocean (~110 pg/m^3) > Chukchi and Beaufort Seas (~93 pg/m^3) > East Asia (~75 pg/m^3) > North Pacific Ocean (~69 pg/m^3).

Xie et al. (2015) carried out a study from September 2011-September 2012 at the German Station in Ny-Ålesund, wherein the concentration of 12 congeners of neutral PFASs in the Arctic were in the range of 6.7–39 pg/m^3 out of which the gaseous PFASs were dominant by 91%; 6:2 FTOH > 10:2 FTOH > 12:2 FTOH varied from 5.6 to 34 pg/m^3; and N- Methyl perfluorobutane sulphonamide (Me-FBSA), N-Methyl perfluorooctane sulphonamide (Me-FOSA) and N-ethyl perfluorooctane sulphonamide (Et-FOSA) were higher than the N-Methyl perfluorobutane sulfonamidoethanol (Me-FBSE), N-Methyl perfluorooctane sulfonamidoethanol (Me-FOSE) and N-ethyl perfluorooctane sulfonamidoethanol (Et-FOSE); neutral PFASs in Arctic snow is in the range of 334–692 pg/L; 8:2 FTOH is dominant with 45% of total concentration (218 = 507 pg/L); and 10:2 FTOH, Me FOSE and 12:2 FTOH were the other abundant PFASs in snow.

Aerosol analyses by Seki et al. (2015) at Alert station revealed deposition of mineral dust, soil organic matter, and baring dust that is long-range transported to higher latitudes due to spring dust storm season in Asia. Levoglucosan is an excellent tracer of biomass burning but in this study the volume in Greenland ice core (north Greenland Eemian Ice Drilling -NEEM) was not correlated to black carbon deposition over the last two centuries due to variations by aeolian processes, glacial processes and total organic carbon.

In the study of Pawlak et al. (2019), twenty samples showed the presence of naphthalene, whereas in six samples, all the PAHs were seen, and in four samples, PAHs were BDL; surface water PAH concentration was <LOD-6210 ng/dm^3 indicating a lower limit contamination; PCB concentrations in 26/30 samples were in the range of <LOD – 273 ng/dm^3 with the largest PCB as PCB153 at 254 ng/dm^3 in a lake sample which is of prime importance to the consumption of stagnant freshwater for

human consumption. α and γ HCH weekly estimates from two the Arctic monitoring stations, i.e. Alert, Nunavut, Canada and Zeppelin Mountain, Svalbard, Norway, indicated that γ HCH/total HCH ration was higher in Zeppelin (proximity to lindane usage areas) as compared to Alert, with increased values during summer and lower concentrations during winter (Becker et al. 2008).

Weekly atmospheric concentrations of OCPs in the Arctic revealed increased summertime use of endosulphan followed by long-range transport (Su et al. 2008).

7.5 OTHER METHODS AND STUDIES

A pilot study for POP assessment was conducted using a combination of gas chromatography mass spectrometer (GC-MS) on air samples collected from polyurethane foam disks inserted in passive air samplers using back trajectory analysis for source determination (Harner et al., 2006). Chirality of POPs determines its deposition into various phases, A study using GC, high performance liquid chromatography (HPLC), capillary electro-chromatography (CEC), micellar electrokinetic chromatography (MEKC), super critical fluid chromatography (SFC) and thin layer chromatography (TLC) to determine chiral pollutants was conducted by Aboul-Enein and Ali (2004).

7.5.1 RESULTS

Harner et al. (2006) depicted concentrations of OCPs between May 2002 and October 2003 at Alert station; mean values include hexachlorocyclohexanes (α 41.33 pg/m^3, γ 8 pg/m^3), chlordanes (*trans* 0.836 pg/m^3, *cis* 2.7 pg/m^3, *trans*-nonachlor 2.2 pg/m^3), dieldrin (10.03 pg/m^3), endosulfan (1–52.33 pg/m^3) and endosulfan-2 (BDL).

Shortcomings of chiral analyses by gas chromatographic techniques by Aboul-Enein and Ali (2004) include inefficiencies due to the volatilisation of some pollutants at high temperatures, reducing the enantiomeric extent analysis process, simultaneous decomposition or racemisation of chiral compounds at high working temperatures. Many GC chiral selectors are unavailable, and the only cyclodextrin-based GC chiral columns are available restricting chiral GC application. Tedious derivatisation experiments are required for environmental samples.

7.6 ICE POPs

Ultra-trace analysis of POPs in Arctic ice was conducted by stir bar sorptive extraction (SBSE) gas chromatography coupled mass spectroscopy (GC-MS) (Lacorte et al. 2009).

7.6.1 RESULTS

GC-quadrupole mass spectrometer with electron impact ionisation was used to detect POPs in ice cores. α-HCH was found to be ~141 pg/L; DDT was found to be ~44 pg/L (out of which 4,4′-DDT values are 4.7–117 pg/L in the Arctic); PCB congeners (118, 101, 28) dominated in the ice cores sampled during the Arctic Ocean Cruise- Antarctic Tourism Opportunity Spectrum (ATOS) campaign in 2007, PCB_3 in ice meltwater was ~20 pg/L; PBDE's in Arctic ice (only 47 out of 99) were detected in low quantities between 0.5 and 2.3 pg/L (Lacorte et al. 2009).

7.7 WATER POPs

Combined solid-phase extraction was carried out with headspace solid-phase micro-extraction (HS-SPME) by Qiu and Cai (2010). To enrich 17 ultra-trace OCPs in water, and they were transferred to gas chromatography capillary column by four steps: SPE, solvent conversion, HS-SPME, and thermal desorption from SPME fibre. An acceptable recovery of 60% was achieved.

7.7.1 RESULTS

Of all the OCPs, hexachlorocyclohexanes (HCHs) are present in abundance in Arctic surface water bodies with a concentration ranging from 0.262 to 3.156 ng/L (Qiu and Cai 2010).

7.8 STATISTICAL STUDIES ON POP's

Evenset et al. (2007a) studied the multivariate relationship between the contaminants in fish and explanatory variables ($\delta^{15}N$, %lipid, length, age) by redundancy analysis followed by a Monte Carlo permutations test and a correspondence analysis to determine the contaminant patterns in water, sediment and invertebrates using CANOCO, ver. 4.5.A statistical software platform SPSS version 19.0.0 for Mac operating system (OS) was employed along with the Hybrid Single Particle Langrangian Integrated Trajectory Model (HYSPLIT) from the Air Resources Laboratory under the National Oceanic and Atmospheric Administration (NOAA) of the United States of America to determine transport and dispersion mechanisms (http://www.arl. noaa.gov/ready.php) (Galbán-Malagón et al. 2012).Rotander et al. (2012) used one-sided ANOVA withTukey comparison of means to assess the variability in sampling and time.

Statistical analysis of time series was performed using the digital filtration technique to study seasonal cycles. Interannual trends with statistical methods and half-lives of the POPs can be ascertained (Hung et al. 2016). One approach to study POP in the Arctic was to create a systematic map of the POP and its impact on the environment and focus on target species categorised on the IUCN red list (Mangano et al. 2017). Endocrine-disrupting compounds (nonylphenols and bisphenol A) were extracted from muscle and liver tissues of Greenland shark using accelerated solvent extraction (ASE) followed by HPLC-fluorescence detection (FLD), and the results were depicted in ng/g wet wt. (Ademollo et al. 2018). A gas chromatography–high-resolution time-of-flight mass spectrometer was used to detect and identify novel and emerging organic pollutants in the Arctic, including those from the USEPA lists (Lebedev et al. 2018).

In-silico screening studies by Muir et al. (2019) assembled a list of 3421 chemicals, out of which 52% were halogenated and 48% were non-halogenated. Subsequently, 'POP scores' were tabulated for transport and accumulation in the Arctic depending on partition coefficients, persistence, transfer efficiency, and bioaccumulation factor.

7.9 PERSISTENT INORGANIC POLLUTANT – MERCURY

Before the disaster of Japan's Minamata Bay, an international treaty titled 'Minamata Convention on Mercury' was signed on 10day of October 2013 and was made effective on 16day August 2017. This treaty outlines the potential risks of anthropogenic activities for mercury poisoning and its bioaccumulation. Articles in this treaty cover mercury in supply sources and trade, mercury-added products, coal use (Lacerda and Marins 1997) manufacturing processes involving mercury, artisanal and small-scale gold mining (Malehase et al. 2016) emissions, releases, environmentally sound storage of mercury other than waste mercury, contaminated sites, etc.

Trace-metal contamination in Canadian Arctic is seen close to emission sources such as automobiles, waste incineration, power plants, etc. (Gorzelska 1989). Party nations are required to eventually fade out mercury in products and related activities by a target year.

Mercury is considered a fatal d-block element of the periodic table with a vapour pressure of 0.18 Pa and exists in multiple forms. The principal oceanic mercury species are as follows: mercuric ion [Hg(II)], elemental Hg [(Hg^0)], monomethyl mercury [MMHG] and dimethyl-Hg [$(CH_3)_2$Hg; DMHg] (Bowman et al. 2020). Mercury has an affinity to form sulphides in soils (Skyllberg et al. 2003).

Braman and Johnson (1974) determined atmospheric particulate mercury, methylmercury (II) chloride-type and mercuric chloride-type compounds, elemental mercury, and dimethylmercury through sensitive sequential, selective absorption tubes to separate and a direct current discharge spectral type detection. Atmospheric depletion mercury events (ADME) at Alert, Nunavut, Canada usually occur three months from mid-March to mid-June (Lu et al. 2001). EDGARv4.tox2 was used to study the global emissions of mercury species (Muntean et al. 2018).

7.10 LONG-RANGE TRANSPORTATION OF MERCURY

Continual monitoring of Hg from Alert and Little Fox Lake sites showed that Asia contributed most ambient Hg to the Canadian Arctic followed by North America, Russia and Europe (Dastoor et al. 2015). Global/Regional Atmospheric Heavy Metals (GRAHM) models (Dastoor and Moran 2010) recorded global total gaseous mercury observations at six Arctic stations, three sub-Arctic stations, and eight stations at mid-latitude, which disclosed Arctic GEM as the highest compared to other source regions and strongest at Barrow, Alert station (Durnford et al. 2010). Air residence times of gaseous oxidized mercury (GOM) over north Atlantic Ocean were higher than those over sub-Arctic north Atlantic Ocean (Fu et al. 2016).

7.11 MERCURY IN THE ATMOSPHERE

Photochemical oxidation of boundary-layer elemental mercury (Hg^0) after polar sunrise by reactive halogens results in deposition of GOM in remote Arctic troposphere which accumulates in the snowpack during polar spring and is bioavailable to bacteria and released back into snowmelt during summer emergence (Lindberg et al. 2002). During summer, TGM in the troposphere at the event of polar sunrise degrades by photochemical means (halogen radicals, i.e. bromine). In contrast, it gets deposited by wet deposition pathways (Cheng and Schroeder 2000). Bromine oxides (BrO_x, Br, BrO) are involved in ozone depletion events (ODE), and BrO_x produces RGM from GEM, and RGM freeloads onto ice and snow (Lu et al. 1998; Schroeder et al. 1998; Steffen et al. 2008). The interaction of ozone and GEM with sea salts, sunlight under stable boundary conditions and cold temperatures have led to heightened levels of mercury in the form of reactive gaseous mercury (RGM) in snow peaks during May month due to springtime deposition at Alert (Steffen et al. 2014a).

A one dimensional simulation model PHANTAS to describe ozone depletion events (ODEs) and ADMEs that occur with snow- air interface due to enriched arctic haze aerosols by Toyota et al. (2014).A chemical mechanism box model was employed for diurnal atmospheric Hg cycling by Ye et al. (2016) conferred that the particulate form of mercury existed as HgBr and the oxidant species were dependent on coastal and inland factors for GOM formation.

In the gaseous phase, Hg (0) converts to Hg (II) and is oxidised by O_3 and OH radicals (Shia et al. 1999). A correlation between the oxidation of carbonyl sulphide (OCS) and Hg(0) as they both decrease in stratosphere suggested OCS as an analogue to Hg(0) depletion as a result of oxidation to Hg (II) and its consequent sedimentation (Lyman and Jaffe 2012).

7.12 MARINE SOURCES OF MERCURY

Mercury in the atmospheric ocean deposition is the primary input of mercury in Arctic waters. The collection of samples from the marine boundary layer has proved to give a good estimate of total Hg, methylmercury involved in wet deposition in surface snow and sea ice (DiMento et al. 2019). The ocean plays a significant role in the cycle of mercury at the poles (Macdonald et al. 2008). Reduction of Hg (II) to Hg (0) takes place in the aqueous phase only (Shia et al. 1999).

Illumina MiSeq shotgun sequencing and metagenomic analysis of marine environments revealed Hg methylation genes *mer*B from mer operon and *hgc*A-like paralogs (*cdh*D genes) showed highest identity matches to more than thousand organisms responsible for the entry of mercury into the Arctic food chains (Podar et al. 2015; Bowman et al. 2020). Most methylmercury is present at a depth of 20–200 metres in the surface water of oceans. The penetration of light ~20 Mg/a[l] is converted in dimethylmercury (17 Mg/a[l]) which escapes into the polar mixed layer (Soerensen et al. 2016).

7.13 MERCURY IN OCEANIC SEDIMENT CORES

Sedimentology studies of cores near the Canadian Arctic Archipelago (CAA) show that the first 10 cm indicate variability of total Hg 5–61 ng/g and a staggering 12–1073 pg/g of monomethyl mercury concentrations (Štrok et al. 2019). Hg accumulates in the soil organic matter (SOM) layer of surface soil (Halbach et al. 2017). Chukchi sea sediment samples gave MMHg ~0.15 ng/g and THg ~0.43 ng/g (Fox et al. 2014).

7.14 MERCURY IN SEA ICE

Open sea ice formations like surface hoars and frost flowers with hypersalinity with prolonged exposure had high mercury (GEM) yields (Douglas et al. 2005).

7.15 TERRESTRIAL SOURCES OF MERCURY

7.15.1 MERCURY IN LAKES

Mercury cycling in the watershed in Amituk Lake, Canada is due to spring freshets (Semkin et al. 2005). Mercury concentrations in EllasjØen have gradually increased since the dawn of the industrial era (Evenset et al. 2007a). Unfiltered total mercury in freshwater across the Arctic ranges from 0.1 to 19.8 ng/L, and ponds, wetlands and lakes have >3 ng/L. Methylmercury in Arctic water bodies is <0.02 ng/L (Chételat et al. 2015). Mackenzie River (second largest river in Canada) has a substantial quantity of particulate (due to basin erosion) total mercury and methyl mercury frozen in its icy portions (Leitch 2006).

It contributes to the mercury budget in marine life in Beaufort Sea (Leitch 2006). Sediment values of total mercury ranged from 5 to 5950 ng/g from bulk samples of sediment sampled from High Arctic islands in Nunavut (Cornwallis, Ellesmere) and Mackenzie River Basin in the northwestern territories (Chételat et al. 2015). The phenomenon of photodemethylation by U.V. ray penetration through water columns led to methylmercury formation in theTundra ecosystems (Inoko 1981). On the other hand, the photochemical oxidation of Hg (II) in aquatic ecosystems occurs by intermediate reactive reductants (O_2/HO_2), of organic matter and through reductive photocatalysed particles such as TiO_2 and photolysis of Hg $(OH)_2$ and Hg (II)-oxalate (Zhang 2006). The Hg (II) reaction with sulphites is significant only in the presence of low hydrochloric concentrations otherwise $HgCl_2$ complex is formed (Shia et al. 1999).

7.16 MERCURY IN SNOW

Glacier snow and firn (ice surviving the summer melts) are adequate Hg sources in terrestrial environments and are sampled at the accumulation and upper sub-surface depths (Gamberg et al. 2015). Gas-phase soluble bromide in the boundary layer and firn air was studied (Dibb et al. 2010). The surface snow had values less than 20 pmol/L for total mercury over a transect across Cornwallis island, NU Canada in which samples over the marine areas had higher mercury than the inland points and increased mercury was also seen over lakes and seas in the interfaces of the snow columns indicating particulate forms of Hg (Poulain, et al., 2007a).

Further indications of reduction of Hg (II) in snow and improved Hg (0) in water suggested dampening effects of snow on mercury reduction (Poulain et al., 2007b). Interstitial air trapped in snowpacks was measured using a Gardis-3 Hg Anlayser with a flow rate of 1.5L/min (Dommergue et al. 2003).

7.17   MERCURY IN MOUNTAINS

Mountain topsoil showed a positive correlation with mercury depositions and elevations and was explained by isotope fractionation in Mt. Leigong by Zhang et al. (2013).

7.18   METHODOLOGY

7.18.1   SAMPLING POINTS

ADME monitoring events were studied at Amderma, Ny-Ålesund, Station Nord, Alert, Churchill, Kuujjuarapik/Whapmagoostui, Resolute and Barrow using times series data collection of GEM concentrations from different locations during the Arctic spring (Steffen et al. 2008). Alternatively, the Pic du Midi (PDM) observatory on Southwest France estimated GEM concentrations over countries located far away (Fu et al. 2016).

Air samples were collected by Steffen et al. (2014a) for 3 hours at Alert Station and snow samples from Alert Nunavut, Canada on Ellesmere Island. AMAP in 1994 encouraged the Norwegian Institute for Air Research (NILU) at Ny-Ålesund, Svalbard, near the western coast of Spitsbergen to conduct air sampling on Zeppelin Mountain (accessible by cable car) (Berg et al. 2003). Offshore air samples within the marine boundary layer were collected from the Western Bering Sea in 2013 (Kalinchuk et al. 2018). U.S. Arctic GEOTRACES cruise to study air-sea Hg flux in 2015 (DiMento et al. 2019).

Soils were collected at sampling sites in Svalbard archipelago between 74°N and 81°N and 10°S and 28°S, i.e. Ny-Ålesund and Adventdalen (periglacial landform) (Halbach et al. 2017), with 18×18 and 5cm thickness using a stainless steel knife followed by removal of mosses, lichens or biotic forms. Between 1991 till 1994, Scheldt Estuary surface waters showed strong complex formation with organic substances in the upper part of the estuary (Leermakers et al. 1995).

Alert (82.5°, 62.3°W) located in northwest Territories of Canada is a sampling site for atmospheric mercury estimations (Cheng and Schroeder 2000), but between 1979 and 1984, it was used as a site for atmospheric chemistry determinations (Barrie and Hoff 1985). Sampling trains were installed at Alert stations for indoor and outdoor total gaseous mercury measurements (Cheng and Schroeder 2000).

Subsurface seawater samples were collected along a transect from the Canada basin across CAA, the Baffin Bay and Barrow Straits into the Labrador Sea (Wang et al. 2018).The US GEOTRACES cruises collected seawater samples for Hg analysis from the Bering Sea, Chukchi Sea, Makarov and Canada basins by Teflon-coated Go-Flo bottles mounted in a trace-metal clean rosette (Bowman et al. 2020).

Snow samples were collected at the High Arctic, Resolute Bay (75°N, 95°W) in low- and high-salinity snow samples in Teflon bottles after exposure to the Sun (Ariya et al. 2004).

7.18.2   MERCURY SPECIATION METHODS

7.18.2.1   Sample Collection

Schroeder et al. (1995) determined the concentrations of aerosol mercury associated with the particulate matter caught onto GF/Fs (Gelman type A/E) in Teflon cassettes by passing air at a rate of 30 L/min or in quartz wool plugs in quartz tubes (6 mm o.d., 4 mm i.d.). Collection of RGM species using KCl-coated annular quartz denuders before the filters has a positive relation with Hg(p) (Landis et al. 2004). RGM was sampled by annular denuders and measured by cold vapour atomic fluorescence spectrometry (CVAFS) (Berg et al. 2003) based on AAS detection (Urba et al. 1995).

Berg et al. (2003) used dual gold cartridges for continuous GEM measurement ~3 m above the ground level on Zeppelin Mountain with high-volume air samplers. GEM and RGM were captured from air using dual gold cartridges and an inline filter (0.2 μm) (Radke et al. 2007).

Atmospheric mercury species were collected in the following manner: GEM on gold traps, RGM on gold denuders and PHg on quartz filters (2.5 μm from impactor set up), wherein GEM was collected continuously with 5-minute intervals (2 traps), while RGM/PHg was collected hourly and analysed over the next hour (Steffen et al. 2013).

Steffen et al. (2014b) drew air into the Tekran instrument through Teflon-coated elutriator and impactor designed to remove particles and snow >2.5 μm at a flow rate of 10 L/min which trapped RGM on a KCL-coated quartz denuder in the 1130 unit. They then passed the PHg over a quartz particulate filter in the 1135 unit, and GEM continues to the 2537 analyser at a flow rate of 1 L/minute. Steffen et al. (2014) sampled aerosol particles from the centre of the flow stream (1000 L/minute diameter in a 10 cm diameter stainless steel manifold) wherein particle size distributions are in the range of 20–500 nm. Steffen et al. (2014b) collected snow of 1 cm depth using PFTE scraper, scooper and gloves on a snow table to discern between fresh and older snowfall depositions.

Fish muscle is partially thawed, and 1 g was excised as per EPA method 7473 (Herger et al. 2016). Fu et al. (2016) used a Tekran2537/1130/1135 system to segregate GEM, RGM and PHg (November 2011–2012) wherein air samples were collected on KCl denuders (GOM), dual gold cartridges (GEM) and regenerable quartz fibre filter (PBM) operated at flow rates of 10 L/minute for GOM and PBM and 1.07 L/minute for GEM.

St. Pierre et al. (2015) collected 70–120 mg of soils and 20–580 mg of lichens to analyse mercury species. Wang et al. (2018) sampled seawater using a Trace-Metal Clean Rosette system in precleaned 12 L Teflon-coated Go-Flo bottles mounted on the system.

7.18.2.2   Pre-Treatment and Pre-Cleaning

Preconcentration of mercury in air was performed by passing air through Carbotrap column (containing graphite black carbon at room temperature) followed by thermal desorption in the U-tube chromatographic column (15% OV-3 on chromosorb WAW_DMSC −196°C in liquid nitrogen) with ramp heating 180°C over 20 minutes to determine mercury speciation by cold vapour atomic fluorescence spectrometry (Bloom and Fitzgerald 1988). For mercury in aqueous samples, methyl mercury was first ethylated with sodium tetraethyl borate (Rapsomanikis and Craig 1991). Krey et al. (2012) freeze-dried 100 mg from the brain tissue samples from Arctic bears prior to the mercury extraction procedure.

A reduction of the total mercury in 100 mL snow melt samples was found useful with 1% $NaBH_4$ (w/v) and 4 M NaOH combined in a glass bubbler (250 mL) and followed by purging (15 minutes from Hg-free generator (Tekran Model 1100)) and final passage over gold filter (at a flow rate of 1 L/minute) (Poulain et al. 2007a).

Krey et al. (2012) placed freeze-dried brain tissue samples in 15 mL Polyethylene terephthalate (PET) centrifuge tubes with extraction mixture (0.1% HCl + 0.1% 2-mercaptoethanol + 0.15% KCl), vortexed, sonicated (30 minutes at 35°C) and centrifuged (3000 rpm at 30°C for 10 minutes). For melted snow samples, Steffen et al. (2013) decanted the sample in precleaned high-density polypropylene (HDPE) bags froze and filtered through nitric acid-washed polypropylene filters (0.45 mm) with frost flowers diluted (1:1000).

The precleaning protocol for snow sample collections by Steffen et al. (2014a) involved sample jars and lids cleaned by soap bath, concentrated HCl, concentrated HNO_3, clean lab air drying and numbered jars were pre-weighed. Acidification of snow samples was a necessity with 0.1 N BrCl to make up 0.4% of total volume (Steffen et al. 2014). Fu et al. (2016) included thermal decomposition of species from collection platforms before measurement of Hg^0 by CVAFS and the inlet of the instrument (Make: Tekran, Model: 1104) was placed in the southwest direction (preset temperature= 70°C and airflow=100 L/minute).

Surface soils were air-dried (duration= 1 month), homogenised (Make: Retsch mill, Model: SM100) and sieved (2mm), weighed (200–500 mg of air-dried organic material) and digested with 6 mL of HNO_3 (50% v/v) whereas, mineral soils were weighed (200–300 mg), digested with 9 mL HNO_3 (50%v/v) and diluted with high purity water to a volume of 60-108mL and transferred to a 15 mL polypropylene (PPE) vial (Halbach et al. 2017).

For the prevention of cross contaminants from chemicals, ultraclean mineral acids were used for Hg analyses (Wang et al. 2018; Bowman et al. 2020). For shipment and storage purposes, Bowman et al. (2020) filtered the seawater samples (0.2 μm) and purged with Hg free Nitrogen gas, transferred to glass bottles (Volume: 250 mL, Make: Borosilicate), acidified with 1% H_2SO_4 and shipped frozen.

Thirty grams of sample was transferred to screw capped glass vials, reduced with BrCl solution (0.3% v/v), left for 24 hours standing, and divided into three aliquots of 10 g each (Krata and Vassileva 2019).

Prior to microwave digestion, decontamination of Teflon CEM vessels was performed by digestion of 10 g HNO_3 followed by Milli Q water rinse (Krata and Vassileva 2019).

The particulate monomethyl mercury (MMHg) was determined on ethylation and acidification of seawater samples neutralized by KOH, buffered with acetic or citric acid supplemented with ascorbic acid, and derivatised with sodium tetraethyl borate pre-concentration (Bowman et al. 2020).

For atmospheric mercury speciation of particulate mercury and reactive gaseous mercury ratios, a mercury analyser (Make: Tekran, Model: 3537A/1130/1135) total mercury was measured followed by step-wise heating of the inline pyrolyser, quartz filter and denuder to convert forms of PHg and RGM into GEM for measurements (Steffen et al. 2014a). St. Pierre et al., (2015) digested lichen sample (200 mg) with a ratio of HF:HNO$_3$ (4:1) at 130°C for 48 hours and further heated to 140°C to completely dry followed by another digestion with acids HCl:HNO$_3$ (1:1) at 130°C for 24 hours and dried in the same manner. To the final 1 mL solution, 0.1 mL of HNO$_3$, 8.8 mL distilled water and internal standards (In, Bi, Sc) was added (St. Pierre et al. 2015).

Pre-concentration is required for fractions of mercury present in minute amounts (parts per quadrillion) namely gaseous oxidized mercury (GOM) and particulate bound mercury (PBM); for GOM a fully heated sample train rejecting large particles by impaction, coated annular denuder for capture of GOM rejecting PBM and GEM, a quartz fibre filter for PBM collection and GEM rejection, to allow continuous detection of ambient air (Fu et al. 2016).

Closed microwave digestion of 0.2 g (marine biota or sediments) and 0.2 g spike solution [Oyster: 4 mL HNO$_3$ (65%) + 1 mL of HCl (36%) and 1 mL H$_2$O$_2$ (30%); Soil: 4 mL of HNO$_3$ (65%) + 1 mL HCl (36%) + 1 mL of H$_2$O$_2$ (30%) + 0.5 mL HF (40%)] with programme one ramp step from 180°C for 30 minutes at controlled pressure of 140 bars and finally evaporated to dryness and diluted to 50 g with 2% HNO$_3$ (Mars-X, CEM Corporation, USA) with 12 Teflon vessel holding carousel (Krata and Vassileva 2019).

7.18.2.3   Extraction

Krey et al. (2012) extracted methylmercury from the freeze-dried brain tissue samples in 15 mL PET tubes on addition of extractant mixture (0.1% HCl + 0.1% 2- mercaptoethanol + 0.15% KCl), vortexed, sonicated (30 minutes at 35°C), centrifuged (3000 rpm at 30°C for 10 minutes). The residue was extracted in 5 mL of extractant mix thereafter supernatants were combined and centrifuged (4000 rpm at 24°C for 10 minutes), filtered (0.45 μm), and made up to the volume (10 mL).

7.18.2.4   Spiking and Recoveries

Berg et al. (2003) developed a methodology with 98% recovery. Krey et al. (2012) achieved overall recoveries between 95% and 102%. Distilled lichen was spiked with MM[201]HgCl (internal standard) and [202]Hg (ambient surrogate), and recovery was 95+/_9% (St. Pierre et al. 2015). Two hundred micrograms of ERM-AE640 (1:100) spike solution was mixed with a sample to be digested (Krata and Vassileva 2019). Spiking for isotope dilution study was done with ERM-AE640 solution (1:100) on [202]Hg isotope (Krata and Vassileva 2019).

According to Fu et al. (2016), the annual mean GEM blank was ~0.04 ng/m^3.

7.18.3   Storage and Transport

Wang et al. (2018) stored collected acidified samples at 4°C until further analysis.

7.19   QUANTIFICATION

Samples were analysed by CVAFS with detection limits at 0.3 pg for Hg and MMHg, 0.4 pg for EEHg and 2.0 pg for MMHg-Cl (Bloom and Fitzgerald 1988). The interference of chloride ions in seawater by CVAFS detection methods of methyl mercury was described by Bloom and Watras (1989). Methyl mercury was detected by quartz furnace atomic absorption spectrometry (Rapsomanikis and Craig 1991). Schroeder et al. (1995) also used a mercury vapour analyser (Make: Tekran, Model: 2537A) with gold-coated beads for entrapment of TGM or quartz sand at 5–30 minutes sampling intervals. Isotope dilution mass spectrometry using inductively coupled plasma mass spectrometer (ICP-MS) for GC analysis has also been performed (Demuth and Heumann 2001). For atmospheric mercury measurements, the arrangement of a gas-phase Mercury analyser (Make: Tekran, Model: 2537A) with a Teflon filter (2 µm) at the inlet of heated sampling line attached to CVAFS (Gardis 1A Hg-Monitor) (λ-253.7 nm) has also been used (Berg et al. 2003). Analysis of snow melts using gas-phase AFS (Make: Tekran, Model: 2537) (flow rate -1.5 L/minute) with a Hg analyser (5 minute data intervals) with volumes (100–350 mL) sent to a bubbler and purged (~20 minutes) using zero air generator (Make: Tekran, Model: 1100), dissolved organic carbon (DOC) autoanalyser (Make: Technicon) on the concept of persulfate UV oxidation followed by conductometric determination of CO_2 has also been employed (Ariya et al. 2004). Water samples were analysed for anions by ion chromatography DIONEX ICS 2000, and cations were analysed using ICP-AES Vista A.X. in the High Canadian Arctic (Poulain et al. 2007a). Tekran 2537A is useful in determining total gaseous mercury (GEM and RGM) in air calibrated by Tekran 2502 Gaseous Mercury Calibrator (Tekran, Inc., Toronto, Canada) (Radke et al. 2007).

High-performance liquid chromatography inductively coupled to plasma mass spectroscopy (HPLC-ICP-MS) was used to detect methyl mercury concentrations in the liver of polar bears versus the total Hg (cross-referenced by mercury analyser NIC MA-2000, Nippon Instruments, College Station, TX, USA) in the brain from Eastern Baffin Island (Krey et al. 2012). Mercury analyser R.A. 915+ (Ltd. 'Lumex', St. Petersburg, Russia) quantified the GEM in the marine boundary layer in summer brought by air masses from the Central Arctic Ocean and the circulation of air which is responsible for the export of mercury to lower altitudes in the Bering Sea (Kalinchuk et al. 2018).

Krey et al. (2012) conducted HPLC-ICP-MS (Make: Agilent 1200 series HPLC system, Agilent Technologies, Canada) with autosampler with a quaternary pump, (loop-100µL, column: Zorbax Eclipse XDB) with a mobile phase of 4% methanol, 0.1% of 2-mercaptoethanol and 0.06 mol/L ammonium acetate attached to the nebuliser of ICP-MS (Make: Agilent, Model: 7500) with overall recoveries between 95% and 102%. Atmospheric mercury was analysed (Make: Tekran, Model: 2537A/1130/1135) measurements of GEM, RGM and PHg at Zeppelin Station, Ny-Ålesund (Steffen et al. 2014a; Wang et al. 2015). Liao et al. (2011) used differential optical absorption spectroscopy (DOAS) for tracing gases such as bromine oxide (BrO) along a long light path.

Steffen et al. (2014a) assessed total Hg by CVAFS after reduction with stannous sulphate (alkaline) with autosampler (Make: Gilson, Model: 222) and AFS (Make: Tekran, Model:2500) using a chromatography interface and software signal capture. St. Pierre et al. (2015) analysed THg by thermal decomposition, pre-concentration and AAS, whereas for non-Hg element, GC-ICP/MS and ICP/OES analyses were carried out. Other meteorological parameters were determined using a 10 m tower in an open field, Setra (pressure), Young model 05103 anemometer (wind speed and direction); released TGM was estimated by CVAFS. Atmospheric CO measured by TEI 48CTL gas filter correlation analyser and ozone concentrations by 49C Ozone analyser (Thermo Environmental Instruments, Inc. U.S.A.) by Fu et al. (2016). EPA method 7473 was used on fish muscle samples (mercury in solids and solutions by thermal decomposition, amalgamation, and AAS, USEPA, 1998).

ICP-MS analyses were used to measure mercury in processed soil samples (Halbach et al. 2017). ICP-MS (Make: Themo scientific), reference materials Soil GBW 07408 and reference material Humus H3 were used for method validation; loss on ignition (LOI) (3–4 g of surface soils were weighed in a crucible and dried at 105°C overnight then in a muffle furnace (550°C)), and pH; (WTW Multi 3430 with pH sensor sentrix 940 combined IDS electrode) experiments were also conducted (Halbach et al. 2017). Wang et al. (2018) used the protocol of acidifying samples to convert di-methyl mercury to monomethyl mercury.

The total Hg was analysed on Hg Analyser (Make: Tekran, Model: 2600) adhering to US EPA Method 1631 using BrCl oxidation, $SnCl_2$ reduction, gold trap preconcentration and measurement by CVAFS. In contrast, MeHg was conducted using a MeHg Analyser (MERX-M, Brooks Rand) following an adapted acidified direct ethylation method using a pre-concentration Tenax trap, gas chromatographic separation, and CVAFS. Continuous flow hydride system (Elemental Scientific, USA) with MP_2 (micro peristaltic pump) for Hg cold vapour generation before ICP-QMS with 5% HCl as carrier fluid and 0.15% $SnCl_2$ in 2% HCl was the reductant with 1-hour Argon purging (Krata and Vassileva 2019) wherein Hg vapour was passed through hydride adapter by ICP-MS nebuliser flow line with isotope dilution calibration. Bowman et al. (2020) used flow injection gas chromatographic cold vapour atomic fluorescence spectrometry (GC-CVAFS) following injection of 2 mol/L HNO_3 at 60°C for 12 hours.

7.20 STATISTICAL ANALYSIS AND MODELLING

Toyota et al. (2014) hypothesised the net transfer of Br_2 from the snowpack to the atmospheric boundary layer (ABL) contrary to the net transfer of HBr from the atmosphere to the snowpack; oxidation of GEM is initiated by Br^- atom giving $HgBr = Hg$ (II) + Br^- with stable products such as $HgBr_2$, $HG(OBr)Br$, Hg^0 and postulated formation of mixed halide complexes such as $HgCl_3Br^{2-}$. Aerosols composed of a mixture of NH_4HSO_4-H_2SO_4-H_2O that serves as a substrate for bromide and Hg (II) species once ozone are depleted in ABL.

Halbach et al. (2017) used ArcMap software version 10.3 to generate element distribution graphs as well as topographic base map databases from Norwegian Polar Institute, along with this statistical software R.2.14.2 was used in which two t-tests,

Shapiro test, Bartlett test, Welch's t-test, Mann-Whitney U test, vector analysis, SPSS Statistics 23 and principal component analysis and rotation by the Varimax method.

7.20.1 RESULTS

Atmospheric concentrations of low GEM values >0.77 ng/m^3 suggested AMDE. The RGM values were lower than PHg with a mean of ~30 pg/m^3 over sea ice; Particles measured over sea ice include the following natural sea salt, aged sea salt and around Arctic Ocean include sea salt sulphate, Arctic haze particles such as non-sea salt sulphate and soot/black carbon increasing the ability of the RGM to get absorbed at PHg; Formation of PHgs is independent of solar radiation (Steffen et al. 2013).

Toyota et al. (2013) developed a model representing gas-phase and aqueous-phase halogenation chemistry to study MADE and ODE from the bottom of snowpack to the ABL in Arctic and Antarctic regions.

Steffen et al. (2014a) declared measurements over 10 years wherein particulate Hg values were low in June through October (median <8.4 pg/m^3), with an increase in November through February (median range <9–42 pg/m^3), and an elevation in PHg values between March and May (median range 21–103 pg/m^3) signified a spring time chemical pattern, whereas, RGM was very low from August to February (median range 0.7–5.3 pg/m^3), moderate in March to July (median 7.4 and 4.6 pg/m^3) and high from April to June (median range 17–100 pg/m^3); transition factors of PHg to RGM were air temperature, sodium nitrate, and sea salts particles; GEM was not a significant component of Arctic snow; Mercury levels in snow increased by day 113 (April end) where PHg and RGM were ~150 and 40 pg/m^3 and were highest between 128 and 131 where PHg and RGM were ~70 and ~150 pg/m^3, respectively.

High levels of particulate Hg and reactive gaseous mercury were seen in a study during springtime with minimal GEM levels (Steffen et al. 2015). Soerensen et al. (2016) explored the Arctic Ocean as a mercury reservoir wherein mercury was trapped in active layers of the benthic shelf and the deep ocean sediment (3900 Mg) and a total concentration in seawater of 2870 Mg with a residence time of 13 years. Soerensen et al. (2016) studied gaseous Hg volatilisation from the water column, which accounted for 44% of Hg loss; sea water Hg was from atmospheric deposition, terrestrial deposition and sea ice erosion.

Soerensen et al. (2016) hypothesised that the stability of methyl mercury in the deep ocean is owed to the fact that 80% of mercury is present in the Arctic Ocean, but the negating factors are the interactions in the PML and subsurface primary productive zones of the ocean

Toyota et al. (2014), reported that BrO radicals exceeded 100 p/mol during daytime dependent on ABL thickness, depth, turbulence, hence, BrO ozone depletion was observed. Aldehydes released from snowpack produced HBr when attacked by Br atom, and HBr further reacted with sulphate aerosols (particulate bromide/filterable bromine). Due to bromine explosion, GEM accumulates in gas phase as GOM in ABL, in which HgBr^{-4} is the dominant portion, i.e. particulate bromine mercury (PBM).

Daytime mixing of bromine radicals is higher in the interface of snowpack and atmosphere than in ambient air; uptake of GOM by liquid surface layer on the grains on the surface of snowpack as Hg (II) increases with time; Hg (0) levels increase due to photo-reduction of Hg (II) which migrates to the lower layers of snowpack; NO_2 causes net oxidation of GEM in the snowpack; the concentration of total mercury is stratified along the snowpack.

7.21   MECHANISM OF ACTION OF MERCURY IN HUMANS

Mercury concentrates from the human samples (such as hair) from the Arctic region are indicative of human dependency on Arctic marine associated diet and the dispersion of emissions of mono-methyl mercury transported via long-range transportation significantly impacted the ecosystem of the Arctic (Kirk et al. 2012). An exposure study was conducted on a group of childbearing-aged individuals of indigenous Arctic Canadians to mercury contaminants through their traditional food consumption patterns was studied. Global-POP-fates and transport model (ACC-Human Bioaccumulation model) revealed PCB exposed women in 2007–2008, cautioned against consumption of traditional foods rich in PCBs and mercury. Hence, reproductive strategies for diet adjustments have been enforced (Binnington et al. 2016). Mercury species (methyl mercury) cross cell membranes on the formation of complexes with thiol molecules, making cysteine complexes on large neutral amino acid carriers and exiting the cell as a complex with reduced glutathione on endogenous carriers (Clarkson et al. 2007).

When primiparous Aboriginals from the Canadian Arctic were biomonitored and compared with Canadian immigrants with multiple births, high concentrations of POPs and mercury were revealed in foreign mothers (Curren et al. 2014). The European Union project ArcRisk is dedicated to study the impacts on human due to climate changes mainly the effects of long range transportation of POPs and mercury on pollutant cycling (Pacyna et al. 2015).

Due to climate change, the Arctic indigenous people's growing food security concerns have laid impetus for methods and strategies through research and policy to provide a solution (Kenny et al. 2019). Monomethyl mercury is a neurotoxin present in contaminated marine life (Bowman et al. 2020).

7.22   MICROPLASTICS: A LOOMING CONCERN

Over the past decade, microplastics (MPs) have gained a lot of attention as a pesky contaminant at the poles (Zhang et al. 2019). The methodology includes sea ice coring during cruise expeditions involved a laborious procedure (use of nitrile gloves, Kovacs corer (9cm diameter), core transfer to LDPE bags and storage at −20°C; core sectioning and fixation of MP onto Anodisc with ethanol (30%); Fourier Transform Infra-red spectrometry (FTIR); Backtracking trajectories with the help of sea ice motion data from satellites (using an algorithm CERSAT/IFREMER) (Peeken et al. 2018) revealed high concentrations of MPs in the upper portion of an ice core from pack ice on sea and on land-fast ice in Fram Strait. Fram Strait is a gateway into the Central Arctic Ocean (Rudels et al. 2000) with constituents such as varnish (a component of cigarette buds) and ethylene-vinyl acetate copolymer (EVA) (ship anti-foulant).

Tyre wear plastics (TWP) and brake wear plastics (BWP) produced due to road traffic have light-absorbing properties that contribute to global warming (Evangeliou et al. 2020).

First microplastics were reported surface waters (depth of 16 cm) microplastics collected using a manta net whereas in subsurface samples (depth of 6 m) through pump enabled filtration by Lusher et al. (2015) in Arctic waters.

7.23 BROMINE AND IODINE

Bromine- and iodine-contaminated codfish from Barents Sea, Norweigian Sea, and North Sea were studied by Sobolev et al. (2019), and partial freshwater fish, Inconnu (Whitefish-Salmon family), showed lower iodine values.

7.24 BLACK CARBON

Black carbon originating from Russian anthropogenic sources (mainly gas flaring) contributed significantly to the Arctic haze (Huang et al. 2015).

7.25 DISCUSSION

7.25.1 Direct Impact

For decades, monitoring of emissions by recognised authoritarian bodies has been of prime importance, and awareness and applications of rules and regulations have reduced emission footprints. However, many legacy and artefact pollutants exist and continue to threaten the Earth's environment.

The studies mentioned above are evidentiary of negative impacts of POP's and Hg to predation in marine, aquatic, benthic and terrestrial pockets in Arctic region highlighting incorporation effects on behavioural changes, interactions, and community dynamics. Furthermore, Arctic ecosystem observations explain the bioaccumulation rate of pollutants in the food web's lower/primary producer strata. Therefore, community structures would inevitably collapse. Human reliance on traditional foods, especially by conventional, aboriginal people, may lead to their eventual harm, extinction, or migration. Making the Arctic devoid of top predation would increase primary productivity which would go unchecked leading to ice melts, increased seawater levels and accelerated release of gaseous pollutants trapped within the ice. A greenhouse effect would be created, leading to the creation of ozone holes or the enlargement of current ozone holes. Exposure scientists need to include a non-target analysis of potential stressors to the environment (POP's and Hg) and to decide policies to be enforced in the Polar Regions to the resident communities to safeguard their health and livelihood (LaKind et al. 2016).

7.25.2 Indirect Impact

Climate change array includes melting of ice as seen most vigorously through the Greenland ice sheet. Mercury and POP compacted in centuries worth of ice traps would be released into the open atmosphere causing a toxic gradient in the Northern Hemisphere. Chemical gradient movements start from a region of higher concentration to a region of lower concentration. The mercury and POPs would thus be distributed over the atmosphere under equilibria.

Atmospheric new particle formation through cloud nucleation was studied for aerosol distribution at various Arctic stations suggesting increased melting, which lead to unique aerosol particle formations (Dall'Osto et al. 2018).

7.26 CONCLUSIONS

In this study, deposition of POPs in the Arctic marine ecosystem's lipid-rich food chains has been researched. POPs are present in very high concentrations in apex predators such as Arctic foxes (*Alopex lagopus*), glaucous gulls, and polar bears. Studies across Nordic countries revealed that POPs alter the integrity of tissues and reproduction of both avian and marine biota. Once mercury (Hg) is biologically transformed by Arctic microbial elements to methylmercury (MeHg) immovable from living tissues .

Neurological toxicity of mercury in apex predators has been well documented in the past. Lastly, ozone depletion by POPs such as chlorofluorocarbons (CFCs) and transformation of inorganic elemental mercury Hg (0) by O_3 have been reported in Arctic aquatic waters. Subsequent studies have confirmed that methylmercury production strongly affects water and aquatic food webs concentrations, including fish. Studies in Remote Canadian Shield Lake revealed that fish size was impeded by mercury bioaccumulation. POPs such as PCBs have been recently found involving in the diabetes mellitus (D.M.) type-2 and insulin resistance developments. Blood samples from Inuit women and their newborn children in north Greenland (1982–1988) revealed an unacceptable high intake of methyl mercury by the mothers. The importance of studying the impacts of these contaminants in this review would establish a relationship between the pollutant and their direct and indirect effects on climate change.

ACKNOWLEDGEMENTS

Authors are thankful to the Secretary,Ministry of Earth Sciences (MoES), New Delhi and Director, National Centre for Polar and Ocean Research (NCPOR), Goa for their support and encouragement. The editor and reviewers of the book are also equally acknowledged for providing an opportunity to write the chapter and reviewing the manuscript.

REFERENCES

Aboul-Enein, H.Y., and Ali, I. (2004) Analysis of the chiral pollutants by chromatography. *Toxicological & Environmental Chemistry* 86: 1–22.
Ademollo, N., Patrolecco, L., Rauseo, J., Nielsen, J., and Corsolini, S. (2018) Bioaccumulation of nonylphenols and bisphenol A in the Greenland shark Somniosus microcephalus from the Greenland seawaters. *Microchemical Journal* 136: 106–112.

Amyot, M., Lean, D.R., Poissant, L., and Doyon, M.-R. (2000) Distribution and transformation of elemental mercury in the St. Lawrence River and Lake Ontario. *Canadian Journal of Fisheries and Aquatic Sciences* 57: 155–163.

Ariya, P.A., Dastroor, A.P., Amyot, M., Schroeder, W.H., Barrie, L., Anlauf, K., et al. (2004) The arctic: a sink for mercury. *Tellus B: Chemical and Physical Meteorology* 56: 397–403.

Balmer, J.E., Hung, H., Vorkamp, K., Letcher, R.J., and Muir, D.C.G. (2019b) Hexachlorobutadiene (HCBD) contamination in the Arctic environment: a review. *Emerging Contaminants* 5: 116–122.

Balmer, J.E., Hung, H., Yu, Y., Letcher, R.J., and Muir, D.C.G. (2019a) Sources and environmental fate of pyrogenic polycyclic aromatic hydrocarbons (PAHs) in the Arctic. *Emerging Contaminants* 5: 128–142.

Barrie, L., and Hoff, R. (1985) Five years of air chemistry observations in the Canadian Arctic. *Atmospheric Environment (1967)* 19: 1995–2010.

Barrie, L.A., Bottenheim, J.W., Schnell, R.C., Crutzen, P.J., and Rasmussen, R.A. (1988) Ozone destruction and photochemical reactions at polar sunrise in the lower Arctic atmosphere. *Nature* 334: 138–141.

Bartlett, P.W., Isaksson, E., and Hermanson, M.H. (2019) 'New' unintentionally produced PCBs in the Arctic. *Emerging Contaminants* 5: 9–14.

Basu, N., Scheuhammer, A.M., Sonne, C., Letcher, R.J., Born, E.W., and Dietz, R. (2009) Is dietary mercury of neurotoxicological concern to wild polar bears (Ursus maritimus)? *Environmental Toxicology and Chemistry* 28: 133.

Becker, S., Halsall, C.J., Tych, W., Kallenborn, R., Su, Y., and Hung, H. (2008) Long-term trends in atmospheric concentrations of α- and γ-HCH in the Arctic provide insight into the effects of legislation and climatic fluctuations on contaminant levels. *Atmospheric Environment* 42: 8225–8233.

Berg, T., Sekkesæter, S., Steinnes, E., Valdal, A.-K., and Wibetoe, G. (2003) Springtime depletion of mercury in the European Arctic as observed at Svalbard. *Science of the Total Environment* 304: 43–51.

Bernhoft, A., Wiig, Ø., and Utne Skaare, J. (1997) Organochlorines in polar bears (Ursus maritimus) at Svalbard. *Environmental Pollution* 95: 159–175.

Bidleman, T.F., Andersson, A., Jantunen, L.M., Kucklick, J.R., Kylin, H., Letcher, R.J., et al. (2019) A review of halogenated natural products in arctic, subarctic and Nordic ecosystems. *Emerging Contaminants* 5: 89–115.

Bidleman, T.F., Laudon, H., Nygren, O., Svanberg, S., and Tysklind, M. (2017) Chlorinated pesticides and natural brominated anisoles in the air at three northern Baltic stations. *Environmental Pollution* 225: 381–389.

Binnington, M.J., Curren, M.S., Chan, H.M., and Wania, F. (2016) Balancing the benefits and costs of traditional food substitution by indigenous arctic women of childbearing age: Impacts on the persistent organic pollutant, mercury, and nutrient intakes. *Environment International* 94: 554–566.

Bloom, N., and Fitzgerald, W.F. (1988) Determination of volatile mercury species at the picogram level by low-temperature gas chromatography with cold-vapour atomic fluorescence detection. *Analytica Chimica Acta* 208: 151–161.

Bloom, N., and Watras, C. (1989) Observations of methylmercury in precipitation. *Science of the Total Environment* 87: 199–207.

Bodaly, R., Rudd, J., Fudge, R., and Kelly, C. (1993) Mercury concentrations in fish related to the size of remote Canadian Shield lakes. *Canadian Journal of Fisheries and Aquatic Sciences* 50: 980–987.

Bogan, J.A., and Bourne, W.R.P. (1972) Organochlorine levels in Atlantic seabirds. *Nature* 240: 358–358.

Borgå, K., Fisk, A.T., Hoekstra, P.F., and Muir, D.C.G. (2004) biological and chemical factors of importance in the bioaccumulation and trophic transfer of persistent organochlorine contaminants in arctic marine food webs. *Environmental Toxicology and Chemistry* 23: 2367.

Borgå, K., Gabrielsen, G.W., and Skaare, J.U. (2001) Biomagnification of organochlorines along a Barents Seafood chain. *Environmental Pollution* 113: 187–198.

Bowes, G.W., and Jonkel, C.J. (1975) Presence and distribution of polychlorinated biphenyls (PCB) in arctic and subarctic marine food chains. *Journal of the Fisheries Research Board of Canada* 32: 2111–2123.

Bowman, K.L., Collins, R.E., Agather, A.M., Lamborg, C.H., Hammerschmidt, C.R., Kaul, D., et al. (2020) Distribution of mercury-cycling genes in the Arctic and equatorial Pacific Oceans and their relationship to mercury speciation. *Limnology and Oceanography* 65.

Braman, R.S., and Johnson, D.L. (1974) Selective absorption tubes and emission technique for determination of ambient forms of mercury in the air. *Environmental Science & Technology* 8: 996–1003.

Carlsson, P., Vrana, B., Sobotka, J., Borgå, K., Bohlin Nizzetto, P., and Varpe, Ø. (2018) New brominated flame retardants and dechlorane plus in the Arctic: local sources and bioaccumulation potential in the marine benthos. *Chemosphere* 211: 1193–1202.

Carlsson, P., Warner, N.A., Hallanger, I.G., Herzke, D., and Kallenborn, R. (2014) Spatial and temporal distribution of chiral pesticides in Calanus spp. from three Arctic fjords. *Environmental Pollution* 192: 154–161.

Carlsson, P.M. (2013) Selective processes for bioaccumulative up-take of persistent organic pollutants (POPs) in Arctic food webs.

Casida, J.E., Holmstead, R.L., Khalifa, S., Knox, J.R., Ohsawa, T., Palmer, K.J., and Wong, R.Y. (1974) Toxaphene insecticide: a complex biodegradable mixture. *Science* 183: 520–521.

Chasar, L.C., Scudder, B.C., Stewart, A.R., Bell, A.H., and Aiken, G.R. (2009) Mercury cycling in stream ecosystems. 3. Trophic dynamics and methylmercury bioaccumulation. *Environmental Science & Technology* 43: 2733–2739.

Cheng, M.-D., and Schroeder, W.H. (2000) Potential atmospheric transport pathways for mercury measured in the Canadian high arctic. *Journal of Atmospheric Chemistry* 35: 101–107.

Chételat, J., Amyot, M., Arp, P., Blais, J.M., Depew, D., Emmerton, C.A., et al. (2015) Mercury in freshwater ecosystems of the Canadian Arctic: recent advances on its cycling and fate. *Science of the Total Environment* 509–510: 41–66.

Choi, S.-D., Baek, S.-Y., Chang, Y.-S., Wania, F., Ikonomou, M.G., Yoon, Y.-J., et al. (2008) Passive air sampling of polychlorinated biphenyls and organochlorine pesticides at the Korean arctic and antarctic research stations: implications for long-range transport and local pollution. *Environmental Science & Technology* 42: 7125–7131.

Clarkson, T.W., Vyas, J.B., and Ballatori, N. (2007) Mechanisms of mercury disposition in the body. *American Journal of Industrial Medicine* 50: 757–764.

Connell, D.W., Miller, G.J., Mortimer, M.R., Shaw, G.R., and Anderson, S.M. (1999) Persistent lipophilic contaminants and other chemical residues in the Southern Hemisphere. *Critical Reviews in Environmental Science and Technology* 29: 47–82.

Curren, M. S., Davis, K., Liang, C. L., Adlard, B., Foster, W. G., Donaldson, S. G., et al. (2014). Comparing plasma concentrations of persistent organic pollutants and metals in primiparous women from northern and southern Canada. *Science of the Total Environment*, 479: 306–318.

Curren, M.S., Liang, C.L., Davis, K., Kandola, K., Brewster, J., Potyrala, M., and Chan, H.M. (2015) Assessing determinants of maternal blood concentrations for persistent organic pollutants and metals in the eastern and western Canadian Arctic. *Science of the Total Environment* 527–528: 150–158.

Dall Osto, M., Geels, C., Beddows, D.C.S., Boertmann, D., Lange, R., Nojgaard, J.K. et al. (2018) Regions of open water and melting sea ice drive new particle formation in North East Greenland. *Scientific Reports* 8: 6109.

Daly, G.L., and Wania, F. (2004) Simulating the influence of snow on the fate of organic compounds. *Environmental Science & Technology* 38: 4176–4186.

Dastoor, A., and Moran, M. (2010) Global/regional atmospheric heavy metals model (GRAHM2005) estimates of atmospheric mercury deposition rates at sites in northern Canada. *Air Quality Research Division, Environment Canada.*

Dastoor, A., Ryzhkov, A., Durnford, D., Lehnherr, I., Steffen, A., and Morrison, H. (2015) Atmospheric mercury in the Canadian Arctic. Part II: Insight from modelling. *Science of the Total Environment* 509–510: 16–27.

Demuth, N., and Heumann, K.G. (2001) Validation of methylmercury determinations in aquatic systems by alkyl derivatization methods for GC analysis using ICP-IDMS. *Analytical Chemistry* 73: 4020–4027.

Derwent, R.G., and Eggleton, A.E.J. (1981) On the validation of one-dimensional CFC–ozone depletion models. *Nature* 293: 387–389.

Diamond, M.L., Bhavsar, S.P., Helm, P.A., Stern, G.A., and Alaee, M. (2005) Fate of organochlorine contaminants in arctic and subarctic lakes estimated by mass balance modelling. *Science of the Total Environment* 342: 245–259.

Dibb, J.E., Ziemba, L.D., Luxford, J., and Beckman, P. (2010) Bromide and other ions in the snow, firn air, and atmospheric boundary layer at Summit during GSHOX. *Atmospheric Chemistry and Physics* 10: 9931–9942.

Dietz, R., Sonne, C., Basu, N., Braune, B., O'Hara, T., Letcher, R.J., et al. (2013) What are the toxicological effects of mercury in arctic biota? *Science of the Total Environment* 443: 775–790.

DiMento, B.P., Mason, R.P., Brooks, S., and Moore, C. (2019) The impact of sea ice on the air-sea exchange of mercury in the Arctic Ocean. *Deep-Sea Research Part I: Oceanographic Research Papers* 144: 28–38.

Dommergue, A., Ferrari, C.P., Gauchard, P.A., Boutron, C.F., Poissant, L., Pilote, M., et al. (2003) The fate of mercury species in a sub-arctic snowpack during snowmelt. *Geophysical Research Letters* 30.

Douglas, T.A., Sturm, M., Simpson, W.R., Brooks, S., Lindberg, S.E., and Perovich, D.K. (2005) Elevated mercury measured in snow and frost flowers near-Arctic sea ice leads: elevated mercury near sea ice leads. *Geophysical Research Letters* 32.

Durnford, D., Dastoor, A., Figueras-Nieto, D., and Ryjkov, A. (2010) Long-range transport of mercury to the Arctic and across Canada. *Atmospheric Chemistry and Physics Discussions* 10: 4673–4717.

Eljarrat, E., Guerra, P., and Barceló, D. (2008) Enantiomeric determination of chiral persistent organic pollutants and their metabolites. *TrAC Trends in Analytical Chemistry* 27: 847–861.

Eriksson, P.G., Banerjee, S., Catuneanu, O., Corcoran, P.L., Eriksson, K.A., Hiatt, E.E., et al. (2013) Secular changes in sedimentation systems and sequence stratigraphy. *Gondwana Research* 24: 468–489.

Evangeliou, N., Grythe, H., Klimont, Z., Heyes, C., Eckhardt, S., Lopez-Aparicio, S., & Stohl, A. (2020). Atmospheric transport is a major pathway of microplastics to remote regions. *Nature Communications*, 11(1): 1–11.

Evenset, A., Carroll, J., Christensen, G.N., Kallenborn, R., Gregor, D., and Gabrielsen, G.W. (2007a) Seabird guano is an efficient conveyer of persistent organic pollutants (POPs) to arctic lake ecosystems. *Environmental Science & Technology* 41: 1173–1179.

Evenset, A., Christensen, G.N., Carroll, J., Zaborska, A., Berger, U., Herzke, D., and Gregor, D. (2007b) Historical trends in persistent organic pollutants and metals recorded in sediment from Lake Ellasjøen, Bjørnøya, Norwegian Arctic. *Environmental Pollution* 146: 196–205.

Evenset, A., Leknes, H., Christensen, G.N., Warner, N., Remberger, M., and Gabrielsen, G.W. (2009) Screening of New Contaminants in Samples from the Norwegian Arctic: Silver, Platinum, Sucralose, Bisphenol A, Tetrabrombisphenol A, Siloxanes, Phtalates (DEHP), Phosphororganic Flame Retardants.

Fenstad, A.A., Jenssen, B.M., Moe, B., Hanssen, S.A., Bingham, C., Herzke, D., et al. (2014) DNA double-strand breaks concerning persistent organic pollutants in a fasting seabird. *Ecotoxicology and Environmental Safety* 106: 68–75.

Ferrari, C.P., Dommergue, A., Boutron, C.F., Skov, H., Goodsite, M., and Jensen, B. (2004) Nighttime production of elemental gaseous mercury in interstitial air of snow at Station Nord, Greenland. *Atmospheric Environment* 38: 2727–2735.

Fisk, A., Hoekstra, P., Gagnon, J., Duffe, J., Norstrom, R., Hobson, K., et al. (2003) Influence of habitat, trophic ecology and lipids on, and spatial trends of, organochlorine contaminants in Arctic marine invertebrates. *Marine Ecology Progress Series* 262: 201–214.

Fox, A. L., Hughes, E. A., Trocine, R. P., Trefry, J. H., Schonberg, S. V., McTigue, N. D., et al. (2014). Mercury in the northeastern Chukchi Sea: Distribution patterns in seawater and sediments and biomagnification in the benthic food web. *Deep Sea Research Part II: Topical Studies in Oceanography*, 102: 56–67.

Frame, G.M. (1997) A collaborative study of 209 PCB congeners and 6 Aroclors on 20 different HRGC columns 2. Semi-quantitative Aroclor congener distributions. *Fresenius' Journal of Analytical Chemistry* 357: 714–722.

Fu, P., Kawamura, K., and Barrie, L.A. (2009a) Photochemical and other sources of organic compounds in the Canadian high Arctic aerosol pollution during winter-spring. *Environmental Science & Technology* 43: 286–292.

Fu, P., Kawamura, K., Chen, J., and Barrie, L.A. (2009b) Isoprene, monoterpene, and sesquiterpene oxidation products in the high Arctic aerosols during late winter to early summer. *Environmental Science & Technology* 43: 4022–4028.

Fu, X., Maruszczak, N., Heimbürger, L.-E., Sauvage, B., Gheusi, F., Prestbo, E.M., and Sonke, J.E. (2016) Atmospheric mercury speciation dynamics at the high-altitude Pic du Midi observatory, southern France. *Atmospheric Chemistry and Physics* 16: 5623–5639.

Gabrielsen, G.W., Skaare, J.U., Polder, A., and Bakken, V. (1995) Chlorinated hydrocarbons in glaucous gulls (Larus hyperboreus) in the southern part of Svalbard. *Science of the Total Environment* 160–161: 337–346.

Galarneau, E., Harner, T., Shoeib, M., Kozma, M., and Lane, D. (2006) A preliminary investigation of sorbent-impregnated filters (SIFs) as an alternative to polyurethane foam (PUF) for sampling gas-phase semivolatile organic compounds in the air. *Atmospheric Environment* 40: 5734–5740.

Galbán-Malagón, C., Berrojalbiz, N., Ojeda, M.-J., and Dachs, J. (2012) The oceanic biological pump modulates the atmospheric transport of persistent organic pollutants to the Arctic. *Nature Communications* 3: 862.

Gamberg, M., Chételat, J., Poulain, A.J., Zdanowicz, C., and Zheng, J. (2015) Mercury in the canadian arctic terrestrial environment: an update. *Science of the Total Environment* 509–510: 28–40.

Gorzelska, K. (1989) Locally generated atmospheric trace metal pollution in Canadian Arctic as reflected by the chemistry of snowpack samples from the MacKenzie delta region. *Atmospheric Environment (1967)* 23: 2729–2737.

Gouin, T., and Wania, F. (2007) Time trends of arctic contamination concerning emission history and chemical persistence and partitioning properties. *Environmental Science & Technology* 41: 5986–5992.

Gustafsson, Ö., Andersson, P., Axelman, J., Bucheli, T.D., Kömp, P., McLachlan, M.S., et al. (2005) Observations of the PCB distribution within and in-between ice, snow, ice-rafted debris, ice-interstitial water, and seawater in the Barents Sea marginal ice zone and the North Pole area. *Science of the Total Environment* 342: 261–279.

Halbach, K., Mikkelsen, Ø., Berg, T., and Steinnes, E. (2017) The presence of mercury and other trace metals in surface soils in the Norwegian Arctic. *Chemosphere* 188: 567–574.

Hallanger, I.G., Ruus, A., Warner, N.A., Herzke, D., Evenset, A., Schøyen, M., et al. (2011) Differences between Arctic and Atlantic fjord systems on bioaccumulation of persistent organic pollutants in zooplankton from Svalbard. *Science of the Total Environment* 409: 2783–2795.

Halsall, C.J. (2004) Investigating the occurrence of persistent organic pollutants (POPs) in the arctic: their atmospheric behaviour and interaction with the seasonal snowpack. *Environmental Pollution* 128: 163–175.

Halsall, C.J., Bailey, R., Stern, G.A., Barrie, L.A., Fellin, P., Muir, D.C.G., et al. (1998) Multi-year observations of organohalogen pesticides in the Arctic atmosphere. *Environmental Pollution* 102: 51–62.

Halsall, C.J., Sweetman, A.J., Barrie, L.A., and Jones, K.C. (2001) Modelling the behaviour of PAHs during atmospheric transport from the UK to the Arctic. *Atmospheric Environment* 35: 255–267.

Hammerschmidt, C.R., and Fitzgerald, W.F. (2006) Methylmercury in Freshwater Fish Linked to Atmospheric Mercury Deposition. *Environmental Science & Technology* 40: 7764–7770.

Hansen, J.C., Tarp, U., and Bohm, J. (1990) Prenatal exposure to methyl mercury among Greenlandic polar Inuits. *Archives of Environmental Health: An International Journal* 45: 355–358.

Hansson, K., Palm Cousins, A., Brorström-Lundén, E., and Leppanen, S. (2006) Atmospheric Concentrations in Air and Deposition Fluxes of POPs at Råö and Pallas Trends and Seasonal and Spatial Variations. In.

Hao, Y., Meng, W., Li, Y., Han, X., Lu, H., Wang, P., et al. (2020) Concentrations and distribution of novel brominated flame retardants in the atmosphere and soil of Ny-Ålesund and London Island, Svalbard, Arctic. *Journal of Environmental Sciences: S1001074220301790.*

Harner, T., Kylin, H., Bidleman, T.F., Halsall, C., Strachan, W.M., Barrie, L.A., and Fellin, P. (1998) Polychlorinated naphthalenes and coplanar polychlorinated biphenyls in Arctic air. *Environmental Science & Technology* 32: 3257–3265.

Harner, T., Pozo, K., Gouin, T., Macdonald, A.-M., Hung, H., Cainey, J., and Peters, A. (2006) Global pilot study for persistent organic pollutants (POPs) using PUF disk passive air samplers. *Environmental Pollution* 144: 445–452.

Haugerud, A.J. (2011) Levels and Effects of Organohalogenated Contaminants on Thyroid Hormone Levels in Glaucous Gulls (Larus hyperboreus) from Kongsfjorden, Svalbard. In: Institutt for Biologi.

Helm, P.A., Bidleman, T.F., Li, H.H., and Fellin, P. (2004) Seasonal and Spatial Variation of Polychlorinated Naphthalenes and Non-/Mono-Ortho-Substituted Polychlorinated Biphenyls in Arctic Air. *Environmental Science & Technology* 38: 5514–5521.

Herbert, B., Halsall, C., Jones, K., and Kallenborn, R. (2006) Field investigation into the diffusion of semi-volatile organic compounds into fresh and aged snow. *Atmospheric Environment* 40: 1385–1393.

Herger, M., van Roye, P., Romney, D.K., Brinkmann-Chen, S., Buller, A.R., and Arnold, F.H. (2016) Synthesis of β-branched tryptophan analogues using an engineered subunit of tryptophan synthase. *Journal of the American Chemical Society* 138: 8388–8391.

Hoferkamp, L., Hermanson, M.H., and Muir, D.C. (2010) Current use pesticides in Arctic media; 2000–2007. *Science of the Total Environment* 408: 2985–2994.

Holden, A.V., and Topping, G. (1972) XIV.—Occurrence of Specific Pollutants in Fish in the Forth and Tay Estuaries. *Proceedings of the Royal Society of Edinburgh Section B Biology* 71: 189–194.

Holdrinet, M.V.H., Frank, R., Thomas, R., and Hetling, L. (1978) Mirex in the sediments of Lake Ontario. *Journal of Great Lakes Research* 4: 69–74.

Holmes, C.D., Jacob, D.J., and Yang, X. (2006) Global lifetime of elemental mercury against oxidation by atomic bromine in the free troposphere. *Geophysical Research Letters* 33: L20808.

Huang, K., Fu, J. S., Prikhodko, V. Y., Storey, J. M., Romanov, A., Hodson, E. L., et al. (2015). Russian anthropogenic black carbon: Emission reconstruction and Arctic black carbon simulation. *Journal of Geophysical Research: Atmospheres*, 120(21): 11–306.

Hung, H., Blanchard, P., Halsall, C.J., Bidleman, T.F., Stern, G.A., Fellin, P., et al. (2005) Temporal and spatial variabilities of atmospheric polychlorinated biphenyls (PCBs), organochlorine (OC) pesticides and polycyclic aromatic hydrocarbons (PAHs) in the Canadian Arctic: Results from a decade of monitoring. *Science of the Total Environment* 342: 119–144.

Hung, H., Kallenborn, R., Breivik, K., Su, Y., Brorström-Lundén, E., Olafsdottir, K., et al. (2010) Atmospheric monitoring of organic pollutants in the Arctic under the Arctic Monitoring and Assessment Programme (AMAP): 1993–2006. *Science of the Total Environment* 408: 2854–2873.

Hung, H., Katsoyiannis, A.A., Brorström-Lundén, E., Olafsdottir, K., Aas, W., Breivik, K., et al. (2016) Temporal trends of Persistent Organic Pollutants (POPs) in arctic air: 20 years of monitoring under the Arctic Monitoring and Assessment Programme (AMAP). *Environmental Pollution* 217: 52–61.

Inoko, M. (1981) Studies on the photochemical decomposition of organomercurials—methylmercury (II) chloride. *Environmental Pollution Series B, Chemical and Physical* 2: 3–10.

Jia, S., Wang, Q., Li, L., Fang, X., Shi, Y., Xu, W., and Hu, J. (2014) Comparative study on PCDD/F pollution in soil from the Antarctic, Arctic, and Tibetan Plateau. *Science of the Total Environment* 497–498: 353–359.

Johansen, S.F. (2019) Riverine Inputs of Polychlorinated Biphenyls (PCBs) and Chlorobenzenes to Isfjorden, Svalbard: Implications for Spatial Distribution and Bioavailability. In: Norwegian University of Life Sciences, Ås.

Kalinchuk, V.V., Mishukov, V.F., and Astakhov, A.S. (2018) Arctic source for elevated atmospheric mercury (Hg 0) in the western Bering Sea in the summer of 2013. *Journal of Environmental Sciences* 68: 114–121.

Kallenborn, R., Hung, H., and Brorström-Lundén, E. (2015) Atmospheric long-range transport of persistent organic pollutants (POPs) into polar regions. In *Comprehensive Analytical Chemistry*. Elsevier, pp. 411–432.

Katsoyiannis, A., and Samara, C. (2004) Persistent organic pollutants (POPs) in the sewage treatment plant of Thessaloniki, northern Greece: occurrence and removal. *Water Research* 38: 2685–2698.

Ke, H., Chen, M., Liu, M., Chen, M., Duan, M., Huang, P., et al. (2017) Fate of polycyclic aromatic hydrocarbons from the North Pacific to the Arctic: field measurements and fugacity model simulation. *Chemosphere* 184: 916–923.

Kelly, B.C., and Gobas, F.A. (2003) An arctic terrestrial food-chain bioaccumulation model for persistent organic pollutants. *Environmental Science & Technology* 37: 2966–2974.

Kenny, T. A. (2019). Climate change, contaminants, and country food: collaborating with communities to promote food security in the Arctic. In *Predicting Future Oceans*. Elsevier, pp. 249–263.

Kirk, J.L., Lehnherr, I., Andersson, M., Braune, B.M., Chan, L., Dastoor, A.P., et al. (2012) Mercury in Arctic marine ecosystems: sources, pathways, and exposure. *Environmental Research* 119: 64–87.

Konoplev, A.V., Volkova, E.F., Kochetkov, A.I., Pervunina, R.I., and Samsonov, D.P. (2012) Monitoring of persistent organic pollutants in the ambient air as an element of the implementation of the Stockholm Convention on persistent organic pollutants. *Russian Journal of Physical Chemistry B* 6: 652–658.

Krata, A.A., and Vassileva, E. (2019) Cold vapour matrix-independent generation and isotope dilution inductively coupled plasma mass spectrometry for reference measurements of Hg in marine environmental samples. *Environmental Science and Pollution Research* 26: 22051–22060.

Krey, A., Kwan, M., and Chan, H.M. (2012) Mercury speciation in brain tissue of polar bears (Ursus maritimus) from the Canadian Arctic. *Environmental Research* 114: 24–30.

Kwok, K.Y., Yamazaki, E., Yamashita, N., Taniyasu, S., Murphy, M.B., Horii, Y., et al. (2013) Transport of Perfluoroalkyl substances (PFAS) from an arctic glacier to downstream locations: Implications for sources. *Science of the Total Environment* 447: 46–55.

Lacerda, L.D., and Marins, R.V. (1997) Anthropogenic mercury emissions to the atmosphere in Brazil: the impact of gold mining. *Journal of Geochemical Exploration* 58: 223–229.

Lacorte, S., Quintana, J., Tauler, R., Ventura, F., Tovar-Sánchez, A., and Duarte, C.M. (2009) Ultra-trace determination of Persistent Organic Pollutants in Arctic ice using stir bar sorptive extraction and gas chromatography coupled to mass spectrometry. *Journal of Chromatography A* 1216: 8581–8589.

LaKind, J.S., Overpeck, J., Breysse, P.N., Backer, L., Richardson, S.D., Sobus, J. et al. (2016) Exposure science in an age of rapidly changing climate: challenges and opportunities. *Journal of Exposure Science and Environmental Epidemiology* 26: 529–538.

Landis, M.S., Keeler, G.J., Al-Wali, K.I., and Stevens, R.K. (2004) Divalent inorganic reactive gaseous mercury emissions from a mercury cell chlor-alkali plant and its impact on the near-field atmospheric dry deposition. *Atmospheric Environment* 38: 613–622.

Larose, C., Prestat, E., Cecillon, S., Berger, S., Malandain, C., Lyon, D., et al. (2013) Interactions between snow chemistry, mercury inputs and microbial population dynamics in an arctic snowpack. *PLoS One* 8: e79972.

Lebedev, A.T., Mazur, D.M., Polyakova, O.V., Kosyakov, D.S., Kozhevnikov, A.Y., Latkin, T.B., et al. (2018) Semi volatile organic compounds in the snow of Russian arctic islands: Archipelago Novaya Zemlya. *Environmental Pollution* 239: 416–427.

Lee, D.-H., Lind, P.M., Jacobs, D.R., Salihovic, S., van Bavel, B., and Lind, L. (2011) Polychlorinated biphenyls and organochlorine pesticides in plasma predict development of type 2 diabetes in the elderly: the prospective investigation of the vasculature in uppsala seniors (PIVUS) study. *Diabetes Care* 34: 1778–1784.

Leermakers, M., Meuleman, C., and Baeyens, W. (1995) Mercury speciation in the Scheldt Estuary. *Water, Air, & Soil Pollution* 80: 641–652.

Leitch, D.R. (2006) Mercury Distribution in Water and Permafrost of the Lower Mackenzie Basin, Their Contribution to the Mercury Contamination in the Beaufort Sea Marine Ecosystem, and Potential Effects of Climate Variation.

Li, Y., Macdonald, R., Ma, J., Hung, H., and Venkatesh, S. (2004) Historical α-HCH budget in the Arctic Ocean: the Arctic Mass Balance Box Model (AMBBM). *Science of the Total Environment* 324: 115–139.

Liao, J., Sihler, H., Huey, L. G., Neuman, J. A., Tanner, D. J., Friess, U., et al. (2011). A comparison of Arctic BrO measurements by chemical ionization mass spectrometry and long path-differential optical absorption spectroscopy. *Journal of Geophysical Research: Atmospheres* 116(D14). https://doi.org/10.1029/2010JD014788

Lindberg, S.E., Brooks, S., Lin, C.-J., Scott, K., Meyers, T., Chambers, L., et al. (2001) *Water, Air, and Soil Pollution: Focus* 1: 295–302.

Lindberg, S.E., Brooks, S., Lin, C.-J., Scott, K.J., Landis, M.S., Stevens, R.K., et al. (2002) Dynamic oxidation of gaseous mercury in the arctic troposphere at polar sunrise. *Environmental Science & Technology* 36: 1245–1256.

Lohmann, R., and Muir, D. (2010) Response to comment on "Global Aquatic Sampling (AQUA-GAPS): using passive samplers to monitor POPs in the waters of the world". *Environmental Science & Technology* 44: 4386–4386.

Lu, J.Y., Schroeder, W.H., Barrie, L.A., Steffen, A., Welch, H.E., Martin, K., et al. (2001) Magnification of atmospheric mercury deposition to polar regions in springtime: the link to tropospheric ozone depletion chemistry. *Geophysical Research Letters* 28: 3219–3222.

Lu, J.Y., Schroeder, W.H., Berg, T., Munthe, J., Schneeberger, D., and Schaedlich, F. (1998) A device for sampling and determination of total particulate mercury in ambient air. *Analytical Chemistry* 70: 2403–2408.

Luo, J., Han, Y., Zhao, Y., Huang, Y., Liu, X., Tao, S., et al. (2020) Effect of northern boreal forest fires on PAH fluctuations across the arctic. *Environmental Pollution* 261: 114186.

Lusher, A. L., Tirelli, V., O'Connor, I., and Officer, R. (2015). Microplastics in Arctic polar waters: the first reported values of particles in surface and sub-surface samples. *Scientific Reports*, 5(1): 1–9.

Lydersen, C., Fisk, A.T., and Kovacs, K.M. (2016) A review of Greenland shark (Somniosus microcephalus) studies in the Kongsfjorden area, Svalbard Norway. *Polar Biology* 39: 2169–2178.

Lyman, S.N., and Jaffe, D.A. (2012) Formation and the fate of oxidized mercury in the upper troposphere and lower stratosphere. *Nature Geoscience* 5: 114–117.

Ma, J., Hung, H., Tian, C., and Kallenborn, R. (2011) Revolatilization of persistent organic pollutants in the arctic induced by climate change. *Nature Climate Change* 1: 255–260.

Ma, Y., Halsall, C.J., Xie, Z., Koetke, D., Mi, W., Ebinghaus, R., and Gao, G. (2017) Polycyclic aromatic hydrocarbons in ocean sediments from the North Pacific to the Arctic Ocean. *Environmental Pollution* 227: 498–504.

Macdonald, R.W., Wang, F., Stern, G., and Outridge, P. (2008) The overlooked role of the ocean in mercury cycling in the arctic. *Marine Pollution Bulletin* 56: 1963–1965.

Malehase, T., Daso, A.P., and Okonkwo, J.O. (2016) Determination of mercury and its fractionation products in samples from legacy use of mercury amalgam in gold processing in Randfontein, South Africa. *Emerging Contaminants* 2: 157–165.

Malmquist, C., Bindler, R., Renberg, I., van Bavel, B., Karlsson, E., Anderson, N.J., and Tysklind, M. (2003) Time trends of selected persistent organic pollutants in lake sediments from Greenland. *Environmental Science & Technology* 37: 4319–4324.

Mangano, M.C., Sarà, G., and Corsolini, S. (2017) Monitoring of persistent organic pollutants in the polar regions: knowledge gaps & gluts through evidence mapping. *Chemosphere* 172: 37–45.

Meng, W., Wang, P., Yang, R., Sun, H., Matsiko, J., Wang, D., et al. (2018) Altitudinal dependence of PCBs and PBDEs in soil along the two sides of Mt. Sygera, southeastern Tibetan Plateau. *Scientific Reports* 8: 1–7.

Möller, A., Xie, Z., Sturm, R., and Ebinghaus, R. (2010) Large-scale distribution of dechlorane plus in air and seawater from the arctic to Antarctica. *Environmental Science & Technology* 44: 8977–8982.

Möller, A., Xie, Z., Sturm, R., and Ebinghaus, R. (2011) Polybrominated diphenyl ethers (PBDEs) and alternative brominated flame retardants in air and seawater of the European arctic. *Environmental Pollution* 159: 1577–1583.

Monitoring, A. (2014) Trends in Stockholm Convention Persistent Organic Pollutants (POPs) in Arctic Air, Human media and Biota. In: Arctic Monitoring and Assessment Programme (AMAP).

Muir, D., and Lohmann, R. (2013). Water as a new matrix for global assessment of hydrophilic POPs. *TrAC Trends in Analytical Chemistry* 46: 162–172.

Muir, D., Zhang, X., de Wit, C.A., Vorkamp, K., and Wilson, S. (2019) Identifying further chemicals of emerging arctic concern based on 'in silico' screening of chemical inventories. *Emerging Contaminants* 5: 201–210.

Muir, D.C.G., and de Wit, C.A. (2010) Trends of legacy and new persistent organic pollutants in the circumpolar arctic: overview, conclusions, and recommendations. *Science of the Total Environment* 408: 3044–3051.

Muir, D.C.G., Wagemann, R., Hargrave, B.T., Thomas, D.J., Peakall, D.B., and Norstrom, R.J. (1992) Arctic marine ecosystem contamination. *Science of the Total Environment* 122: 75–134.

Mullins, M.D., Pochini, C.M., McCrindle, S., Romkes, M., Safe, S.H., and Safe, L.M. (1984) High-resolution PCB analysis: synthesis and chromatographic properties of all 209 PCB congeners. *Environmental Science & Technology* 18: 468–476.

Muntean, M., Janssens-Maenhout, G., Song, S., Giang, A., Selin, N.E., Zhong, H., et al. (2018) Evaluating EDGARv4.tox2 speciated mercury emissions ex-post scenarios and their impacts on modelled global and regional wet deposition patterns. *Atmospheric Environment* 184: 56–68.

Murvoll, K.M., Skaare, J.U., Moe, B., Anderssen, E., and Jenssen, B.M. (2006) spatial trends and associated biological responses of organochlorines and brominated flame retardants in hatchlings of north Atlantic kittiwakes (Rissa tridactyla). *Environmental Toxicology and Chemistry* 25: 1648.

Oehme, M., and Stray, H. (1982) Quantitative determination of ultra-traces of chlorinated compounds in high-volume air samples from the Arctic using polyurethane foam as collection medium. *Fresenius' Zeitschrift für Analytische Chemie* 311: 665–673.

Ortiz, J.M.C. (2016) Levels of Persistent Organic Pollutants (POPs) and Metals in Breeding Kittiwakes (Rissa tridactyla) from Kongsfjorden, Svalbard. In: Universitat de Barcelona (UB) & the Norwegian University of Science and Technology: 1–46

Overland, J.E., and Wang, M. (2010) Large-scale atmospheric circulation changes are associated with the recent loss of arctic sea ice. *Tellus A: Dynamic Meteorology and Oceanography* 62: 1–9.

Pacyna, J., and Keeler, G.J. (1995) Sources of Mercury in the Arctic. In *Mercury as a Global Pollutant*. Springer, pp. 621–632.

Pacyna, J. M., Cousins, I. T., Halsall, C., Rautio, A., Pawlak, J., Pacyna, E. G., et al. (2015). Impacts on human health in the Arctic owing to climate-induced changes in contaminant cycling–The EU ArcRisk project policy outcome. *Environmental Science & Policy* 50: 200–213.

Pawlak, F., Kozioł, K., Ruman, M., and Polkowska, Ż. (2019) Persistent organic pollutants (POPs) as an indicator of surface water quality in the vicinity of the Polish Polar Station, Horsund. *Monatshefte für Chemie - Chemical Monthly* 150: 1573–1578.

Pearce, J., Caracciolo, J., Greig, R., Wenzloff, D., and Steimle Jr, F. (1979) Benthic fauna and heavy metal burdens in marine organisms and sediments of a continental slope dumpsite off the northeast coast of the United States (Deepwater Dumpsite 106). *Ambio Special Report* 101–104.

Peeken, I., Bergmann, M., Gerdts, G., Katlein, C., Krumpen, T., Primpke, S., and Tekman, M. B. (2018). Microplastics in the Marine Realms of the Arctic with special emphasis on sea ice. *Arctic Report Card* 2018: 88.

Perron, T., Chételat, J., Gunn, J., Beisner, B.E., and Amyot, M. (2014) Effects of experimental thermocline and oxycline deepening on methylmercury bioaccumulation in a Canadian shield lake. *Environmental Science & Technology* 48: 2626–2634.

Pidwirny, M. (2006) Introduction to the Oceans. Retrieved from: http://www physicalgeography net/fundamentals/8o html.

Podar, M., Gilmour, C.C., Brandt, C.C., Soren, A., Brown, S.D., Crable, B.R., et al. (2015) Global prevalence and distribution of genes and microorganisms involved in mercury methylation. *Science Advances* 1: e1500675.

Pouch, A., Zaborska, A., and Pazdro, K. (2017) Concentrations and origin of polychlorinated biphenyls (PCBs) and polycyclic aromatic hydrocarbons (PAHs) in sediments of western Spitsbergen fjords (Kongsfjorden, Hornsund, and Adventfjorden). *Environmental Monitoring and Assessment* 189: 175.

Poulain, A.J., Garcia, E., Amyot, M., Campbell, P.G.C., and Ariya, P.A. (2007a) Mercury distribution, partitioning and speciation in coastal vs. inland high arctic snow. *Geochimica et Cosmochimica Acta* 71: 3419–3431.

Poulain, A.J., Garcia, E., Amyot, M., Campbell, P.G.C., Raofie, F., and Ariya, P.A. (2007b) Biological and chemical redox transformations of mercury in fresh and salt waters of the high arctic during spring and summer. *Environmental Science & Technology* 41: 1883–1888.

Prather, M.J., and Watson, R.T. (1990) Stratospheric ozone depletion and future levels of atmospheric chlorine and bromine. *Nature* 344: 729–734.

Pućko, M., Stern, G.A., Burt, A.E., Jantunen, L.M., Bidleman, T.F., Macdonald, R.W., et al. (2017) Current use pesticide and legacy organochlorine pesticide dynamics at the ocean-sea ice-atmosphere interface in a resolute passage, Canadian arctic, during the winter-summer transition. *Science of the Total Environment* 580: 1460–1469.

Pućko, M., Stern, G.A., Macdonald, R.W., Barber, D.G., Rosenberg, B., and Walkusz, W. (2013) When will α-HCH disappear from the western arctic ocean? *Journal of Marine Systems* 127: 88–100.

Pućko, M., Stern, G.A., Macdonald, R.W., Jantunen, L.M., Bidleman, T.F., Wong, F., et al. (2015) The delivery of organic contaminants to the arctic food web: why sea ice matters. *Science of the Total Environment* 506–507: 444–452.

Qiu, C., and Cai, M. (2010) Ultra trace analysis of 17 organochlorine pesticides in water samples from the Arctic based on the combination of solid-phase extraction and headspace solid-phase microextraction–gas chromatography-electron-capture detector. *Journal of chromatography A* 1217: 1191–1202.

Radke, L.F., Friedli, H.R., and Heikes, B.G. (2007) Atmospheric mercury over the NE Pacific during spring 2002: gradients, residence time, upper troposphere lower stratosphere loss, and long-range transport. *Journal of Geophysical Research* 112: D19305.

Ramlal, P.S., Kelly, C.A., Rudd, J.W.M., and Furutani, A. (1993) Sites of methyl mercury production in remote Canadian shield. *Canadian Journal of Fisheries and Aquatic Sciences* 50: 972–979.

Rapsomanikis, S., and Craig, P. (1991) Speciation of mercury and methylmercury compounds in aqueous samples by chromatography-atomic absorption spectrometry after ethylation with sodium tetraethyl borate. *Analytica Chimica Acta* 248: 563–567.

Ratcliffe, D.A. (1967) Decrease in eggshell weight in certain birds of prey. *Nature* 215: 208–210.

Ratcliffe, D.A. (1970) Changes attributable to pesticides in egg breakage frequency and eggshell thickness in some British birds. *The Journal of Applied Ecology* 7: 67.

Rayne, S., and Forest, K. (2009) Perfluoroalkyl sulfonic and carboxylic acids: a critical review of physicochemical properties, levels and patterns in waters and wastewaters, and treatment methods. *Journal of Environmental Science and Health Part A* 44: 1145–1199.

Reppas-Chrysovitsinos, E., Sobek, A., and MacLeod, M. (2017) Screening-level exposure-based prioritization to identify potential POPs, vPvBs, and planetary boundary threats among Arctic contaminants. *Emerging Contaminants* 3: 85–94.

Rigét, F., Bignert, A., Braune, B., Stow, J., and Wilson, S. (2010) Temporal trends of legacy POPs in arctic biota, an update. *Science of the Total Environment* 408: 2874–2884.

Rotander, A., van Bavel, B., Polder, A., Rigét, F., Auðunsson, G.A., Gabrielsen, G.W., et al. (2012) Polybrominated diphenyl ethers (PBDEs) in marine mammals from arctic and north Atlantic regions, 1986–2009. *Environment International* 40: 102–109.

Rowland, F.S. (1990) Stratospheric ozone depletion by chlorofluorocarbons. *Ambio* 281–292.

Rudels, B., Muench, R. D., Gunn, J., Schauer, U., and Friedrich, H. J. (2000). Evolution of the Arctic Ocean boundary current north of the Siberian shelves. *Journal of Marine Systems* 25(1), 77–99.

Sakin, A.E., Esen, F., and Tasdemir, Y. (2017) Effects of sampling interval on the passive air sampling of atmospheric PCBs levels. *Journal of Environmental Science and Health, Part A* 52: 673–679.

Sanei, H., Outridge, P.M., Goodarzi, F., Wang, F., Armstrong, D., Warren, K., and Fishback, L. (2010) Wet deposition mercury fluxes in the Canadian sub-arctic and southern Alberta, measured using an automated precipitation collector adapted to cold regions. *Atmospheric Environment* 44: 1672–1681.

Savinov, V., Muir, D.C.G., Svetochev, V., Svetocheva, O., Belikov, S., Boltunov, A., et al. (2011) Persistent organic pollutants in ringed seals from the Russian arctic. *Science of the Total Environment* 409: 2734–2745.

Scheuhammer, A.M., Meyer, M.W., Sandheinrich, M.B., and Murray, M.W. (2007) Effects of environmental methylmercury on the health of wild birds, mammals, and fish. *AMBIO: A Journal of the Human Environment* 36: 12–19.

Schroeder, W., Anlauf, K., Barrie, L., Steffen, A., and Lu, J. (1999) Depletion of mercury vapour in the arctic troposphere after polar sunrise. *WIT Transactions on Ecology and the Environment* 36.

Schroeder, W.H., Anlauf, K., Barrie, L., Lu, J., Steffen, A., Schneeberger, D., and Berg, T. (1998) Arctic springtime depletion of mercury. *Nature* 394: 331–332.

Schroeder, W.H., Keeler, G., Kock, H., Roussel, P., Schneeberger, D., and Schaedlich, F. (1995) International Field Intercomparison of Atmospheric Mercury Measurement Methods. In *Mercury as a Global Pollutant*. Porcella, D.B., Huckabee, J.W., and Wheatley, B. (eds.), Dordrecht: Springer Netherlands, pp. 611–620.

Seki, O., Kawamura, K., Bendle, J.A., Izawa, Y., Suzuki, I., Shiraiwa, T., and Fujii, Y. (2015) Carbonaceous aerosol tracers in ice-cores record multi-decadal climate oscillations. *Scientific Reports* 5: 14450.

Semkin, R.G., Mierle, G., and Neureuther, R.J. (2005) Hydrochemistry and mercury cycling in a high arctic watershed. *Science of the Total Environment* 342: 199–221.

Shia, R.-L., Seigneur, C., Pai, P., Ko, M., and Sze, N.D. (1999) Global simulation of atmospheric mercury concentrations and deposition fluxes. *Journal of Geophysical Research: Atmospheres* 104: 23747–23760.

Shiraiwa, M., Li, Y., Tsimpidi, A.P., Karydis, V.A., Berkemeier, T., Pandis, S.N., et al. (2017a) Global distribution of particle phase state in atmospheric secondary organic aerosols. *Nature Communications* 8: 1–7.

Shiraiwa, M., Ueda, K., Pozzer, A., Lammel, G., Kampf, C.J., Fushimi, A., et al. (2017b) Aerosol health effects from molecular to global scales. *Environmental Science & Technology* 51: 13545–13567.

Shrestha, R.A., Pham, T.D., and Sillanpää, M. (2009) Effect of ultrasound on removal of persistent organic pollutants (POPs) from different types of soils. *Journal of Hazardous Materials* 170: 871–875.

Skov, D.S., Andersen, J., Olsen, J., Jacobsen, B., Knudsen, M., Jansen, J., et al. (2020) Constraints from cosmogenic nuclides on the glaciation and erosion history of Dove Bugt, northeast Greenland. *Geological Society of America Bulletin* 132: 2282–2294.

Skyllberg, U., Qian, J., Frech, W., Xia, K., and Bleam, W.F. (2003) Distribution of mercury, methyl mercury and organic sulphur species in soil, soil solution and stream of a boreal forest catchment. *Biogeochemistry* 64: 53–76.

Sobolev, N., Aksenov, A., Sorokina, T., Chashchin, V., Ellingsen, D.G., Nieboer, E., et al. (2019) Essential and non-essential trace elements in fish consumed by indigenous peoples of the European Russian arctic. *Environmental Pollution* 253: 966–973.

Soerensen, A.L., Jacob, D.J., Schartup, A.T., Fisher, J.A., Lehnherr, I., St. Louis, V.L., et al. (2016) A mass budget for mercury and methylmercury in the arctic ocean: arctic ocean hg and mehg mass budget. *Global Biogeochemical Cycles* 30: 560–575.

Sonne, C. (2010) Health effects from long-range transported contaminants in arctic top predators: An integrated review based on studies of polar bears and relevant model species. *Environment International* 36: 461–491.

St. Pierre, K., St. Louis, V., Kirk, J., Lehnherr, I., Wang, S., and La Farge, C. (2015) Importance of open marine waters to the enrichment of total mercury and monomethyl mercury in lichens in the Canadian high arctic. *Environmental Science & Technology* 49: 5930–5938.

Steffen, A., Bottenheim, J., Cole, A., Ebinghaus, R., Lawson, G., and Leaitch, W.R. (2014a) Atmospheric mercury speciation and mercury in snow over time at alert, Canada. *Atmospheric Chemistry and Physics* 14: 2219–2231.

Steffen, A., Bottenheim, J., Cole, A., Ebinghaus, R., Lawson, G., and Leaitch, W. (2014b) Atmospheric mercury speciation and mercury in snow over time at alert, Canada. *Atmospheric Chemistry and Physics* 14: 2219.

Steffen, A., Douglas, T., Amyot, M., Ariya, P., Aspmo, K., Berg, T., et al. (2008) A synthesis of atmospheric mercury depletion event chemistry in the atmosphere and snow. *Atmospheric Chemistry and Physics* 8: 1445–1482.

Steffen, A., Lehnherr, I., Cole, A., Ariya, P., Dastoor, A., Durnford, D., et al. (2015) Atmospheric mercury in the Canadian arctic. Part I: a review of recent field measurements. *Science of the Total Environment* 509–510: 3–15.

Steffen, K., Emery, C.A., Romiti, M., Kang, J., Bizzini, M., Dvorak, J., et al. (2013) High adherence to a neuromuscular injury prevention programme (FIFA 11+) improves functional balance and reduces injury risk in Canadian youth female football players: a cluster randomised trial. *British Journal of Sports Medicine* 47: 794–802.

Stern, G.A., Halsall, C.J., Barrie, L.A., Muir, D.C.G., Fellin, P., Rosenberg, B., et al. (1997) Polychlorinated biphenyls in arctic air. 1. Temporal and spatial trends: 1992–1994. *Environmental Science & Technology* 31: 3619–3628.

Stohl, A. (2006) Characteristics of atmospheric transport into the arctic troposphere. *Journal of Geophysical Research* 111: D11306.

Štrok, M., Baya, P.A., Dietrich, D., Dimock, B., and Hintelmann, H. (2019) Mercury speciation and mercury stable isotope composition in sediments from the Canadian arctic archipelago. *Science of the Total Environment* 671: 655–665.

Su, Y., Hung, H., Blanchard, P., Patton, G.W., Kallenborn, R., Konoplev, A., et al. (2008) A circumpolar perspective of atmospheric organochlorine pesticides (OCPs): results from six arctic monitoring stations in 2000–2003. *Atmospheric Environment* 42: 4682–4698.

Sweetman, A.J., Dalla Valle, M., Prevedouros, K., and Jones, K.C. (2005) The role of soil organic carbon in the global cycling of persistent organic pollutants (POPs): interpreting and modelling field data. *Chemosphere* 60: 959–972.

Szczybelski, A.S., van den Heuvel-Greve, M.J., Kampen, T., Wang, C., van den Brink, N.W., and Koelmans, A.A. (2016) Bioaccumulation of polycyclic aromatic hydrocarbons, polychlorinated biphenyls and hexachlorobenzene by three Arctic benthic species from Kongsfjorden (Svalbard, Norway). *Marine Pollution Bulletin* 112: 65–74.

Temme, C., Einax, J.W., Ebinghaus, R., and Schroeder, W.H. (2003) Measurements of atmospheric mercury species at a coastal site in the Antarctic and over the South Atlantic Ocean during polar summer. *Environmental Science & Technology* 37: 22–31.

Toyota, K., Dastoor, A.P., and Ryzhkov, A. (2013) Air-snowpack exchange of bromine, ozone, and mercury in the springtime arctic simulated by the 1-D model PHANTAS – Part 2: mercury and its speciation. *Atmospheric Chemistry and Physics Discussions* 13: 22151–22220.

Toyota, K., McConnell, J., Staebler, R., and Dastoor, A. (2014) Air–snowpack exchange of bromine, ozone, and mercury in the springtime arctic simulated by the 1-D model PHANTAS–Part 1: in-snow bromine activation and its impact on ozone. *Atmospheric Chemistry and Physics* 14: 4101–4133.

Turk, M., Jakšić, J., Vojinović Miloradov, M., and Klanova, J. (2007) Post-war levels of persistent organic pollutants (POPs) in the air from Serbia determined by active and passive sampling methods. *Environmental Chemistry Letters* 5: 109–113.

Ubl, S., Scheringer, M., Stohl, A., Burkhart, J.F., and Hungerbuhler, K. (2012) Primary source regions of polychlorinated biphenyls (PCBs) measured in the Arctic. *Atmospheric Environment* 62: 391–399.

Urba, A., Kvietkus, K., Sakalys, J., Xiao, Z., and Lindqvist, O. (1995) A new sensitive and portable mercury vapor analyzer Gardis-1A. In *Mercury as a Global Pollutant*. Springer, pp. 1305–1309.

Vieweg, I., Hop, H., Gabrielsen, G.W., Brey, T., and Huber, S. (2010) Persistent organic pollutants in four bivalve species from Kongsfjorden and Liefdefjorden, Svalbard. In.

Vorkamp, K., Balmer, J., Hung, H., Letcher, R.J., and Rigét, F.F. (2019a) A review of chlorinated paraffin contamination in arctic ecosystems. *Emerging Contaminants* 5: 219–231.

Vorkamp, K., Balmer, J., Hung, H., Letcher, R.J., Rigét, F.F., and de Wit, C.A. (2019b) Current-use halogenated and organophosphorous flame retardants: a review of their presence in arctic ecosystems. *Emerging Contaminants* 5: 179–200.

Waite, D.T., Hunter (Retired), F.G., and Wiens, B.J. (2005) Atmospheric transport of lindane (γ-hexachlorocyclohexane) from the Canadian prairies—a possible source for the Canadian Great Lakes, Arctic, and Rocky mountains. *Atmospheric Environment* 39: 275–282.

Wang, K., Munson, K.M., Beaupré-Laperrière, A., Mucci, A., Macdonald, R.W., and Wang, F. (2018) Subsurface seawater methylmercury maximum explains biotic mercury concentrations in the Canadian Arctic. *Scientific Reports* 8: 14465.

Wang, Z., Na, G., Ma, X., Ge, L., Lin, Z., and Yao, Z. (2015) Characterizing the distribution of selected PBDEs in soil, moss, and reindeer dung at Ny-Ålesund of the arctic. *Chemosphere* 137: 9–13.

Wang-Andersen, G., Utne Skaare, J., Prestrud, P., and Steinnes, E. (1993) Levels and congener pattern of PCBs in arctic fox, Alopex lagopus, in Svalbard. *Environmental Pollution* 82: 269–275.

Wania, F., Hoff, J., Jia, C., and Mackay, D. (1998) The effects of snow and ice on the environmental behaviour of hydrophobic organic chemicals. *Environmental Pollution* 102: 25–41.

Warner, N.A., Evenset, A., Christensen, G., Gabrielsen, G.W., Borgå, K., and Leknes, H. (2010) Volatile siloxanes in the European arctic: assessment of sources and spatial distribution. *Environmental Science & Technology* 44: 7705–7710.

Wiener, J.G. (2013) Mercury exposed: advances in environmental analysis and ecotoxicology of a highly toxic metal: advances in environmental mercury research. *Environmental Toxicology and Chemistry* 32: 2175–2178.

Wittsiepe, J., Schrey, P., Lemm, F., Eberwein, G., and Wilhelm, M. (2008) Polychlorinated dibenzo-p-dioxins/polychlorinated dibenzofurans (PCDD/Fs), polychlorinated biphenyls (PCBs), and organochlorine pesticides in human blood of pregnant women from Germany. *Journal of Toxicology and Environmental Health, Part A* 71: 703–709.

Wong, F., Jantunen, L.M., Pućko, M., Papakyriakou, T., Staebler, R.M., Stern, G.A., and Bidleman, T.F. (2011) Air–water exchange of anthropogenic and natural organohalogens on International Polar Year (IPY) expeditions in the Canadian arctic. *Environmental Science & Technology* 45: 876–881.

Wu, X., Lam, J.C.W., Xia, C., Kang, H., Xie, Z., and Lam, P.K.S. (2014) Atmospheric hexa-chlorobenzene determined during the third China arctic research expedition: sources and environmental fate. *Atmospheric Pollution Research* 5: 477–483.

Xiao, H., Shen, L., Su, Y., Barresi, E., DeJong, M., Hung, H., et al. (2012) Atmospheric con-centrations of halogenated flame retardants at two remote locations: the Canadian High Arctic and the Tibetan Plateau. *Environmental Pollution* 61: 154–161.

Xie, Z., Wang, Z., Mi, W., Möller, A., Wolschke, H., and Ebinghaus, R. (2015) Neutral poly-/perfluoroalkyl substances in air and snow from the arctic. *Scientific Reports* 5: 8912.

Ye, Z., Mao, H., Lin, C. J., and Kim, S. Y. (2016) Investigation of processes controlling sum-mertime gaseous elemental mercury oxidation at midlatitudinal marine, coastal, and inland sites. *Atmospheric Chemistry and Physics* 16(13): 8461–8478.

Zaborska, A., Carroll, J., Pazdro, K., and Pempkowiak, J. (2011) Spatio-temporal patterns of PAHs, PCBs, and HCB in sediments of the western Barents Sea. *Oceanologia* 53: 1005–1026.

Zepp, R., Wolfe, N., Azarraga, L., Cox, R., and Pape, C. (1977) Photochemical transforma-tion of the DDT and methoxychlor degradation products, DDE and DMDE, by sunlight. *Archives of Environmental Contamination and Toxicology* 6: 305–314.

Zhang, H. (2006) Photochemical Redox Reactions of Mercury. In *Recent Developments in Mercury Science*. Atwood, D.A. (ed.), Berlin/Heidelberg: Springer-Verlag, pp. 37–79.

Zhang, H., Yin, R.-S., Feng, X.-B., Sommar, J., Anderson, C.W., Sapkota, A., et al. (2013) Atmospheric mercury inputs in montane soils increase with elevation: evidence from mercury isotope signatures. *Scientific Reports* 3: 1–8.

Zhang, Y., Gao, T., Kang, S., and Sillanpää, M. (2019). Importance of atmospheric transport for microplastics deposited in remote areas. *Environmental Pollution*, 254: 112953.

8 Fate and Transport of Mercury in the Arctic Environmental Matrices under Varying Climatic Conditions

Gopikrishna VG and Kannan VM
Mahatma Gandhi University

Krishnan KP
National Centre for Polar and Ocean Research

Mahesh Mohan
Mahatma Gandhi University

CONTENTS

DOI: 10.1201/9781003265177-8

8.1   INTRODUCTION

The Arctic region extends across northern Europe, northern Asia and northern North America (Bennett 2014). This region is delineated by the Arctic Circle (66°32′N), which is the approximate southern boundary of the mid-night sun (Nevitt and Percival 2018). Based on climate, vegetation, marine and geographical coverage, the Arctic Monitoring and Assessment Programme (AMAP) defined it as terrestrial and marine areas, north of the Arctic Circle (62°32′N) and north of 62°N in Asia and 60°N in North America. The polar ecosystem exists under extreme climatic conditions such as cold temperatures, large seasonal fluctuations with short growing seasons, extensive snow and ice cover (Callaghan and Jonasson 1995). These conditions affect productivity, species diversity and the food chain in the Arctic ecosystems.

The Arctic is undergoing rapid environmental changes, and the entire world is more conscious about the current climatic fluctuations, which will affect the rest of the world by the rise in sea level, biodiversity, various climatic conditions and socio-economic aspects (Hwang et al. 2020; Moerlein and Carothers 2012; Kofinas et al. 2005). Melting of Arctic sea ice and the thawing of permafrost are increasing day by day (Yadav et al. 2020; Stroeve et al. 2014; Nghiem et al. 2007; Jorgenson et al. 2006), and the warming is more than twice the global average, known as Arctic amplification, which causes sea ice loss (Pistone et al. 2014; Serreze and Barry 2011). Over the past three decades, the decline in sea ice is a major impact of global climate change (Yadav et al. 2020; IPCC 2013). The reduction in the extent and thickness of the Arctic sea ice and later freeze-up and earlier break-up may disturb the marine environment (Kovacs et al. 2011). During the last 50 years, the annual Arctic surface air temperature has risen by 2.7°C (AMAP 2019). This temperature variation and the rise in surface and ocean temperature led to the high melting of snow and ice in the Arctic. Moreover, the increase of sea ice melting may result in a rise in the lower atmosphere temperature in the Arctic. Furthermore, changes in precipitation patterns such as frequency, intensity and distribution influenced freshwater flow into the Arctic Ocean, affecting the nutrient level, acidification, biological productivity and circulation. These precipitation changes alter the soil moisture, and the higher rate of atmospheric humidity contributes to the increasing rate of warming and snowmelt in the Arctic (AMAP 2019).

The pristine environment of the Arctic is getting contaminated with toxic pollutants such as metals, persistent organic pollutants (POPs) and macro as well as microplastic (Warner et al. 2019; Skaar et al. 2019; Iannilli et al. 2019). Most of these are capable of long-range transport by atmospheric circulations and oceanic currents, i.e., the global distillation process (Letcher et al. 2010; Kallenborn and Herzke 2001). It mainly depends on the physical properties and chemical characteristics of the pollutants. These pollutants are deposited in the marine and terrestrial environments of the Arctic region. Environmental changes are caused by climatic variations, and pollutants are re-emitted into the atmosphere, impacting transport and transformation (Semeena and Lammel 2005; Gouin et al. 2004). This makes the Arctic a source of contaminants (Dastoor and Durnford 2013). For example, the increase in permafrost temperature (1°C–2°C during the past three decades) enhances thawing, which releases deposited pollutants like mercury (Stern et al. 2012; Walker 2007; MacDonald et al. 2005).

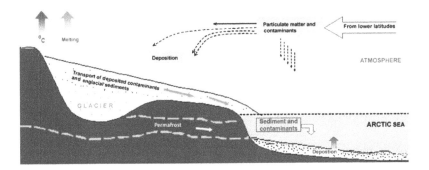

FIGURE 8.1   Conceptual diagram of the transport of contaminants into the Arctic. (Modified from Mohan et al. 2018.)

8.1.1   MERCURY IN THE ARCTIC

Long-range transport of contaminants, along with local sources, enhanced the number of toxic metals in the Arctic environment (Figure 8.1). Emission of metals such as cadmium from natural sources is significant; however, the anthropogenic emissions are estimated to be higher than two to three times the natural sources (Tabelin et al. 2018; AMAP 2005, 1998). One of the major toxic pollutants reaching the Arctic is mercury (Hg).

Hg occurs naturally in the Earth's crust in very low concentration (0.05 mg/kg) and is ranked third in the list of the world's most toxic elements for human health (UNEP 2002; Budnik and Casteleyn 2019). Because of its unique chemical and physical characteristics, mercury is found in all environment compartments and is the only metal in liquid form at room temperature. The naturally occurring Hg exists in several forms such as elemental, inorganic and organic. Elemental Hg is the metallic, pure form of mercury, a shiny, silver-white metal existing as a liquid at room temperature. Mercury can exist in three oxidation states such as 0, +1 and +2, and in the atmosphere, it occurs mainly as elemental gaseous Hg, with an oxidation state of 0 (Hg0).

8.1.2   SOURCES AND TRANSPORT OF MERCURY

Mercury is emitted into the environment by both natural sources and anthropogenic activities. The naturally existing Hg in the Earth's crust is emitted via weathering of rocks and geological movements such as volcanoes and geothermal activities. Changes in land-use patterns, meteorological conditions, biomass burning and exchange mechanisms of gaseous Hg at topsoil, air-water, snow–ice interfaces result in the re-emission of mercury. The present-day anthropogenic Hg emission rate is higher compared to the primary geological sources. As per the global mercury assessment report (GMA 2018), artisanal and small-scale gold mining is the major source of Hg pollution (Figure 8.2). Mercury emission into the environment has increased, and it is estimated that the emission of Hg is around 2000–2500 tonnes per year so far in the 21st century (GMA 2018). About 37.7% (838 tonnes) of mercury

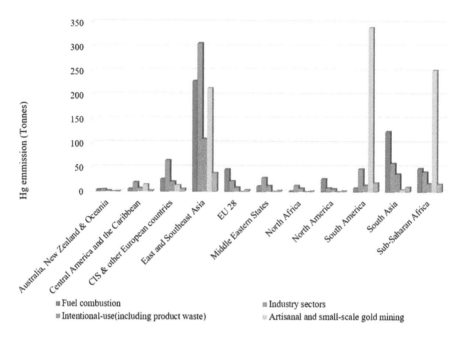

FIGURE 8.2 Emission of mercury from different sectors in different regions (GMA 2018).

is emitted from the artisanal and small-scale gold mining industries, followed by stationary combustion of coal in power plants (13.1%, 292 tonnes). From 2010 to 2015, global Hg emission increased by 1.8% per year (Streets et al. 2019), and the rate of emission has grown in Asia, Africa and other industrial regions. The natural and anthropogenic emissions of mercury can travel to the Arctic and remain deposited there. Disasters like forest fires also contribute to polar Hg concentrations (Witherow and Lyons 2008). The physical and chemical properties of Hg render it as a transboundary pollutant. The major long-range transport mechanism taking Hg to the polar environment is the grasshopper effect (O'Driscoll et al. 2005).

8.1.3 FATE OF MERCURY

The fate of mercury in any ecosystem depends on the physical environment, biological activities, chemical transformations, Hg content in the environment, etc. (Lyman et al. 2020; Selin 2009). The variation in any of these may enhance or decline the rate of transport of Hg in the environmental matrices and subsequently influence its biogeochemical cycle (Boening 2000). The physical factors mainly support the biological or chemical actions and may enhance the release of Hg into the atmosphere through degassing and vaporisation. The photochemical reactions (both in air and water) and density can lead to the deposition of Hg into the terrestrial and aquatic environment (Pirrone and Mason 2009). The anaerobic biological and chemical reactions in aquatic and terrestrial areas can result in chemical transformations of Hg (Kim and Zoh 2012), such as the biological transformation into highly toxic and biomagnifying methyl mercury, as well as other unstable organic Hg forms, which may transport to another matrix. The different forms of Hg in the aquatic and terrestrial environment can accumulate in biota (aquatic organisms and vegetation) and sediment or soil. Bioaccumulation will have health impacts through biomagnification. In anoxic conditions of sediments, Hg can bind with sulphur and form stable compounds (Ravichandran 2004).

The polar environment is highly vulnerable to climate change and will significantly influence Hg circulation due to an increase in temperature (Moritz et al. 2002). The increasing temperature will release both the recently deposited and the historically deposited Hg due to glacier melting and permafrost thawing. Such a sudden release of mercury remains a significant addition to global Hg sources, which will decide the rate of its transformations and transport in the environment.

The deposition of Hg to the Arctic is by multiple processes such as atmospheric mercury depletion events (AMDEs) during springtime and other atmospheric deposition processes like wet and dry deposition (Sprovieri et al. 2005). These Hg forms can be either deposited on snow and ice or in the land and freshwater surfaces (Dommergue et al. 2010). The deposited Hg may enter the fjords, lake, river and snow environments, or it remains in the soils or permafrost (Muir et al. 2009). The major outputs are gaseous Hg^0 re-emitted from the snow and ice surfaces, mainly due to the photochemical reduction of Hg(II) or microbial activities and the elusion of dissolved gaseous Hg^0 (DGM) from freshwater or marine water. In seawater, Hg (II) is photochemically or microbially reduced into Hg^0 (Sommar et al. 2004). However, sea ice prevents the release of Hg^0 back into the atmosphere. Moreover, the photodemethylation process converts MHg into DGM.

Melting of snow is the major seasonal transition in the Arctic from winter to spring. During the snowmelt, the inorganic Hg (II) present in the snow makes complexes with OH, Cl and Br. Moreover, there is a significant chance for the formation of Hg (II) bound to organic matter (Poulain et al. 2007). This may enhance the concentration of methyl mercury (MHg) in the snowpack. However, there is no evidence regarding the amount of MHg formed in the snow compared to the amount of Hg delivered during the polar sunrise (Constant et al. 2007). Finally, the deposited THg and MHg in the snow will melt and subsequently reach rivers, lakes and oceans, though it is re-emitted back to the atmosphere to some extent (Larose et al. 2010). The increase of THg in the freshwater sediments indicated that the Hg deposited by the atmospheric process affected mercury levels in the lake sediments, which affects the freshwater Hg budget in the Arctic (St. Louis et al. 2005).

Mercury concentration in the permafrost peat ranges from 20 to 100 ng/g (Givelet et al. 2004; Leitch 2006). Generally, mercury concentration in the underlying permafrost is lower than that in the active layer. The flux of Hg released from thawing permafrost depends on its concentration in the permafrost, the thawing and the erosion rate (Stern et al. 2012). The release of Hg from thawing permafrost in the high Arctic region has already been demonstrated by Klaminder et al. (2008).

The earlier studies indicated that snowmelt water is a major source of methyl mercury in the Arctic ecosystem (Loseto et al. 2004). The total mercury (THg) and methyl mercury (MHg) were identified from the freshwater ecosystem of the Arctic, mainly in the water and sediment of the lakes (Jiang et al. 2011; Semkin et al. 2005; Riget et al. 2000; Outridge et al. 2007). By its absorption into the water ecosystem,

Hg can pass from solution or suspension into the sediments (Riget et al. 2000). The studies found a significant level of Hg in the tissues of freshwater fishes, waterfowl and game birds, mammals such as beaver, river otter and mink of the Arctic (Hallangeret al. 2019; Aubail et al. 2012; Fortin et al. 2001; Hui et al. 1998).

Various mercury forms are present in the marine environment of the Arctic (Kirk et al. 2005). Mercury can absorb and accumulate in the biota through bioconcentration, bioaccumulation and biomagnification processes (Fisk et al. 2003). Mercury's presence in the marine biota reveals its trophic level of transfer (Fisk et al. 2003). Marine invertebrates, sea birds and mammals such as walrus, belugas, ringed seals and polar bears also show a higher concentration of Hg in their tissues (Fisk et al. 2003; Dietz et al. 2000). Biomagnification is the major reason behind the higher concentration of contaminants in the higher trophic level animals, even though the water, air and soil have a low concentration of Hg or other contaminants (AMAP 1997).

The changing climate may influence the mercury cycle the same as the other elemental cycles. It may influence the chemical and biological transformations and transport of mercury in the ecosystem. The present study assesses the contents of THg and MHg in the soil, sediment, vegetation of the high Arctic region and various mercury fractions in the sediments of marine and freshwater ecosystems.

8.2 MATERIALS AND METHODS

8.2.1 STUDY AREA AND SAMPLE COLLECTION

The samples were collected from Ny-Ålesund, the International Arctic research base located in the Spitsbergen Island of the Svalbard archipelago (Figure 8.3). The period of sampling was four years from 2015 to 2018. The freshwater lakes, meltwater stream, fjord, and tundra soils and vegetation were sampled during the summer seasons of this period (Figure 8.3).

Surface sediment samples were collected from three shallow, summer-active freshwater lakes such as Knudsenheia, Storvatnet and Tvillingvatnet. The lakes were fed by meltwater from glaciers, snowdrift and local precipitation. Additionally, surface sediment samples were collected from the Bayelva stream, which originated from the Austre and Vestre Broggerbreens. A Van-Veen grab is used for collecting sediment samples from Kongsfjorden, the marine glacial fjord system in Ny-Ålesund with a length of 20 km and width varying from 4 to 10 km (Svendsen et al. 2002).

The soil sample locations were fixed based on the distance from the residential area and fjord and coverage of surface soil with vegetation and organic matter. The areas near the closed mining sites were excluded because of the mining debris in that region. The soil samples were collected by cutting squares with an 18×18 cm area, with 5 cm thickness, using a stainless-steel knife. The vegetation cover was removed with the knife, and the samples were transferred to the laboratory and kept at $-20°C$.

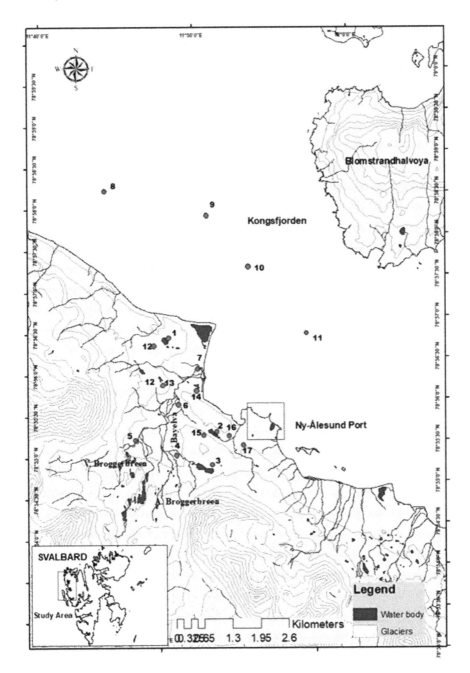

FIGURE 8.3 Study area showing sampling locations.

Vegetation such as moss campion (*Silene acaulis (L.) Jacq*), bog saxifrage (*Saxifraga hirculus L.*) and tufted saxifrage (*Saxifraga cespitosa L.*) were selected for the analysis of THg. These three species are common all over Svalbard and are collected from Ny-Ålesund randomly. All the collected samples were kept at −20°C and cold shipped to India for analysis.

8.2.2 SAMPLE ANALYSIS

The geochemical parameters, such as total organic carbon (TOC) and pH, were analysed using standard methods (Maiti 2003). Mercury detection was carried out using Direct Mercury Analyser (DMA-80, Milestone srl, Italy). The THg concentration in the soil, sediments and biota were analysed using the method USEPA 7473e. Methyl mercury analysis was carried out by Maggie et al. (2009), and Bloom's five-step sequential extraction procedures were used to fractionate mercury in the sediment (Bloom et al. 2003; Ramasamy et al. 2012).

8.2.3 QUALITY ASSURANCE

The estimation of mercury was validated with the certified reference material 'Estuarine Sediment (ERM CC 580)'. The percentage recovery of total mercury was 98.07 ± 0.21.

8.3 RESULTS

8.3.1 TOTAL MERCURY AND METHYL MERCURY

The mercury reaching into the Arctic atmosphere is deposited in terrestrial, freshwater and marine environments during spring by AMDE. The terrestrial ecosystems such as tundra, snow/glaciers and freshwater ecosystem receive the deposited mercury during AMDE. The lake sediments were slightly alkaline with a pH of 7.49, and the TOC was found to be 1.69%. The mean concentration of THg in the freshwater lake sediments varied from 18.01 to 27.3 ng/g, with an average of 22.23 ng/g. The MHg mean concentration was 0.41 ng/g (Table 8.1).

The average pH of Bayelva stream sediment was 7.88, depicting the slightly alkaline nature of the sediment, and the TOC showed an average of 0.17%. The THg concentration remained high at the confluence point of the Bayelva (31.97 ng/g). The sample collected from the glacier front of Austre Brøggerbreen showed a higher level of THg (28.21 ng/g) compared to Vestre Brøggerbreen forefront (22.45 ng/g). The mean MHg concentration was 0.14 ng/g in the stream sediment.

The mean TOC content in the fjord increased from the outer fjord (sample 8) to the middle part. Liu et al. (2013) observed that the average organic carbon content in the Kongsfjorden was 0.62%, which is lower than the present study (1.73%). The THg concentration in the surface sediments varied from 24.8 to 72.6 ng/g, with a mean value of 47.43 ng/g. The highest concentration of THg was observed for sample 8, which is considered as the outer part of the fjord.

TABLE 8.1
THg (ng/g) in Soil/Sediment from Different Ecosystems during the Period of 2015–2018

Sample No.	Type	Region	THg (ng/g)
1	Lake sediment	Knudsenheia	25.14
2	Lake sediment	Storvatnet	21.63
3	Lake sediment	Tvillingvatnet	19.93
4	Stream sediment	Near Austre Brøggerbreen	28.21
5	Stream sediment	Near Vestre Brøggerbreen	22.45
6	Stream sediment	The middle part of Bayelva	29.3
7	Stream sediment	The confluence point of Bayelva	31.97
8	Marine sediment	ᵃKongsfjorden – outer	68.86
9	Marine sediment	ᵃKongsfjorden – middle	29.23
10	Marine sediment	ᵃKongsfjorden – middle	47.43
11	Marine sediment	ᵃKongsfjorden – inner	44.2
12	Terrestrial soil	Ny-Ålesund	25
13	Terrestrial soil	Ny-Ålesund	33.33
14	Terrestrial soil	Ny-Ålesund	43.33
15	Terrestrial soil	Ny-Ålesund	20.83
16	Terrestrial soil	Ny-Ålesund	49.99
17	Terrestrial soil	Ny-Ålesund	75.83

ᵃPeriod 2015–2017.

In the terrestrial soils, THg varied from 20.83 to 75.83 ng/g, and the maximum concentration was observed near the Bayelva stream (S6), followed by Ny-Ålesund town (S5). The mean THg concentration varied from 22.08 to 187.62 ng/g in the vegetation, with an average of 83.45 ng/g (Table 8.2). The concentration of THg was higher in vegetation in 2016, and the maximum concentration was observed in the tufted saxifrage (Figure 8.4). There was no significant difference between the THg concentration and the year ($F = 0.380$; $p > 0.05$).

TABLE 8.2
THg Concentration in the Vegetation (ng/g)

		THg (ng/g)							
Common Name	**Scientific Name**	**2016**					**2017**		
Moss Campion	*Silene acaulis* (L.) Jacq	141.22	30.61	ND	73.47	ND	20.08	26.14	61.28
Bog Saxifrage	*Saxifraga hirculus* L.	43.16	26.06	118.27	28.77	162.15	43.66	22.59	ND
Tufted Saxifrage	*Saxifraga cespitosa* L.	114.85	173.35	282.39	117.08	86.82	187.62	122.41	179.52

ND, No data.

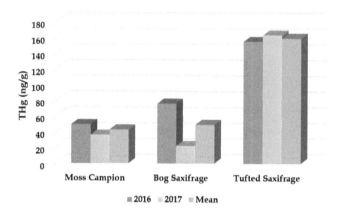

FIGURE 8.4   THg concentration (ng/g) in the vegetation.

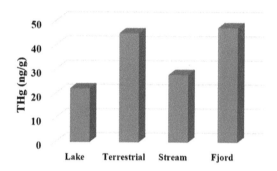

FIGURE 8.5   Mean THg concentration (ng/g) in various environmental matrices.

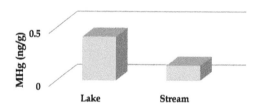

FIGURE 8.6   Mean MHg concentration (ng/g) in lake and stream.

The total mercury concentration was high in fjord sediments, followed by soil, stream and lake (Figure 8.5). However, the MHg content was high in lake sediments compared to stream sediments (Figure 8.6). The presence of MHg in stream sediments (0.14 ng/g) and lake sediments (0.41 ng/g) indicates the potential for methylation in the Arctic freshwater ecosystems. Hence, there are high chances of conversion of more quantity of less toxic inorganic form to highly toxic organic forms if more Hg reaches the stream sediment. All the reactions in water and sediment will enhance rising temperature and subsequently affect the normal cycling of Hg in the terrestrial environment.

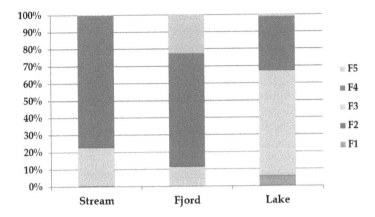

FIGURE 8.7 Mean content of various mercury fractions in different ecosystems.

8.3.2 FRACTIONATION OF MERCURY

The mercury fractionation has shown that the organo-chelated (F3) and elemental (F4) Hg fractions owed the major portions (Figure 8.7). It is indicated that the Hg present in these systems is available for methylation reactions. The residual fractions observed only in the fjord sediments might be due to the presence of anaerobic conditions and availability of sulphur and geological conditions. The lake sample has shown the presence of all the fractions, and a very small fraction is bioavailable (F1 and F2). The fifth fraction is negligible in the lake sediments. It indicated that the mercury in lake and stream sediments is mainly of anthropogenic origin, which came from long-range transport and subsequent AMDE. Otherwise, the concentration of the residual fraction might be more. The proglacial streams are seasonal and hence the natural Hg content in the sediments is below the detectable limit of zero. The F3 and F4 fractions are available for further transformation reactions, and they are high in fjord systems against the natural F5 content. Hence, the sediment Hg content in the fjord and streams is related, and the sources might be the long-range transport and deposition of mercury. The maximum bioavailable fractions are observed in the lake system.

8.4 DISCUSSION

Rainwater is a major source of water for lakes. The sources of Hg in the freshwater lakes are the same compared to meltwater streams. Moreover, erosion of permafrost during the summer acts as an important source of Hg to lakes. The amount of Hg received through snow melting is very low. Hence, the Hg concentration observed in lake sediment was slightly lower than in the other ecosystems. The mercury content is influenced by the catchment inputs of dissolved organic carbon (DOC), and photoreduction enhances the retention of Hg in the lakes. The productivity of the lakes strongly affects the retention of Hg by increasing the scavenging rate of mercury from the water column. After that, it is accumulated in the sediments (Outridge et al. 2007). The retreat of glaciers might have influenced the thickening of the active layer of permafrost, which added to the risk of contaminant flow into the lakes.

The deposited mercury (Hg-II) can accumulate in the soil or snow; however, it may melt away during summer. The mercury accumulated in the active layer of permafrost may be released into nearby environmental matrices. The increase in the thickness of the active layer is highly significant as the previously deposited Hg may be re-released into the environment. The soil Hg can accumulate in the vegetation. The changing climate may enhance the growth of tundra vegetation, which can support more accumulation in the above-ground biomass. This remains significant as the lichen, moss and vascular plants in the Arctic terrestrial ecosystem are the main diet for the tundra animals (Gauthier et al. 1989) and may lead to bioaccumulation and biomagnifications across the food web. The primary and secondary uptake of Hg by flora and fauna leads to bioaccumulation and can accumulate into organisms such as moose, caribou, reindeer and wolf (Larter and Nagy 2000; AMAP 1997).

The study conducted by Ford et al. (1995) observed that the moss species *Hylocomium splendens* (glittering woodmoss) have an average THg concentration of 44 ng/g, ranging from 35 to 54 ng/g. *Racomitrium lanuginosum* moss found in the tundra showed THg concentration from 88 to 137 ng/g from Nuuk, Greenland, with an average of 91 ng/g (Riget et al. 2000). The present study also reported comparable values. The THg observed was slightly higher than the earlier studies carried out in the tundra vegetation. The rise in temperature and changes in the weather pattern in various seasons might have enhanced the growth rate of flora, which might be attributed to the increased bioaccumulation.

The rise in temperature affects the geochemistry of soil mercury and may cause the emission of Hg back into the atmosphere. The emission can be physical volatilisation and biological transformation. Photochemical transformation of Hg(II) to Hg^0 in soils supports the transport of mercury into the atmosphere. The prevalence of Hg-tolerant bacteria in the soils of tundra regions of Ny-Ålesund has been observed in an early study (Binish et al. 2017). It indicates the occurrence of Hg transformation in the tundra soils and the subsequent emission into the surrounding environment, which may have increased amid climate change. This may be the reason behind the declining trend of Hg content in the soil in the recent past. These changes could affect the geochemistry and transport of Hg in the tundra region.

Generally, mosses and lichens do not have a vascular system, minimising the possibility to uptake the trace metals from substrates. Moreover, the contaminants are accumulated mainly from the atmosphere. This might be the reason for the low concentration of Hg in mosses. Along with this accumulation, geographical and ecological factors make mosses a useful bio-monitor of atmospheric deposition (Steinnes 1995).

Climate change has already affected seasonal snow deposition and melting (Stern et al. 2012). The changing climate has adversely affected the glacier mass balance. The melting of snow produces a high quantity of meltwater, which contains metals and other toxic elements that have been deposited in the recent past. In this scenario, the Hg deposited during AMDE cannot hold more in the snow. Mercury can reach out through the meltwater into sediments of streams and fjords. Hence, the deposition time and adsorption of Hg in the soils will be reduced. The thickening of the active permafrost layer will have a synergic effect on the deposition of Hg in the soils. It can be concluded that the opportunity for deposition and residence time of Hg in soils is declining, which transports it faster to other environmental matrices such as air, water and sediments.

The mercury released from the soils and snowmelt reaches meltwater streams or proglacial streams, where it can accumulate in the sediments or be transported to the nearby fjord system. The mercury entered into the sedimentary environment of streams or fjord may undergo chemical and biological transformations (Figure 8.8). It can transform into more toxic forms like methyl mercury and less stable ethyl or phenyl mercury. Moreover, the Hg(II) in water can undergo photoreduction and form Hg^0, which can be emitted into the air. However, the photooxidation of dissolved gaseous mercury (Hg^0) to Hg(II) by halogens, main chlorides in the Arctic coastal streams and ponds and marine waters, provides more mercury for further chemical or biological transformation (Whalin et al. 2007). The rate of photoreduction and photooxidation processes will be balanced in lakes. However, the presence of DOC can support the photoreduction on the surface of water, which results in the retention of mercury in lakes/streams (O'Driscoll et al. 2018). The mercury in water can also be attached to the organic matter (both dissolved and particulate) and transported to fjords.

It is confirmed that the AMDE and Hg deposition is the major source of lake sediment mercury. However, a very small percentage of residual mercury is present in the lakes, which can be attributed to the presence of Hg in the geological material. The available Hg component in the stream sediments confirms its long-range transport. The fractionation results indicated the independent character of the lake system. The mercury accumulated in the surface sediments of proglacial streams is of anthropogenic origin, as the deposition has been mainly linked with the melting of glaciers. It has been proved that the glacier melt has caused for release of Hg and other metals into the fjord sediments during the last few decades (Mohan et al. 2018). The core sediments of the Kongsfjorden have shown an increased sedimentation rate and contaminant deposition. The major reason for this is the changing climate and rise in temperature. The increased melting of glaciers has released metals and other toxic pollutants, along with the englacial sediments, into the nearby fjords on account of increased sedimentation and deposition of metals and other pollutants in the tidewater glacier fjords (Mohan et al. 2018). The difference in accumulation or retention of Hg content in different ecosystems has been shown a significant difference ($p < 0.01$). It indicates that the sources, deposition and transformation in the matrices of the Arctic environment depend on geographical locations, glacial melting and permafrost thawing (Figure 8.8).

8.5   CONCLUSIONS

Mercury transformation and transport in the Arctic terrestrial environment such as soil, fresh water and fjord ecosystems are shown in Figure 8.8. It can be concluded that climate change significantly influences the cycling of Hg in the Arctic environment by impacting physical conditions and geochemical parameters. This, in turn, affects Hg transformation and transport in the terrestrial environment. The changes in Hg cycling in the Arctic terrestrial environment may be mainly due to:

1. Rise in temperature
2. Changes in factors affecting photochemical reactions – oxidation and reduction

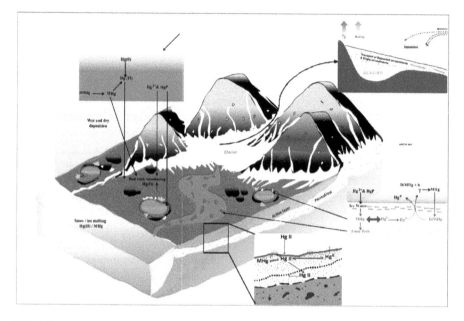

FIGURE 8.8   Schematic diagram representing fate and transport of Hg in the Arctic tundra.

3. Increase in snowmelt and permafrost thawing as well as thickening of the active layer
4. Glacier melting.

The fate of Hg is dependent on specific factors. Overall, the rise in temperature due to global climate change has influenced the rate of transformation in the matrix and movement of Hg from one matrix to another. The major variation occurring in the Hg cycle is the less retention time of Hg in the Arctic environment, especially in the snow and the release of deposited mercury. It may enhance the emission of mercury than deposition, which converts the Arctic into the source of Hg, whereas earlier it was considered as a sink of mercury.

ACKNOWLEDGEMENTS

We would like to thank the Ministry of Earth Sciences (MoES), Department of Science and Technology (DST) PURSE & FIST programmes for their financial support and DST-INSPIRE for the research fellowship to the first author. Moreover, we are grateful to the National Centre for Polar and Ocean Research (NCPOR), Earth System Science Organisation of the Ministry of Earth Sciences, Government of India, for its logistical and financial support through the Arctic Expedition and PACER-POP Project.

REFERENCES

AMAP, 1997. AMAP Assessment Report: Heavy Metals. In *Arctic Monitoring and Assessment Programme (AMAP)*, Oslo, Norway.

AMAP, 1998. AMAP Assessment Report: Arctic Pollution Issues. In *Arctic Monitoring and Assessment Programme (AMAP)*, Oslo, Norway.

AMAP, 2005. AMAP Assessment: Heavy Metals in the Arctic. In *Arctic Monitoring and Assessment Programme (AMAP)*, Oslo, Norway.

AMAP, 2019. Arctic Climate Change Update 2019. In *Arctic Monitoring and Assessment Programme (AMAP)*, Oslo, Norway.

Aubail, A., Dietz, R., Rigét, F., Sonne, C., Wiig, Ø., & Caurant, F. (2012). Temporal trend of mercury in polar bears (Ursus maritimus) from Svalbard using teeth as a biomonitoring tissue. *Journal of Environmental Monitoring* 14(1), 56–63.

Bennett, M.M. (2014). North by Northeast: toward an Asian-Arctic region. *Eurasian Geography and Economics* 55(1), 71–93.

Bloom, N.S., Preus, E., Katon, J., & Hiltner, M. (2003). Selective extractions to assess the biogeochemically relevant fractionation of inorganic mercury in sediments and soils. *Analytica Chimica Acta* 479(2), 233–248.

Boening, D.W. (2000). Ecological effects, transport, and the fate of mercury: a general review. *Chemosphere* 40(12), 1335–1351.

Budnik, L.T., & Casteleyn, L. (2019). Mercury pollution in modern times and socio-medical consequences. *Science of the Total Environment* 654, 720–734.

Callaghan, T.V., & Jonasson, S. (1995). Arctic terrestrial ecosystems and environmental change. *Philosophical Transactions of the Royal Society of London. Series A: Physical and Engineering Sciences* 352(1699), 259–276.

Constant, P., Poissant, L., Villemur, R., Yumvihoze, E., & Lean, D. (2007). The fate of inorganic mercury and methyl mercury within the snow cover in the low arctic tundra on the shore of Hudson Bay (Québec, Canada). *Journal of Geophysical Research: Atmospheres* 112(D8).

Dastoor, A.P., & Durnford, D.A. (2013). Arctic Ocean: is it a sink or a source of atmospheric mercury? *Environmental Science & Technology* 48(3), 1707–1717.

Dietz, R., Riget, F., & Born, E.W. (2000). Geographical differences of zinc, cadmium, mercury and selenium in polar bears (Ursus maritimus) from Greenland. *Science of the Total Environment* 245(1–3), 25–47.

Dommergue, A., Larose, C., Faïn, X., Clarisse, O., Foucher, D., Hintelmann, H., Schneider, D., & Ferrari, C.P. (2010). Deposition of mercury species in the Ny-Ålesund area (79 N) and their transfer during snowmelt. *Environmental Science & Technology* 44(3), 901–907.

Epa, U. (1998). Method 7473 (SW-846): mercury in solids and solutions by thermal decomposition, amalgamation, and atomic absorption spectrophotometry, Revision 0. Washington, DC, p. 17.

Fisk, A.T., Hobbs, K., & Muir, D.C.G. (2003). *Contaminant levels, trends and effects in the biological environment. Canadian Arctic Contaminants Assessment Report II*. Ottawa: Indian and Northern Affairs Canada.

Ford, J., Landers, D., Kugler, D., Lasorsa, B., Allen-Gil, S., Crecelius, E., & Martinson, J. (1995). Inorganic contaminants in Arctic Alaskan ecosystems: long-range atmospheric transport or local point sources? *Science of the Total Environment* 160, 323–335.

Fortin, C., Beauchamp, G., Dansereau, M., Lariviere, N., & Belanger, D. (2001). Spatial variation in mercury concentrations in wild mink and river otter carcasses from the James Bay Territory, Quebec, Canada. *Archives of Environmental Contamination and Toxicology* 40(1), 121–127.

Gauthier, L., Nault, R., & Crête, M. (1989). Seasonal variation in the diet of George River Caribou Herd, Québec. *Naturaliste Canadien* 116, 101–112.

Givelet, N., Roos-Barraclough, F., Goodsite, M.E., Cheburkin, A.K., & Shotyk, W. (2004). Atmospheric mercury accumulation rates between 5900 and 800 calibrated years BP in the high arctic of Canada recorded by peat hummocks. *Environmental Science & Technology* 38(19), 4964–4972.

GMA, (2018). Global Mercury Assessment, 2018. UN-Environment Programme, Chemicals and Health Branch, Geneva, Switzerland. p. 59.

Gouin, T., Mackay, D., Jones, K.C., Harner, T., & Meijer, S.N. (2004). Evidence for the "grasshopper" effect and fractionation during long-range atmospheric transport of organic contaminants. *Environmental Pollution* 128(1–2), 139–148.

Hallanger, I.G., Fuglei, E., Yoccoz, N.G., Pedersen, Å.Ø., König, M., & Routti, H. (2019). Temporal trend of mercury in relation to feeding habits and food availability in arctic foxes (Vulpes Lagopus) from Svalbard, Norway. *Science of the Total Environment* 670, 1125–1132.

Hui, A., Takekawa, J.Y., Baranyuk, V.V., & Litvin, K.V. (1998). Trace element concentrations in two subpopulations of lesser snow geese from Wrangel Island, Russia. *Archives of Environmental Contamination and Toxicology* 34(2), 197–203.

Hwang, B., Aksenov, Y., Blockley, E., Tsamados, M., Brown, T., Landy, J., Stevens, D., & Wilkinson, J. (2020). Impacts of climate change on arctic sea ice. *MCCIP Science Review* 2020, 208–227.

Iannilli, V., Pasquali, V., Setini, A., & Corami, F. (2019). The first evidence of microplastics ingestion in benthic amphipods from Svalbard. *Environmental Research* 179, 108811.

Intergovernmental Panel on Climate Change (2013). Climate change 2013: The physical science basis. Working Group I Contribution to the Fifth Assessment Report of the Intergovernmental Panel on Climate Change. Cambridge: Cambridge University Press.

Jiang, S., Liu, X., & Chen, Q. (2011). Distribution of total mercury and methylmercury in lake sediments in Arctic Ny-Ålesund. *Chemosphere* 83(8), 1108–1116.

Jorgenson, M.T., Shur, Y.L., & Pullman, E.R. (2006). The abrupt increase in permafrost degradation in Arctic Alaska. *Geophysical Research Letters* 33(2).

Kallenborn, R., & Herzke, D. (2001). Long-Range Transport of persistent pollutants in the northern hemisphere. *Environmental Science and Pollution Research* 8(3), 215–215.

Kim, M.K., & Zoh, K.D. (2012). Fate and transport of mercury in environmental media and human exposure. *Journal of Preventive Medicine and Public Health* 45(6), 335.

Kirk, J.L., St. Louis, V.L., & Hintelmann, H. (2005). *What Is the Source of Methyl Mercury in the Canadian Arctic Oceans? ArcticNet Annual Scientific Meeting.* Banff, Alberta.

Klaminder, J., Yoo, K., Rydberg, J., & Giesler, R. (2008). An explorative study of mercury export from a thawing palsa mire. *Journal of Geophysical Research: Biogeosciences* 113(G4).

Kofinas, G., Forbes, B., Beach, H., Berkes, F., Berman, M., Chapin, T., Csonka, Y., Kjell, D., Osherenko, G., Semenova, T., & Tetlichi, J. (2005). A research plan for the study of rapid change, resilience, and vulnerability in social-ecological systems of the Arctic. *The Common Property Resource Digest* 73, 1–10.

Kovacs, K. M., Lydersen, C., Overland, J. E., & Moore, S. E. (2011). Impacts of changing sea-ice conditions on Arctic marine mammals. *Marine Biodiversity* 41(1), 181–194.

Larose, C., Dommergue, A., De Angelis, M., Cossa, D., Averty, B., Marusczak, N., Soumis, N., Schneider, D., & Ferrari, C. (2010). Springtime changes in snow chemistry lead to new insights into mercury methylation in the Arctic. *Geochimica et Cosmochimica Acta* 74(22), 6263–6275.

Larter, N.C., & Nagy, J.A. (2004). Seasonal changes in the composition of the diets of Peary caribou and muskoxen on Banks Island. *Polar Research* 23(2), 131–140.

Leitch, D.R. (2006). Mercury distribution in water and permafrost of the lower Mackenzie Basin, their contribution to the mercury contamination in the Beaufort Sea marine ecosystem, and potential effects of climate variation.

Letcher, R.J., Bustnes, J.O., Dietz, R., Jenssen, B.M., Jørgensen, E.H., Sonne, C., Verreault, J., Vijayan, M.M., & Gabrielsen, G.W. (2010). Exposure and effects assessment of persistent organohalogen contaminants in arctic wildlife and fish. *Science of the Total Environment* 408(15), 2995–3043.

Loseto, L.L., Siciliano, S.D., & Lean, D.R. (2004). Methylmercury production in High Arctic wetlands. *Environmental Toxicology and Chemistry: An International Journal* 23(1), 17–23.

Lyman, S.N., Cheng, I., Gratz, L. E., Weiss-Penzias, P., & Zhang, L. (2020). An updated review of atmospheric mercury. *Science of the Total Environment* 707, 135575.

Macdonald, R.W., Harner, T., & Fyfe, J. (2005). Recent climate change in the Arctic and its impact on contaminant pathways and interpretation of temporal trend data. *Science of the Total Environment* 342(1–3), 5–86.

Maggi, C., Berducci, M.T., Bianchi, J., Giani, M., & Campanella, L. (2009). Methylmercury determination in marine sediment and organisms by direct mercury analyzer. *Analytica Chimica Acta* 641(1–2), 32–36.

Maiti, S.K. (2003). *Handbook of Methods in Environmental Studies*, Vol. 2: Air, Noise, Soil, Overburden, Solid Waste and Ecology. Japur: ABD Publishers.

Mechirackal Balan, B., Shini, S., Krishnan, K.P., & Mohan, M. (2018). Mercury tolerance and biosorption in bacteria isolated from Ny-Ålesund, Svalbard, arctic. *Journal of Basic Microbiology* 58(4), 286–295.

Moerlein, K.J., & Carothers, C. (2012). Total environment of change: impacts of climate change and social transitions on subsistence fisheries in northwest Alaska. *Ecology and Society* 17(1).

Mohan, M., Sreelakshmi, U., Vishnu, S., Gopikrishna, V.G., Pandit, G.G., Sahu, S.K., Tiwari, M., Ajmal, P.Y., Kannan, V.M., Abdul Shukur, M., & Krishnan, K.P. (2018). Metal contamination profile and sediment accumulation rate of arctic fjords: implications from a sediment core. Kongsfjorden. Svalbard. *Marine Pollution Bulletin* 42, 131.

Moritz, R.E., Bitz, C.M., & Steig, E.J. (2002). Dynamics of recent climate change in the arctic. *Science* 297(5586), 1497–1502.

Muir, D.C.G., Wang, X., Yang, F., Nguyen, N., Jackson, T.A., Evans, M.S., Douglas, M., Kock, G., Lamoureux, S., Pienitz, R., & Smol, J.P. (2009). Spatial trends and historical deposition of mercury in eastern and northern Canada inferred from lake sediment cores. *Environmental Science & Technology* 43(13), 4802–4809.

Nevitt, M.P., & Percival, R.V. (2018). Polar opposites: assessing the state of environmental law in the world's polar regions. *BCL Review* 59, 1655.

Nghiem, S.V., Rigor, I.G., Perovich, D.K., Clemente-Colón, P., Weatherly, J.W., & Neumann, G. (2007). Rapid reduction of Arctic perennial sea ice. *Geophysical Research Letters* 34(19).

O'Driscoll, N.J., Rencz, A., & Lean, D.R. (2005). The biogeochemistry and fate of mercury in the environment. *Metal Ions in Biological Systems* 43, 221–238.

O'Driscoll, N.J., Vost, E., Mann, E., Klapstein, S., Tordon, R., & Lukeman, M. (2018). Mercury photoreduction and photooxidation in lakes: effects of filtration and dissolved organic carbon concentration. *Journal of Environmental Sciences* 68, 151–159.

Outridge, P.M., Sanei, H., Stern, G.A., Hamilton, P.B., & Goodarzi, F. (2007). Evidence for control of mercury accumulation rates in Canadian High Arctic lake sediments by variations of aquatic primary productivity. *Environmental science & technology* 41(15), 5259–5265.

Pirrone, N., & Mason, R. (2009). *Mercury Fate and Transport in the Global Atmosphere.* Dordrecht, The Netherlands: Springer. 1–532.

Pistone, K., Eisenman, I., & Ramanathan, V. (2014). Observational determination of albedo decrease caused by vanishing Arctic sea ice. *Proceedings of the National Academy of Sciences* 111(9), 3322–3326.

Poulain, A.J., Garcia, E., Amyot, M., Campbell, P.G., & Ariya, P.A. (2007). Mercury distribution, partitioning and speciation in coastal vs. inland High Arctic snow. *Geochimica et Cosmochimica Acta* 71(14), 3419–3431.

Ramasamy, E. V., Toms, A., Shylesh, C. M. S., Jayasooryan, K. K., & Mahesh, M. (2012). Mercury fractionation in the sediments of Vembanad wetland, west coast of India. *Environmental Geochemistry and Health* 34(5), 575–586.

Ravichandran, M. (2004). Interactions between mercury and dissolved organic matter—a review. *Chemosphere* 55(3), 319–331.

Riget, F., Dietz, R., Johansen, P., & Asmund, G. (2000). Lead, cadmium, mercury and selenium in Greenland marine biota and sediments during AMAP phase 1. *Science of the Total Environment* 245(1–3), 3–14.

Selin, N.E. (2009). Global biogeochemical cycling of mercury: a review. *Annual Review of Environment and Resources* 34, 43–63.

Semeena, V.S., & Lammel, G. (2005). The significance of the grasshopper effect on the atmospheric distribution of persistent organic substances. *Geophysical Research Letters* 32(7).

Semkin, R.G., Mierle, G., & Neureuther, R.J. (2005). Hydrochemistry and mercury cycling in a High Arctic watershed. *Science of the Total Environment* 342(1–3), 199–221.

Serreze, M.C., & Barry, R.G. (2011). Processes and impacts of Arctic amplification: A research synthesis. *Global and Planetary Change* 77(1–2), 85–96.

Skaar, J.S., Ræder, E.M., Lyche, J.L., Ahrens, L., & Kallenborn, R. (2019). Elucidation of contamination sources for poly-and perfluoroalkyl substances (PFASs) on Svalbard (Norwegian Arctic). *Environmental Science and Pollution Research* 26(8), 7356–7363.

Sommar, J., Wängberg, I., Berg, T., Gårdfeldt, K., Munthe, J., Richter, A., Urba, A., Wittrock, F., & Schroeder, W.H. (2004). Circumpolar transport and air-surface exchange of atmospheric mercury at Ny-Ålesund (79 N), Svalbard, Spring 2002. *Atmospheric Chemistry and Physics* 7(1), 151–166.

Sprovieri, F., Pirrone, N., Landis, M.S., & Stevens, R.K. (2005). Atmospheric mercury behavior at different altitudes at Ny Alesund during Spring 2003. *Atmospheric Environment* 39(39), 7646–7656.

St. Louis, V.L., Sharp, M.J., Steffen, A., May, A., Barker, J., Kirk, J.L., Kelly, D.J., Arnott, S.E., Keatley, B., & Smol, J.P. (2005). Some sources and sinks of monomethyl and inorganic mercury on Ellesmere Island in the Canadian High Arctic. *Environmental Science & Technology* 39(8), 2686–2701.

Steinnes, E. (1995). A critical evaluation of the use of naturally growing moss to monitor the deposition of atmospheric metals. *Science of the Total Environment* 160, 243–249.

Stern, G.A., Macdonald, R.W., Outridge, P.M., Wilson, S., Chetelat, J., Cole, A., Hintelmann, H., Loseto, L.L., Steffen, A., Wang, F., & Zdanowicz, C. (2012). How does climate change influence arctic mercury? *Science of the Total Environment* 414, 22–42.

Streets, D.G., Horowitz, H.M., Lu, Z., Levin, L., Thackray, C.P., & Sunderland, E.M. (2019). Global and regional trends in mercury emissions and concentrations, 2010–2015. *Atmospheric Environment* 201, 417–427.

Stroeve, J.C., Markus, T., Boisvert, L., Miller, J., & Barrett, A. (2014). Changes in arctic melt season and implications for sea ice loss. *Geophysical Research Letters* 41(4), 1216–1225.

Svendsen, H., Beszczynska-Møller, A., Hagen, J.O., Lefauconnier, B., Tverberg, V., Gerland, S., BørreØrbæk, J., Bischof, K., Papucci, C., Zajaczkowski, M., & Azzolini, R. (2002). The physical environment of Kongsfjorden–Krossfjorden, an Arctic fjord system in Svalbard. *Polar Research* 21(1), 133–166.

Tabelin, C.B., Igarashi, T., Villacorte-Tabelin, M., Park, I., Opiso, E.M., Ito, M., & Hiroyoshi, N. (2018). Arsenic, selenium, boron, lead, cadmium, copper, and zinc in naturally contaminated rocks: a review of their sources, modes of enrichment, mechanisms of release, and mitigation strategies. *Science of the Total Environment* 645, 1522–1553.

UNEP, 2002. *Global Mercury Assessment 2002*. United Nations Environment Programme (UNEP), Geneva, Switzerland.

Walker, G. (2007). Climate change 2007: a world melting from the top down. *Nature* 446, 718–721.

Warner, N.A., Sagerup, K., Kristoffersen, S., Herzke, D., Gabrielsen, G.W., & Jenssen, B.M. (2019). Snow buntings (Plectrophenax nivealis) as bio-indicators for exposure differences to legacy and emerging persistent organic pollutants from the Arctic terrestrial environment on Svalbard. *Science of the Total Environment* 667, 638–647.

Whalin, L., Kim, E.H., & Mason, R. (2007). Factors influencing the oxidation, reduction, methylation and demethylation of mercury species in coastal waters. *Marine Chemistry* 107(3), 278–294.

Witherow, R.A., & Lyons, W.B. (2008). Mercury deposition in a polar desert ecosystem. *Environmental Science & Technology* 42(13), 4710–4716.

Yadav, J., Kumar, A., & Mohan, R. (2020). The dramatic decline of arctic sea ice linked to global warming. *Natural Hazards: Journal of the International Society for the Prevention and Mitigation of Natural Hazards* 103, 1–5.

9 Increasing Presence of Non-Polar Isolates in the Tundra and Fjord Environment – A Pointer towards Warming Trends in the Arctic

A. A. Mohamed Hatha
Cochin University of Science and Technology

Krishnan KP
National Centre for Polar and Ocean Research

CONTENTS

DOI: 10.1201/9781003265177-9

9.1 INTRODUCTION

It is well documented that the earth is warming as a result of an increase in greenhouse gases (GHGs). However, the warming is not uniform across the globe and the Arctic is warming at more than double the rate of warming elsewhere ($2.5°C$–$3°C$ compared to $0.8°C$ global average) (Overland et al. 2014). This has resulted in an increasing loss of sea ice cover (Figure 9.1, which is based on NASA image), thawing of permafrost over tundra and incursion of warm Atlantic water (Figure 9.2) into the Arctic fjord. Ice-free summer in the Arctic Ocean with far-reaching consequences is predicted as a result of rapid and pronounced warming (Wang 2009). The Arctic Ocean is also particularly sensitive to ocean acidification as the colder waters allow more dissolution of the CO_2, while the loss of sea ice exposes more and more ocean surface for potential diffusion of atmospheric CO_2 into the sea (National Research Council 2015). While the impact of global warming and related changes in the macro-level life in the Arctic and elsewhere is relatively better studied (Renaud et al. 2008), the changes happening to microbial communities are more or less unknown. Microbes as a group are the most successful organisms on planet earth and play a pivotal role in several key processes, including biogeochemical cycling, which is essential for the maintenance of life on earth.

9.2 SCOPE OF THIS CHAPTER

The scope of this book chapter is to highlight the changes in microbial communities as a proxy to the climate-induced changes that are happening in the Årctic. We have considered two major drivers of change such as the melting of sea ice (Figure 9.1) and the resultant incursion of warm Atlantic water into the fjords, especially Kongsfjorden, our study area (Figure 9.2) and the loss of ice cover over the tundra. While the warm Atlantic water mass can bring in a kind of microflora that is non-polar in origin, the expanding tundra cover provides excellent foraging ground for migratory birds, that come with a lot of allochthonous (foreign) microflora from their wintering grounds. Considerable quantities of meltwater from the tundra also reach the fjords in Arctic summer. Both these processes can bring in several allochthonous microbes into the Arctic fjords, which might be transient seasonally or some of them may adapt and become part of the autochthonous (native) microflora of the fjords over some time. The objectives of our research are to understand the diversity of bacteria, both culturable and unculturable, in the Kongsfjorden water and sediment and to understand the presence of non-polar isolates, especially the health significant mesophilic bacteria in the Arctic fjord environment. Another objective was to study the survival of the mesophilic health of significant bacteria such as *Escherichia coli* to understand whether they are adapting to the changing Arctic environment.

9.3 STUDY AREA AND COLLECTION OF SAMPLES

Kongsfjorden, located off the west coast of Spitsbergen, is considered as a model site for climate-induced changes in the Arctic (Piontek et al. 2011; Iverson and Seuthe 2012). Kongsfjorden is a glacial fjord system with a 20 km length and a varying width

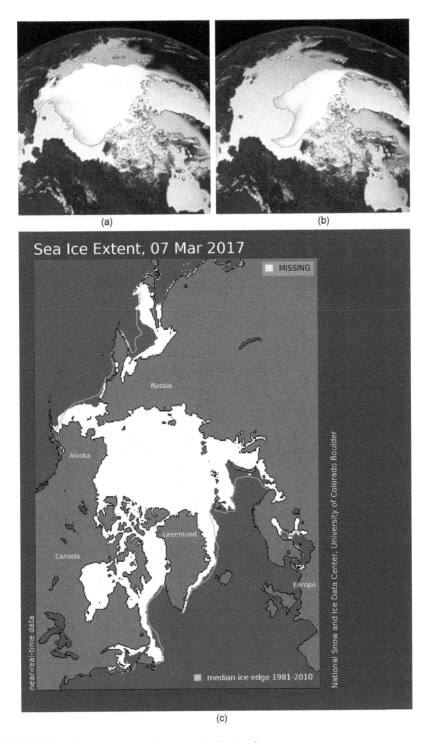

FIGURE 9.1   Fast retreating sea ice cover in the Arctic.

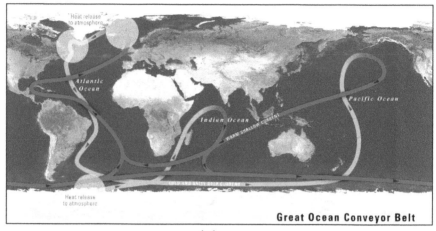

(a)

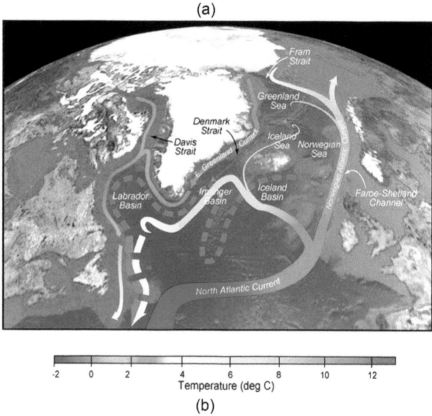

(b)

FIGURE 9.2 The great ocean conveyor belt and the extension of warm Atlantic water into the Spitsbergen region.

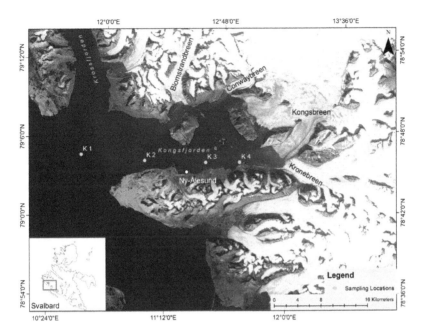

FIGURE 9.3 Map of the Kongsfjorden, Svalbard with sampling locations. The map was created using Landsat 8 OLI (path/row: 217/003) satellite data in ArcGIS 10.3.

of 4–10 km (Figure 9.3). The total volume of the fjord is estimated to be 29.4 km^3 (Ito and Kudoh 1997). Three large tidal glaciers, namely Kongsbreen, Conwaybreen and Blomstrandbreen, terminate in the Kongsfjorden. The major water masses of this fjord are – surface waters originating from glacier meltwater outflows, Atlantic water coming from the shelf of Spitsbergen, local water and bottom water formed due to deep convection during the winter (Svendsen et al. 2002).

Water and sediment samples were collected onboard RV Teisten. Water samples were collected using a Niskin sampler, and surface sediments were collected using a grab sampler (Van-Veen grab). Water samples for bacteriology were collected aseptically into a sterile container, and the surface samples were aseptically scooped out and transferred to a sterile poly bag. Samples were brought to the shore lab within 4 hours of collection. Geographical coordinates and physicochemical characteristics of the water and sediment samples are given in Table 9.1.

9.4 OUR APPROACH TO STUDIES ON BACTERIAL DIVERSITY

Our approach to studying microbial diversity was two-pronged. One is the classical culture-based method where we have used a range of culture media and dilutions to cultivate as many possible varieties as possible that grew on the culture medium. Since it is a known fact that we can never recreate the natural conditions on the culture medium, the number of bacteria recovered on any culture medium is hardly 0.1%–1%. Hence, we have also attempted a metagenome-based approach where the

TABLE 9.1

Geographical Coordinates of the Sampling Stations and Physicochemical Parameters of Water and Sediment Samples

Sampling Stations	Geographical Coordinates	Depth (m)	Surface Water Temp. (°C)	Bottom Water Temp. (°C)	Water Salinity (ppt)	Sediment Temp. (°C)	Sediment Particle Size (%)
1	79°02′36″ N 11°19′14″ E	320	6.13	0.8	33.69	1.0	Clay (13.7) Silt (82.6) Sand (3.7)
2	79°01′22″ N 11°43′06″ E	185	6.51	2.3	33.56	1.5	Clay (15.0) Silt (80.5) Sand (4.5)
3	79°00′01″ N 12°02′50″ E	110	3.61	1.2	33.11	−0.8	Clay (23.1) Silt (75.8) Sand (1.1)
4	78°57′29″ N 12°19′27″ E	45	4.14	0.8	33.13	−0.6	Clay (12.3) Silt (67.6) Sand (20.1)

water- and sediment-bound DNA was extracted and the bacterial community was selectively amplified using PCR with the help of 16srRNA gene primer.

Samples were taken from outer and inner fjord stations and also from the stations kept for comprehensive Kongsfjorden monitoring. India has also established moored observatory in the Kongsfjorden, Arctic, for real-time observations on the important physicochemical parameters of the fjord environment.

9.5 CULTURING, ISOLATION AND IDENTIFICATION OF THE BACTERIA

The samples upon reaching the shore lab (Kingsbay Marine Laboratory at Ny-Ålesund International Arctic Research Station) were processed for bacteriology. Water and sediment samples were diluted using isotonic saline, and 0.2 mL of appropriate dilutions were plated on various selective and non-selective media. We have used various strengths of Zobell Marine Agar (ZMA, Full strength, half strength and quarter strength), nutrient agar and thiosulphate citrate bile salts sucrose (TCBS) agar. The plates were incubated at 5°C and 20°C for 7–14 days. Plates with counts ranging from 25 to 250 were taken for estimation of the bacterial load, which is expressed as cfu/ mL and cfu/gm for water and sediment, respectively.

After counting and estimating bacterial load in the samples, well-separated colonies were aseptically transferred to sterile ZMA slants using a sterile inoculation loop. The purity of the isolates was ensured by restreaking on the media, and the purified isolates were numbered and stored on sterile ZMA slants for further characterisation as per Bergey's manual of determinative bacteriology.

For molecular characterisation, total bacterial genomic DNA was extracted using the bacterial genomic DNA (prep) kit (Chromus Biotech, India). 16srRNA genes of these isolates were amplified by PCR using universal primers 27f and 1492r (Bosshard and Santini 2000). PCR products were checked by electrophoresis, purified using PCR cleaned up kit and subjected to sequencing using an ABi 3730 XL genetic analyser (Applied Biosystem, USA). Sequencing of the PCR products was done at Scigenome, Kochi. The obtained DNA sequences were subjected to sequence similarity search at NCBI (http://blast.ncbi.nlm.nih.gov/BLAST) to identify the nearest taxa. The obtained sequences were aligned using CLUSTAL W, and phylogenetic trees were constructed by tree-making algorithms such as maximum likelihood (ML) and neighbourhood joining (NJ) methods using MEGA Version 5 (Tamura et al. 2011).

9.6 DETECTION OF MESOPHILIC HEALTH SIGNIFICANT BACTERIA INTO THE ARCTIC ENVIRONMENT

As a result of accelerated warming in the Arctic region, ice cover on the tundra is fast disappearing. This, in a way, opens up new foraging ground for seasonal migratory birds in the Arctic. We presume that these migratory birds, especially the Svalbard population of barnacle goose (*Branta leucopis*) (Figure 9.3), which do wintering in the Solway Firth of Scotland-England border bring in a lot of allochthonous bacteria into the pristine Arctic environment, including health significant mesophilic bacteria. To test this hypothesis, we have collected the faecal pellets of *B. leucopis* from the tundra around the international Arctic research station at Ny-Ålesund for isolation of the typical faecal indicator bacteria *Escherichia coli*.

9.7 SURVIVAL OF MESOPHILIC BACTERIA IN THE ARCTIC FJORDS

Any allochthonous bacteria, when introduced into a foreign environment, are subject to various physical, chemical and biological factors that affect their survival in that environment. Almost all natural waters have great self-purifying capacity thanks to various physical, chemical and biological factors operating in that environment. While major physical factors include dilution and mixing in the natural waters, this could not be studied in the lab. However, the effect of another physical factor such as light (both UV and visible fractions) could be well studied in the laboratory. Similarly, various biological factors such as competition, predation, antibiosis and the chemical composition of the receiving water body have a significant effect on the survival of an allochthonous intruder.

We have designed microcosm experiments to study the survival of *Escherichia coli* in the Kongsfjorden water during the summer and early winter. The following microcosms were used to study the effect of various self-purifying factors:

- **Biotic Factors**: Raw Kongsfjorden water with all its biotic components was used. Incubation is done in the dark to avoid the effect of the light factor.
- **Chemical Factors**: Filtered and autoclaved Kongsfjorden water was used to neutralise the effect of all biotic factors. Incubation is done in the dark.
- **Light**: Filtered and autoclaved Kongsfjorden water was used. Incubation is done under illumination. Natural 24-hour day cycle lighting was given to understand the effect of the summer sun.

Microcosm studies were conducted as per the methodology detailed in Chandran and Hatha (2005). Briefly, microcosms (100 mL), designed to study each of the factors described above, were prepared and inoculated with *E. coli* isolated from *B. leucopis* at an initial inoculation density of 1.5×10^7 cells/mL. The inoculated microcosms were incubated as described above for 16 days with sampling after 1, 3, 5, 7, 10, 14 and 16 days. Percentages of survival and injury were evaluated using a double-layer agar method (Chandran and Hatha 2005).

9.8   OBSERVATIONS AND DISCUSSION

Physicochemical characteristics of the water and sediment from various sampling stations are given in Table 9.1. Water temperature found to vary from 3°C to 6°C depending on the closeness of the stations to glaciers. While stations close to the open ocean recorded more than 6°C, the stations from the inner fjord ranged between 3°C and 4°C. There could be two reasons: the effect of an influx of warm Atlantic water is felt more at the outer fjord, and the meltwater from three major glaciers such as Kongsbreen, Conwaybreen and Blomstrandbreen considerably cools down the water in the inner fjord. However, the salinity found to be more or stable across the stations ranging from 33.11 to 33.69 ppt, with a minor reduction in water's salinity from inner fjord stations. The summer season in the Arctic witnessed a free exchange of Atlantic water with the fjord water; the possible reduction in salinity due to the glacier melt is more or less offset (Walkusz and Kwasniewski 2009). The influence of warm North Atlantic current on the coastal waters of western Svalbard has also been documented earlier (Svendson et al. 2002).

Sediment characteristics reveal that all of them were silty clay in nature, though the fraction of clay and sand varied depending on the station location. Inner fjord sediment samples, especially the one closer to the glacier, had a considerable quantity of clay fraction. While the outer fjord water was more or less clear blue waters, the inner fjord waters were reddish-brown due to the large-scale land run-off and influx of terrigenous material from the tundra. Such observations in the Kongsfjorden were made by earlier researchers (Zajaczkowski 2008; Piwosz et al. 2009). Sediment samples from each station varied considerably in their colour, which is attributed to the amount of organic content in them and the level of oxygenation at the bottom (Jorgensen et al. 2005).

9.9   RETRIEVABLE BACTERIAL LOAD IN THE FJORD WATER AND SEDIMENT SAMPLES

The retrievable bacterial load of the water samples ranged from 1.22×10^3 to 1.68×10^3 cfu/mL and that of sediment sample from Kongsfjorden varied from 1.35×10^5 to 8.45×10^6 cfu/gm. A similar level of retrievable bacterial load in Kongsfjordenwater was reported by Sinha et al. (2017). They have also taken the total counts with epiflourescence microscopy using 4′6-diamidino-2-phenylindole (DAPI) stain. As expected, the total counts ranged from 10^7 to 10^9 cells/mL.

In general, the retrievable bacterial load was nearly 2 logs high in the sediment samples. Sediment, in general, provides a much stable environment and nutrients, and acts as a repository of bacteria. In areas where active dredging or bottom trawls are happening, this sediment-borne bacteria could be further re-suspended to the water column. However, such activities are non-existent in the study region. Compared to sediment, water is highly dynamic, and the bacterial load is more or less transient. Increased melting could result in more freshwater input to the fjord environment, which could, in turn, injure the waterborne bacteria. Injured bacteria might not even show up on the media which we use for cultivation. While land run-off in summer could bring in a lot of terrigenous material, including organic matter, most of it is recalcitrant and do not favour the nutrition of bacteria in a significant manner (Kirchman et al. 2009). We have also encountered a slightly higher retrievable bacterial load in the inner fjord samples, which is similar to previous observations (Zheng et al. 2009). The bacterial communities in general and benthic ones in particular play key role in the remineralisation of complex organic polymers in the sediment and make it available to production through benthic-pelagic coupling (Teske et al. 2011) and regulate the concentration of biogenic elements such as C, N, P, Fe, O and S (Kirchman et al. 2009).

9.10 BACTERIAL FLORA OF THE KONGSFJORDEN WATER

Observations on culturable bacteria encountered during the summers of 2009 and 2011 are given in Table 9.2 and Figures 9.4 and 9.5. The composition of various bacterial species in the summer months (June to September) of 2011 is represented in Figure 9.4. Proteobacteria (both α and γ) and Actinobacteria contributed a major fraction in almost all the months, while the contribution of firmicutes and bacteroidetes varied depending on the month of sampling. The dominance of γ-proteobacteria is also observed in summer 2009 samples. Prevalence of α and γ proteobacteria and firmicutes was reported previously from the Arctic environment (Piquet et al. 2010). In general, the peak summer month of August revealed a better diversity of all groups. The Kongsfjorden hydrography remains highly dynamic during peak summer (Hop et al. 2002) with a considerable influx of Atlantic water from the ocean side (outer fjord) and significant freshwater influx from the glacier side (inner fjord). This will result in a mix-up of communities resulting in better diversity during mid-summertime.

Various genera that contributed to each phylum encountered in the summer of 2009 are given in Figure 9.3 and that in 2011 summer are given in Table 9.2. Results reveal good diversity of genera in the α-proteobacteria and γ-proteobacteria. This includes several non-polar genera, suggesting their invasion possibly through the warm Atlantic water. *Halomonastitanicae* is one such genus that is identified with Atlantic water mass. Microbial signatures associated with specific water masses have been reported earlier (Agouge et al. 2011). Their study, with the help of massively parallel sequencing, reveals that deepwater masses act as bio-oceanographic islands for bacterioplankton leading to water mass-specific bacterial communities in the deep waters of the Atlantic.

Results of 2009 summer samplings (Figure 9.5) also revealed potential mesophilic forms such as *Enterobacter ludwigii* and *Stenotrophomonas maltophilia* contributing a significant percentage to the total number of organisms from both sediment

TABLE 9.2
Percentage of Incidence of Various Genera of Heterotrophic Bacteria under Each Phylum during the Summer of 2011

Name of Phylum	Name of Genus and sp.	% Prevalence During		
		Jun 2011	Aug 2011	Sep 2011
Firmicutes	Bacillus subtilis	7	0.5	0
	Paenibacillus barcinonencis	0	2.7	0
	Staphylococcus haemolyticus	0	5.4	0
	Staphylococcus pasteuri	0	4.8	0
	Paenibacillus sp.	0	0.7	0
	Bacillus idriensis	0	0	2
	Paenibacillus urinalis	0	0	1
Actinobacteria	Agrococcus baldri	17	4.8	0
	Rhodococcus fasciens	0	32.3	9
	Micrococcus yunnanensis	0	1.2	0
	Nocardiodus basaltis	0	0	12
α-Proteobacteria	Erythrobacter citreus	2	2.4	0
	Sphingopyxis baekryungensis	2.5	2.7	10
	Brevundimonas mediterranea	0	1.1	0
	Phenylobacterium falsum	0	3.6	0
	Sphingopyxis flavimarus	0	1.9	0
	Alterythrobacter sp.	0	0.3	0
	Paracoccus marinus	0	0	3
	Brevundimonas variabilis	0	0	15
	Brevundimonas sp.	0	0	3
γ-Proteobacteria	Psychrobacter nivimaris	9	4.8	6
	Acinetobacter iwoffii	0	1.6	0
	Halomonas titanicae	0	4	16
	Janibacter limosus	0	0.8	0
	Psychrobacter fozii	0	2	0
	Psychrobacter glanicola	0	5.6	0
	Psychrobacter okhotskensis	0	4.4	0
	Psychrobacter piscatorii	0	0	11
Bacteroidetes	Leewenhoekiella aqueoria	40	12	12

Source: Modified from Sinha et al. (2017).

and water from Kongsfjorden. *Stenotrophomonas maltophilia* is getting increasing importance as an emerging opportunistic pathogen (Brooke 2012). Nosocomial infections caused by this organism has been reported (Kwa et al. 2008; Paez et al. 2008), which are often difficult to treat (Orcutt et al. 2011). Though the presence of *S. maltophilia* is reported from the marine realm (Zheng et al. 2009, 2011), its presence

FIGURE 9.4 Barnacle geese feeding in the tundra around international research base in Ny-Ålesund.

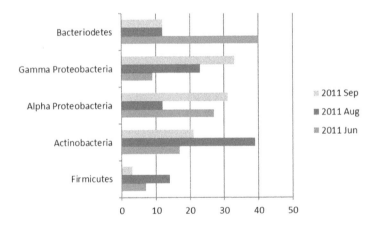

FIGURE 9.5 Percentage composition of various bacterial phyla in the Kongsfjorden water during the summer of 2011. (Adopted from Rupesh Sinha et al., 2017, *Brazilian Journal of Microbiology*, 48: 51–61.)

in the Arctic environment is uncommon. The effect of fast retreating ice cover and resultant opening up of shipping channels as well as mixing up of water masses from lower latitudes need to be studied to clearly understand the presence of such organisms in the Arctic fjord environments. *Enterobacter ludwigii* is also recognised as a novel Enterbacter species, which is gaining clinical significance (Hoffman et al. 2005).

Among the firmicutes, *Bacillus thuringiensis* was the only one encountered in the water samples from Kongsfjorden, whereas *B. thuringiensis* and *B. flexus* were isolated from the sediment samples. We could also recover *Staphylococcus cohnii* from

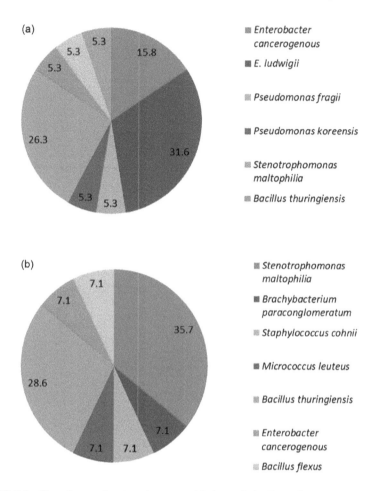

FIGURE 9.6 Prevalence of various heterotrophic bacteria in Kongsfjorden water and sediment during 2009 summer.

the sediment samples of Kongsfjorden. Srinivas et al. (2009) reported the presence of *B. thuringiensis* from the fjord samples. Perrault et al. (2008) observed the presence of *B. thuringiensis* in the glacier meltwater and sediments from the Arctic, while Gourdieva (2004) reported the presence of *Staphylococcus* in the spring samples of the high Arctic. However, as of now, there are no reports on the presence of *S. maltophilia* and *E. ludwigii* from the Arctic samples. Since both these organisms are currently considered as emerging opportunistic pathogens, further studies are required to understand the virulence markers and antibiotic resistance among them.

Figure 9.7 represents the distribution of polar and non-polar bacterial isolates (depending on the reports on the presence of these bacteria) in the Kongsfjorden water during the 2011 summer. This indicates the presence of more diverse non-polar isolates during the mid-summer months when the incursion of Atlantic water is maximum. Certain phytopathogens such as *Rhodococcus fasians* who's dominating presence is noticed during the summer bloom of phytoplankton in the Arctic fjord environment. However, most of the reports on the association of *R. fascians* are related to terrestrial plants (Goethals et al. 2001; Vandeputte et al. 2005). Some of the other non-polar genera include *R. yunnanensis, Micrococcus stanieri, Erythrobacter citreus,* and *H. boliviensis,* to name a few.

9.11 MESOPHILIC BACTERIA IN THE ARCTIC ENVIRONMENT

Specific observation on the presence of mesophiles such as *Escherichia coli* associated with migratory bird *B. leucopis* is given in Table 9.3. Serotyping of the *E. coli* strains at National Salmonella and Escherichia Centre, Central Research Institute, Kasauli, Himachal Pradesh revealed the presence of nine different serotypes, including pathogenic ones such as O_2. O_2 serotype of *E. coli* constituted nearly 10% of the total isolates, which is reported to possess virulence factors such as alpha-hemolysin (hly), cytotoxic necrotising factor type 1 (CNF-1) and expressed P-fimbriae or mannose resistant haemagglutinin (MRHA). *E. coli* serotypes O149 (24%) and O24 (17%) were the most frequently encountered ones. Table 9.3 also represents the phenotyping results of *E. coli* isolates from *B. leucopis*. Phenotyping is done as per the methodology described by Clermont et al. (2000). Pathogenic phylotype B2 constituted nearly 32%, while B1 phylotype was dominant with a 41% share.

Dissemination of pathogens, especially ultidrug resistant (MDR) ones through migratory birds, is gaining a lot of significance. Migratory birds act as long-distance dissemination tool of the above organisms, when they mix and feed with a local population of birds, at recovery points during their long-distance migration. During such mingling, they can pick up potentially pathogenic and multidrug-resistant bacteria when they feed in polluted environments such as solid waste dumps and such polluted environments. These organisms carry with themselves and distribute to hitherto pristine environments such as the Arctic. The role of Arctic birds in the dissemination

TABLE 9.3
Diversity of *Escherichia coli* in the Droppings of the Svalbard Population of Barnacle Goose (*Branta leucopis*)

The Serotype of *E. coli*	% of Incidence	Phylotype of *E. coli*	Percentage of Incidence
O149	24.39	A	17.07
O24	17.07	B1	41.46
O148	12.17	B2	31.7
O2	9.75	D	9.75
O21	2.43		
O91	2.43		
O130	2.43		
O32	2.43		
O37	2.43		
Rough/untypable	24.39		

Source: Adopted from Hatha et al. (2013) *Current Science* 104(8): 1078–1080.

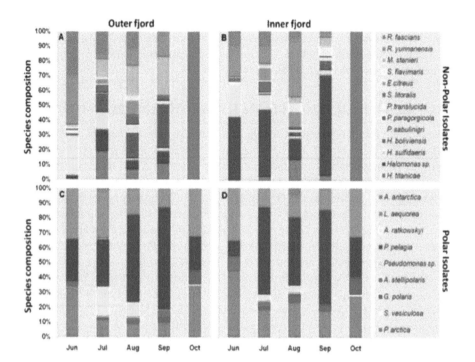

FIGURE 9.7   Distribution of polar and non-polar isolates in the inner and outer fjords (Kongsforden) during June through October 2011.

of multidrug-resistant bacteria into Siberia and Greenland regions of the Arctic has been reported previously (Sjolund et al. 2008). The migratory birds which we have analysed in this study is a Svalbard population of barnacle goose (*Branta leucopis*, Figure 9.6), which does wintering in the Solway Firth on the England/ Scotland border. Some of the wintering areas of migratory birds are frequented by birds from six different continents (Winker et al. 2007). Such environments provide a potential concoction, where MDR pathogens could be picked up by these birds and disseminate it too far and wide locations.

9.12   SURVIVAL OF *ESCHERICHIA COLI* IN THE FJORD ENVIRONMENT

We have studied the survival of *Escherichia coli* strain (isolated from *B. leucopis*) using microcosms to evaluate the power self-purifying factors contained in the fjord environment. This includes biotic factors (competition, antibiosis, predation, bacteriophage mediated lysis, etc.), chemical factors (chemical composition of the fjord water) and sunlight (both UV and visible fractions contained in the sunlight) that impinge on the Kongsfjorden water during summer (24-hour daylight cycle). We could observe that (results unpublished) chemical composition of the water does not result in a drastic decline of the cells, and the *E. coli* cells were able to survive up to 8 days in this microcosm. Biotic factors exerted a strong neutralising effect followed by sunlight. *E. coli* cells could survive only up to 3 days in the microcosms designed to study the effect of biotic factors and light. The results suggest that the self-purifying factors in the fjord environment can inactivate the contaminant *E. coli* cells in a matter of 3–7 days.

9.13 CONCLUSIONS

It is a well-known fact that the Arctic region is warming twice the rate of warming elsewhere on earth. This has implications for the entire fauna and flora in the Arctic region. The movement of sub-Arctic fauna and flora to the Arctic region and expanding tundra are quite visible. Indirect effects through changes in salinity, turbidity and sedimentation will also have a considerable effect on benthic organisms, including microbenthos. The swampy nature of the tundra could also act as an ideal environment for the methanogens to flourish with the resulting release of considerable quantities of methane into the environment. While bioinvasions through warm water influx and migratory birds into the Arctic will considerably alter the microbial diversity in the Arctic, accelerated melting of glaciers might even unlock the microorganisms, including potential pathogens, trapped deeper in the ice along with dead polar animals such as reindeer and polar bear. Elevated water temperatures could also have a telling effect on the biogeochemical processes mediated by the microbes as the polar oceans have a considerable amount of structurally stable polysaccharides stored in these environments. Ice-free summers could result in the opening up of shipping channels and increased trade and commerce in the Arctic region which could considerably impact the environment. The spread of multidrug-resistant pathogens and virulence features into the pristine Arctic region might also have serious negative consequences. Along with the efforts to study the changes in macro-level life, sufficient funding should be made to study the changes in the microbial communities and their processes in the Arctic which play crucial roles in this ecosystem.

ACKNOWLEDGEMENTS

A. A. Mohamed Hatha is thankful to the Ministry of Earth Sciences, Govt. of India and Director, National Centre for Polar and Ocean Research, Goa for providing field access to the Arctic. He is also thankful to the Registrar, Cochin University of Science and Technology for giving the necessary permissions to engage in this research.

REFERENCES

Agouge, H., Lamy, D., Neal, P.R., Sogin, M.L. and Herndl, G.J. (2011) Water mass specificity of bacterial communities in the North Atlantic revealed by massively parallel sequencing. *Molecular Ecology* 20(2): 258–274.

Bosshard, P.P., Santini, Y., Gruter, D., Stettler, R. and Bachofen, R. (2000) Bacterial diversity and composition in the chemocline of the meromictic alpine lake Cadagnoa reveal by 16S rDNA analysis. *FEMS Microbial Ecology* 31: 173–182.

Brooke, J.S. (2012) *Stenotrophomonas maltophilia*: an emerging global opportunistic pathogen. *Clinical Microbiology Reviews* 25(1): 2–41.

Chandran, A. and Hatha, A.A.M. (2005) Relative survival of *Escherichia coli* and *Salmonella typhimurium* in a tropical estuary. *WaterResearch* 39: 1397–1403.

Goethals, K., Vereecke, D., Jaziri, M., Van Montagu, M. and Holsters, M. (2001) Leafy gall formation by *Rhodococcusfascians*. *Annual Reviews in Phytopathology* 39: 27–52.

Hoffman, H., Stindle, S., Stumpf, A., Mehlen, A., Monget, D., Heeseman, J., Sleifer, K.H. and Roggenkamp, A. (2005) Description of Enterobacterludwigii sp. Nov., a novel Enterobacter species of clinical significance. *Systematic and Applied Bacteriology* 28(3): 206–212.

Hop, H., Pearson, T., Hegseth, E.N., Kovacs, K.M., Wiencke, C., Kwaniewski, S., Eiane, K., Mehlum, F., Gulliksen, B., Wlodarsk-Kowalczuk, M., Lydersen, C., Weslawski, J.M., Cochrane, S., Gabrielsen, G.W., Leakey, R.J.G., Lonee, O.J., Zajaczkowski, M., Falk-Petersen, S., Kendall, M., Wnagberg, S.A., Bischof, K., Voronkov, A.Y., Kovaltchouk, N.A., Wiktor, J., Poltermann, M., di Prisco, G., Papucci, C. and Gerland, S. (2002) The marine ecosystem of Kongsfjorden, Svalbard. *Polar Research* 21: 167–208.

Ito, H. and Kudoh, S. (1997) Characteristics of water in Kongsfjorden, Svalbard. Proceedings of NIPR Symposium. *Polar Meteorology and Glaciology* 11: 211–232.

Jorgensen, B.B., Glud, R.N. and Holby, O. (2005) Oxygen distribution and bioirrigation in Arctic fjord sediments (Svalbard, Barents Sea). *Marine Ecology Progressive Series* 292: 85–95.

Kirchman, D. L., Moran, X. A.G. and Ducklow, H. (2009) Microbial growth in polar oceans – the role of temperature and potential impact of climate change. *Nature Reviews in Microbiology* 7: 451–459.

Kwa, A.L., Low, J.G., Lim, T.P., Leow, P.C., Kurup, A. and Tam, V.H. (2008) Independent predictors for mortality in patient with positive *Stenotrophomonasmaltophilia* cultures. *Annals of Academy of Medicine Singapore* 37(10): 826–830.

National Research Council (2015) *Arctic Matters: The Global Connection to Changes in the Arctic.* Washington DC. The National Academic Press, https://doi.org/10.17226/21717.

Orcutt, B.N., Sylvan, J.B., Knab, N.J. and Edwards, K.J. (2011) Microbial Ecology of the dark ocean above, at, and below the seafloor. *Microbiology and Molecular Biology Reviews* 75: 361–422.

Overland, J.E., Wang, M., Walsh, J.E. and Stroeve, J.C. (2014) Future arctic climate changes: Adaptation and mitigation time scales. *Earths Future* 2, 68–74.

Paez, J.I., Tengan, F.M., Barone, A.A., Levine, A.S. and Costa, F.S. (2008) Factors associated with mortality in patients with bloodstream infection and pneumonia due to *Stenotrophomonasmaltophilia*. *European Journal of Clinical Microbiology and Infectious Diseases* 27(10): 901–906.

Perreault, N.N., Greer, C.W., Andersen, D.T., Tille, S., Couloume, G.L., Lollar, B.S. and Whyte, L.G. (2008) Heterotrophic and autotrophic microbial populations in cold perennial springs of the high Arctic. *Applied and Environmental Microbiology* 74: 898–907.

Piontek, J., Borchard, C., Sperling, M., Schulz, K.G., Riebesell, U and Engel, A (2012) Response of bacterioplankton activity in an Arctic fjord system to elevated pCO_2: results from a mesocosm perturbation study. *BiogeosciencesDiscussion* 9: 10467–10511.

Piquet, A.M.-T., Scheepens, J.F., Bolhuis, H., Wiencke, C. and Buma, A.G.J. (2010) Variability of protistan and bacterial communities in two Arctic fjords (Spitsbergen). *Polar Biology* 33:1521–1536.

Piwosz, K., Walkusz, W., Hapter, R., Wieczorek, P., Hop, H. and Wiktor, J. (2009) Comparison of productivity and phytoplankton in a warm (Kongsfjorden) and a cold (Hornsund) Spitzbergen fjord in mid summer 2002. *Polar Biology* 32: 549–559.

Renaud, P.E., Carrol, M.L. and Ambrose, W.G. Jr. (2008) Effect of global warming on Arctic seafloor communities and its consequences for higher trophic levels. In *Impacts of Global Warming on Polar Ecosystems.* C.M. Duarte (Ed.). Fundacion BBVA, Spain. ISBN 978–84–96515-74-1. pp. 1–39.

Sinha, R.K., Krishnan, K.P., Hatha, A.A.M., Rahiman, K.M.M., Thresyamma, D.D. and Kerkar, S. (2017) Diversity of retrievable heterotrophic bacteria in Kongsfjorden, an Arctic fjord. *Brazilian Journal of Microbiology* 48: 51–61.

Sjolund, M., Bonnedhal, J., Hernandez, J., Bengtsson, S., Cederbrant, G., Pinhassi, J., Kahlmeter, G. and Olssen, B. (2008) Dissemination of multidrug-resistant bacteria into the Arctic. *Emerging Infectious Diseases* 14, 70–72.

Srinivas, T.N.R., Nageswara Rao, S.S.S., Reddy, P.V.V., Pratibha, M.S., Sailaja, B., Kavya, B., Kishore, K.H., Begum, Z., Singh, S.M. and Shivaji, S. (2009) Bacterial diversity and bioprospecting for cold-active lipases, amylases and proteases, from culturable bacteria of Kongsfjorden and Ny-Alesund, Svalbard, Arctic. *CurrentMicrobiology* 59: 537–547.

Svendson, H., Beszczynska-Møller, A., Hagen, J. O., Lefauconnier, B., Tverberg, V. and Gerland, S. (2002) The physical environment of Kongsfjorden--Krossfjorden, an Arctic fjord system in Svalbard. *Polar Research* 21, 133–166.

Tamura, K., Peterson, D., Peterson, N., Stecher, G., Nei M. and Kumar, S. (2011) MEGA5: Molecular evolutionary genetics analysis using maximum likelihood, evolutionary distance and maximum parsimony methods. *Molecular Biology and Evolution* 28(10): 2731–2739.

Teske, A., Durbin, A., Ziervogel, K., Cox, C. and Arnosti, C. (2011) Microbial community composition and function in permanently cold seawater and sediments from an Arctic fjord of Svalbard. *Applied Environmental Microbiology* 77(6): 2008–2018.

Vandeputte, O., Oden, S., Mol, A., et al. (2005) Biosynthesis of auxin by the gram-positive phytopathogen *Rhodococcusfasciansis* controlled by compounds specific to infected plant tissues. *Applied and Environmental Microbiology* 71(3): 1169–1177.

Walkusz, W., Kwaniewski, S., Falk-Petersen, S., Hop, H., Tverberg, V., Wieczorek, P. and Weslawski, J.M. (2009) Seasonal and spatial changes in the zooplankton community in Kongsfjorden, Svalbard. *Polar Research* 28: 254–281.

Wang, M. and Overland, J.E. (2009) A sea ice-free summer in the Arctic within 30 years? *Geophysical Research Letters* 36. doi: 10.1029/2009GL037820.

Winker, K., McCracken, K.G., Gibson, D.D., Pruett, C.L., Meier, R., Huettmann, F., Wege, M., Kulikova, I.V., Zhuravlev I.N., Perdue, M.L., Spackman, E., Suarez, D.L. and Swayne, D.E. (2007) Movement of birds and avian influenza from Asia into Alaska. *Emerging Infectious Diseases* 13: 547–552.

Zajaczkowski, M. (2008) Sediment supply and fluxes in glacial and outwash fjords, Kongsfjorden and Adventfjorden, Svalbard. *Polar Research* 1: 59–72.

Zeng, Y., Zou, Y., Grebmeier, J.M., He, J. and Zheng, T. (2009) Culture-independent and dependent methods to investigate the diversity of planktonic bacteria in the northern Bering Sea. *Polar Biology* 35: 117–129.

Zheng, Y., Zheng, T. and Li, H. (2009) Community composition of the marine bacterio-plankton in Kongsfjorden (Spitzbergen) as revealed by 16S rRNA gene analysis. *Polar Biology* 32: 1447–1460.

Zheng, Y., Zou, Y., Chen, B., Grebmeir, J.M., Li, H., Yu, Y. and Zheng, T. (2011) Phylogenetic diversity of sediment bacteria in the northern Bering Sea. *Polar Biology* 34: 907–919.

10 Zooplankton of the Past, Present and Future
Arctic Marine Ecosystem

Jasmine Purushothaman,
Haritha Prasad, and Kailash Chandra
Zoological Survey of India

CONTENTS

10.1 INTRODUCTION

The Arctic marine milieu is changing, reflecting a combination of increasing land and ocean temperature, shrinking of sea ice, freshwater runoff and associated hydrographical variations (Haakstaad et al.1994), advection of species inhabiting the south (Wassmann et al. 2015) and migration of Arctic species to the northwards (Vihtakari et al. 2018). All these factors contribute to the 'borealization' process of the Arctic ecosystems, also referred to as 'atlantification', thereby causing a temperate state shift of the Arctic environment (Weydmann et al. 2018; Vihtakari et al. 2018).

Zooplanktons are an important indicator of these climates variations in the Arctic marine ecosystem due to the sudden expansion or reduction of their bioclimatic ranges as a result of alterations in the ocean circulation as well as temperature (Taylor et al. 2002). Arctic zooplankton communities are usually larger with lesser diversity (Gluchowska et al. 2017). They represent the Arctic versus Atlantic waters' influence in the Arctic seas (Hop et al. 2019). Copepods mainly dominate the mesozooplankton community in the Arctic seas; of those, the smaller-sized *Oithona similis*, *Microcalanus* spp. and *Pseudocalanus* spp., and the larger-sized *Calanus finmarchicus* and *Calanus glacialis* were most common. Other notable groups were amphipods (e.g. *Themisto libellula* and *Themisto abyssorum*), euphausiids

DOI: 10.1201/9781003265177-10

(e.g. *Thysanoessa raschii, T. longicaudata* and *T. inermis*), pteropods (e.g. *Limacina helicina*) and other crustacea (Hop et al. 2019).

Several studies have notified the warm water-associated northward extension of Arctic zooplankton communities in the recent past. This results in a considerable decline in the population of zooplankton species inhabiting colderwaters (Dalpadado et al. 2012; Haug et al. 2017). Furthermore, the increasing dominance of warm water copepods was also recorded from the high Arctic seas in the past decade with the advection of Atlantic warm water (Walkusz et al. 2009; Weydmann et al. 2014; Hop et al. 2019).

Apart from the climate-triggered alterations in the distribution, biomass and the abundance of Arctic zooplankton species, changes in their body size is also an important indicator of changes in their ecosystem properties (Trudnowska et al. 2014). According to a model put forward by Slagstad et al. (2011), the overall rate of secondary production in zooplankton communities is expected to increase with an increase in the temperature. Later, Garzke et al. (2016) documented the probability of a reduction in the total zooplankton biomass in the upcoming years. This reduction in the body sizes and life cycle of the Arctic zooplanktons can be ascribed to prolonged open-water conditions in the warm Arctic.

Among the Arctic zooplankton, *Calanus* species are known for their larger lipid storages and for their larger size than their boreal counterparts, which can be transferred to higher energy levels in the marine food web. Continued warming and the changes associated with it have largely affected the primary production systems of the high Arctic seas. This has led to a shift from *Calanus hyperboreus* and *Calanus glacialis* (large-sized, lipid-rich Arctic copepods) to *Calanus finmarchicus* (small-sized boreal copepod) in the Arctic environment (Moller and Nielsen 2020). These variabilities in the assemblages of planktonic communities and planktonic size structure can cause severe consequences in the structural composition of the pelagic food web, thereby affecting higher trophic level species (planktivorous seabirds and polar cod). All of these changes show a gradual occurrence. Hence, long-term surveillance and multiyear data are required for predicting the possible implications of these climate-driven alterations in the Arctic ecosystems (Kwasniewski et al. 2012).The present chapter aims at revisiting the diverse zooplankton studies conducted along the Arctic Seas from the past, through the present along with an analysis of the future knowledge gaps.

10.2   METHODOLOGY

Arctic zooplankton studies dated back to 1893 were collected for the preparation of this chapter. We mainly focused on literature dealing with the arctic zooplankton secondary productivity (dated since 1969), diel vertical migration and biological pump. Unpublished documents, newspaper reports and popular articles were excluded during the preparation of this chapter to eliminate any sources of uncertainty. Global ocean temperature anomalies data (°C) was collected from https://www.ncdc.noaa.gov/cag/ from 1969-2020. In addition to this, zooplankton biomass data based on our collections from the high Arctic fjord of Svalbard during the years 2018 and 2019, using standard collection techniques, were also employed to delineate the current temperature associated trends in the Arctic pelagic ecosystems.

10.3 STUDIES ON ARCTIC ZOOPLANKTON

10.3.1 PAST RESEARCH

The legendary Nansen's *Fram* expedition (1893–96) by Fridtjof Nansen, a Norwegian explorer, marked the beginning of zooplankton research in the central Arctic. During the initial 80 years, zooplankton was collected from the ships found frozen in the ice, namely, the Russian ice-breaker Georgiy Sedov and the Norwegian *Fram*, or from the drifting ice platforms. During this timeframe of sporadic data accumulation, baseline information on seasonal dynamics, as well as various structural parameters of the zooplankton communities present in the Arctic Ocean, was accumulated (Pautzke 1979; Dawson 1978; Kosobokova 1978, 1982, 1989; Hughes 1968; Brodsky and Pavshtiks 1976; Hopkins 1969a, b; Johnson 1963; Virketis 1959; Brodsky and Nikitin 1955).

Kielhorn (1952) described the biology of surface layer zooplanktons regarding the zooplankton community composition, seasonal progression, fluctuations and pertinent physical factors. Grainger (1965) also studied zooplankton in the Arctic Ocean, northwest Canadian coastal waters and also from the Beaufort Sea with special emphasis on the physical features of the water sampled.

Various researches on the estimation of zooplankton biomass also took place during this phase of time (Kosobokova 1981, 1982; Hopkins 1969a, b; Minoda 1967). However, few of these assessments were found to be difficult to compare because of incomplete sampling works done (Minoda 1967), or due to differences in the methodologies employed (Hopkins 1969a, b). A study on the estimation of zooplankton standing crop of the Arctic Basin by Hopkins (1969a) found copepods as the most dominant group, accounting for about 80% of the total biomass. According to this study, the total biomass (dry weight) of zooplankton in the central Arctic was $1\text{-}2 \times 10^6$ metric tonnes. The early 1980s witnessed the next phase in zooplankton research marked by the arrival of sophisticated ice-breakers which facilitated interdisciplinary research and also allowed improved sampling design. This interdisciplinary research, addressing the various chemical, physical and biogeochemical parameters, brought a major break-through in elucidating relationships between the pelagic community structure, environmental factors and hydrographic processes. Several studies on zooplankton biomass distribution (Auel and Hagen 2002; Kosobokova and Hirche 2000; Mumm et al. 1998; Mumm 1993; Hirche and Mumm 1992) showed the distribution of biomass across the Eurasian basins to have undergone considerable regional variability (Kosobokova and Hirche 2000; Mumm et al. 1998; Mumm 1993; Hirche and Mumm 1992). Regional variability was found to be chiefly related to the water circulation pattern. Apart from this, local food availability also has a significant effect on regional variability, reported in zooplankton biomass distribution (Kosobokova and Hirche 2009).

Due to the increasing number of studies in the Arctic Ocean, it is not anymore seen as an ice desert with no sprouts of green (Hopcroft et al. 2005; Ashjian et al. 2003; Kosobokova and Hirche 2000; Thibault et al. 1999; Mumm et al. 1998; Vinogradov and Melnikov 1980; Wheeler et al. 1996). The increased changes in global climate have been sighted as the cause of recent interest in understanding the structural complexity and function of the Arctic ecosystem and also in quantifying various biological processes (Ashjian et al. 2003; Kosobokova and Hirche 2000, 2001; Sherr et al. 2003; Lane et al. 2008; Hirche and Kosobokova 2003; Melnikov and Kolosova 2001; Olli et al. 2007; Auel and Hagen 2002; Hopcroft et al. 2005). The high rate of global warming has induced glacial melting, increase in Atlantic water temperatures and its circulation pattern (Perovich et al. 2008; Comiso et al. 2008; Maslanik et al. 2007; Grebmeier et al. 2006; Carmack et al. 2005; Rudels et al. 2000; Schauer et al. 1997; Carmack et al. 1997). These changes have inflicted the stability of the Arctic ecosystem. Recently, several studies have been conducted to provide well-grounded data on the effect of climate change on the current structure, composition and distribution of the zooplankton community from the Arctic (Trudnowska et al. 2020; Koplin 2020; Kasyan 2020; Trudnowska et al. 2020; Abe et al. 2020).

Vital information regarding the diversity and productivity of the Arctic pelagic community is essential for the detection and quantification of possible changes in the same as a result of climate change. Raskoff et al. (2005) studied gelatinous zooplankton of the Arctic Ocean with special reference to their vertical distributions, and the observed four main groups are cnidarians, ctenophores, pelagic tunicates and chaetognaths. Detailed data on the distribution of these soft-bodied animals help to bridge the gap in understanding the connectivity between primary productivity and secondary productivity in the Arctic Ocean.

Hopcroft et al. (2005, 2010) studied the community patterns of zooplankton in the Chukchi Sea and the Arctic's Canada Basin, respectively. The latter had given special emphasis on contributions by smaller taxa. Copepods are the most diverse group and contribute the bulk of the total holo-zooplankton community abundance and biomass at most stations. According to these studies, an increase in climate change may cause changes in the boundaries, size spectra, extent of penetration and productivity of these communities.

Global warming and climate change will provide increased opportunities for the southern organisms to get entrenched in the Arctic Ocean. About this, Nelson et al. (2009) employed molecular techniques to identify the population genetic structuring of zooplankton from the Pacific into the western Arctic Ocean.

The possible risks of biomagnification in the Arctic food webs, energy adaptation of the zooplankton community, the role of biological pump, and diel vertical migration in Arctic ecosystem are also extensively studied in the recent past. Several studies reported that PCBs and hydrophobic organic contaminants (HOCs) cause biomagnification in aquatic Arctic food webs (Fisk et al. 2001; Hop et al. 2002; Borga et al. 2002). A study on persistent organic pollutants (POPs) present in Arctic zooplankton by Fisk et al. (2001) revealed that hydrophobic POP concentrations present in zooplankton are more likely to reflect water concentrations. This study also suggested that POPs do not undergo biomagnification in small, herbivorous zooplankton. The occurrence of polychlorinated biphenyls (PCBs) in the Arctic is also well documented. But these studies mainly addressed the pollutant transfer upward in the food web from zooplankton and, thus, did not focus much on the potential biomagnification taking place from phytoplankton to zooplankton. A study conducted by Sobek et al. (2006) supported equilibrium partitioning of PCBs between zooplankton of the size range from 20 to >500 μm and Arctic seawater. The results suggest minimal uncertainty in the modelling of trophic transfer in the aquatic food webs, and hence, significant biomagnification does not have to be considered in zooplanktons. Characteristics of organic matter and its consequences on the sorption of organic pollutants are the study areas which demand future research efforts to understand both spatial and seasonal variability in partition coefficients.A study by Riser et al. (2008) clearly showed the importance of zooplankton for vertical flux regulation. Zooplanktons ingest POC or particulate organic carbon (about 22%-44% of the daily primary production). However, through the production of fast-sinking faecal pellets, they can accelerate vertical flux. Miquel et al. (2015) also highlighted the significance of the zooplankton community in transforming carbon, issued as a result of primary production, and its transition from the productive surface layer to the interior of the Arctic Ocean. Furthermore, Turner (2015) discussed the role of zooplankton in driving the 'biological pump' which is the exporting of photosynthetically produced carbon from the surface to the ocean interiors. Zooplankton faecal pellets are essential components in the biological pump.

Vertical migration by zooplankton enhances the efficiency of the biological pump by translocation of nitrogen (N) and carbon (C) beneath the mixed layer due to respiration and excretion at depth. According to a study by Darnis et al. (2017), euphausiids (mainly *Thysanoessa* spp.) make up more than 90% biomass of the diel migrants. Certain areas of the Atlantic side of the Arctic Ocean have euphausiids dominating the zooplankton community. In these areas, the role of diel vertical migration in the operation of the biological pump is increased due to the same. This study also addressed the biogeochemical role of zooplanktons present in Kongsfjorden. Vertical migration patterns throughout the polar night are mainly due to *Thysanoessa* spp.; however, copepods and chaetognaths may be co-responsible for the same (Grenvald et al. 2016).

Darnis and Fortier (2012) detailed zooplankton respiration and also provided data on the export of carbon in the Arctic Ocean. This study emphasizes the significance of incorporating active transport in carbon budgets by large migrant zooplanktons. During spring-summer, zooplankton migrants present in the surface layer of the Arctic seas accumulate lipids, and during winter, they respire these lipids at depth. In late summer, large zooplanktons undergo downward migration, and it coincides with a steady decrease in specific respiration rates. This signals the onset of diapause and also initiates endogenous fuelling of metabolic processes.

Grainger (1989) described the seasonal vertical migration of zooplankton groups in the central Arctic Ocean and reported that this habit might help maintain the group in the Arctic Ocean. Diel vertical migration or DVM is a global phenomenon exhibited by zooplanktons. At high latitudes, DVM patterns have been documented. Numerous studies were done with a major focus on the adaptive significance of DVM (Hays et al. 2001; Lampert 1989; Ohman 1990) and its ecosystem consequences (Buesseler et al. 2007). Several studies documented the frequent occurrence of this behaviour during autumn and spring when there is a pronounced day-night cycle (Wallace et al. 2010; Falk-Petersen et al. 2008; Cottier et al. 2006).

In contrast, many other studies have failed to observe any coordinated vertical migration at times of continuous light (Blachowiak-Samolyk et al. 2006; Fisher and Visbeck 1993). Berge et al. (2014) showed that thermal properties of water masses are the most essential factor for the vertical positioning of migrants during the daytime. At night, phytoplankton biomass is thought to be the major determining factor.

Recently, Last et al. (2016) studied moonlight-driven vertical migration of Arctic zooplankton during winter. This Lunar Vertical Migration or LVM may help in monthly pulses of carbon remineralization. The extent of LVM indicates adaptive and highly conserved behaviour. Therefore, LVM can be considered a 'baseline' activity exhibited by zooplanktons in a changing Arctic Ocean. The study also infers the role of moonlight in carbon sequestration and predator–prey interactions during the Arctic winter.

The presence of large amounts of lipids, either wax esters or triglycerides is a characteristic feature of Arctic zooplanktons addressed in research works carried out by Lee (1975), Hagen and Auel (2001) and Conover and Siferd (1993). According to these studies, the predominant energy storage in copepods is a lipid, primarily wax esters. Triglycerides are the main reserve fuel reported in amphipods. Wax esters were also found in amphipods but at a much lower rate. The life-history traits of oceanic zooplanktons were found to be much diversified. Polar oceans are known for extreme fluctuations in primary production. Hence, variations in the seasonal availability of food are considered to be the major determining factor of their life strategies.

The process of lipid biosynthesis in animals such as carnivores and omnivorous is not too elaborate as it is the least affected by seasonal food shortages. Herbivores do not utilize much of their lipid reserves for overwintering. Instead, they use it for reproductive processes during late winter or early spring. High latitudes call for critical reproductive timings because of the short production period, and reserved lipid allow early spawning without being dependent on external resources. Extensive vertical migrations, trophic flexibility, dormancy and change in the mode of life are the prominent dark season survival life strategies that supplement this energetic adaptation (lipid storage) (Hagen and Auel 2001; Conover and Siferd 1993; Hirche and Kosobokova 2011).

Recently, Schmid et al. (2016) developed an underwater imaging system Lightframe On-sight Keyspecies Investigation (LOKI) system, and also an automated zooplankton identification model for 114 taxonomic categories to identify LOKI images of zooplanktons. This model successfully distinguished species and stages. LOKI's image quality helps to develop automatic identification models with very high taxonomic detail and a high level of accuracy. These models will play a prime role in future studies on zooplankton dynamics and the coupling of zooplanktons with other trophic levels. Further, this model also provides insights into the vertical distribution of taxa, on a finer scale.

Several types of research have been done to enhance our understanding of the seasonal processes in the fjord. In the recent past, Kongsfjorden has received considerable research attention and will be a significant Arctic monitoring site in the coming decades. This current interest in Kongsfjorden is because, with an influx of Atlantic water and tidal glacier melting, it is a suitable site for exploring the influences of possible climate change. The benthic ecosystem is more sensitive to variations in hydrography, glacial runoff and sedimentation, whereas the pelagic ecosystem is more affected by the Arctic vs Atlantic influence.

Walkusz et al. (2009) found that seasonal changes in hydrographical regime and seasonal variability in zooplankton abundance are related. This study also addressed the importance of various factors such as advection, presence of local front, decrease in the depth of inner fjord basin and flow pattern in shaping the zooplankton community. Willis et al. (2006) found the influence of advection on the community composition of zooplankton in the Arctics. A study conducted by Basedow et al. (2004) qualified the impact of advection on local planktonic populations in an Arctic fjord, Kingsfjorden. The present changes in carbon flux and composition and trophodynamics of Arctic pelagic ecosystems can be attributed to the colonization of the Arctic waters by Atlantic species (Hirche and Kosobokova 2007). Ormańczyk et al. (2017) also reported the possible consequences of increasing influence that the Atlantic waters have on the zooplankton communities present in the marine pelagic ecosystem of the Arctic fjord. Furthermore, the zooplankton characteristics had undergone significant changes over the years in terms of biogeographic structure and abundance, which corresponds to the changes in the fjords' physical environment.

Basedow et al. (2018) also studied seasonal variations in zooplankton transport through the Atlantic gateway into the Arctic Basin. The detailed data obtained from this study revealed the importance of species phenology for the amount of advected biomass. In the Arctic Basin, advective input of zooplankton carbon is found to be important in all seasons. Further, the advective input of zooplankton is of great importance to mesopelagic planktivorous predators, especially during winter.

Several studies addressing the zooplanktons in Kongsfjorden (e.g. Hop et al. 2002, 2019) and its distribution (e.g. Kwasniewski et al. 2003; Feng et al. 2014) have also been done. Hop et al. (2019) found that the Arctic versus Atlantic water masses majorly contribute to the zooplankton species of Kongsfjorden. These communities are changing as a result of temperature rise, melting of sea ice and the volume of Atlantic water inflow. Since 1996, mesozooplankton species have shown significant changes in their secondary productionand in the proportion of Arctic as well as Atlantic species. These inter-annual variations are found to be closely related to hydrographical changes of Kongsfjorden. Higher abundance of Atlantic species such as *Calanus finmarchicus*, *Thysanoessa longicaudata*, *Oithona atlantica* and *Themisto abyssorum* characterizes the 'Warm years' in Kongsfjorden, whereas the 'Cold years' are marked by increased abundance of *Themisto libellula*.

Some of the landmark publications discussing the trends in biomass of Arctic planktonic community include Hopkins (1969a, 1969b), Hirche and Mumm (1992), Richter (1994), Astthorsson and Gislason (1995), Mumm et al. (1998), Gislason and Astthorsson (1998), Thibault et al. (1999), Kosobokova and Hirche (2000), Hop et al. (2002), Dalpadado et al. (2003), Hopcroft et al. (2005), Lane et al. (2008), Kosobokova and Hirche (2009), Hirche and Kosobokova (2011), Weydmann et al. (2013), Kwasniewski et al. (2013), Ormańczyk et al. (2017) and Hop et al. (2019). Studies have also revealed that certain species had their abundance increase as a result of increasing temperature. An abundance of Atlantic species in the pelagic realm of Kongsfjorden is the one to show this particular trend. Also, an increase in the total biomass of the zooplanktons was noticed in the fjord, which can further result in potentially greater secondary production (Hop et al. 2019). Slagstad et al. (2011) reported an uneven spatial distribution of secondary production in the copepods inhabiting Arctic Ocean.). Furthermore, in 2016, Garzke et al. showed the possibilities of reduction in biomass of Arctic zooplankton species in the coming years. Based on several of the published literature, these uncertainities in the zooplankton secondary production may be attributed to rising temperature (Global ocean temperature anomalies data available from

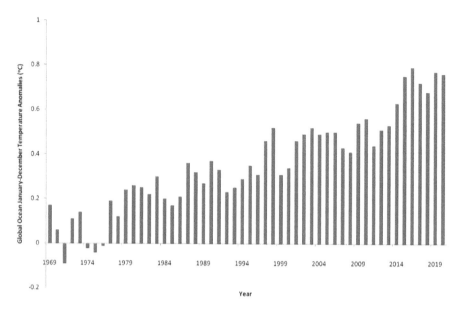

FIGURE 10.1   Global ocean temperature anomalies data (°C) from 1969-2020. Temperature data was obtained from https://www.ncdc.noaa.gov/cag/.

Figure 10.1). However, the direction of these changes remains largely undefined, and demands more systematic studies in the future.

10.3.2 PRESENT SCENARIO

In this chapter, apart from reviewal and analysis of the previous research studies carried out on Arctic zooplankton, we have also shared the biomass studies of our zooplankton collections taken during the years 2018 and 2019 from Kongsfjord. This glacial fjord situated in the West Spitsbergen region of Svalbard Archipelago always receives freshwater from glacial melting and are also prone to fluctuating water temperature (Aagaard and Carmack 1989). Zooplankton collections of 2018 from the surface layers of Kongsfjord waters had larger biomass values (Figure 10.2) than those collected during the 2019 expedition (Figure 10.3). In 2018, the biomass values range from 0.086 to 5.312 g/m³ with stations K6 and 4A having the maximum and minimum reported biomass values, respectively. During 2019, the maximum value of 0.273 g/m³ was recorded from the station KF9A whereas the minimum value of 0.042 g/m³ was from the station KF10A. The above-mentioned observations of declining biomass in recent years are following the findings of Garzke et al. (2016), where lengthened open waters in the warm Arctic are the pivotal reason behind the decline in biomass. Ellingsen et al. (2007) reported a significant decrease in the zooplankton biomass ranging about 50% using an ocean model. Based on the studies carried out by Kordas et al. (2011), copepods (dominant zooplankton group) are largely affected by increasing temperature in terms of their egg production and causing earlier reproduction. Warmer conditions also cause higher

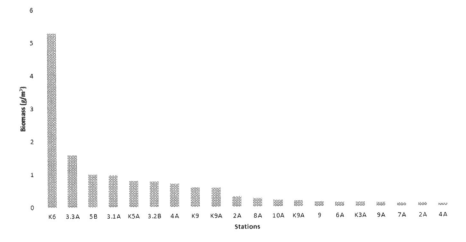

FIGURE 10.2 Biomass of Arctic zooplanktons collected from the surface level of 20 stations in Kongsfjord during the 2018 Indian Arctic Expedition.

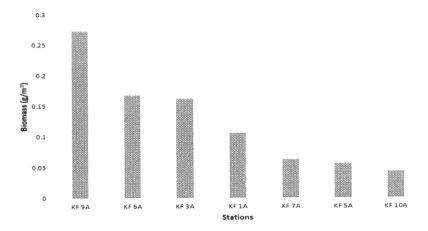

FIGURE 10.3 Biomass of Arctic zooplanktons collected from the surface level of seven stations in Kongsfjord during the 2019 Indian Arctic Expedition.

rates of mortality in copepods (Klein Breteler and Schogt 1994). These factors, considerably affecting the copepod populations, can also affect the overall zooplankton biomass of the warmer Arctic.

Furthermore, Dam and Baumann (2017) have also reported the reduction of body size and biomass in the Arctic zooplanktons. Apart from these observations, there are also studies reporting an increase in biomass of zooplanktons with the increasing ocean temperatures. Hence, climate change-driven alterations in the Arctic zooplankton biomass can be considered as the much-debated area which requires systematic studies for deciphering the relationship between biomass and temperature.

10.3.3 FUTURE RESEARCH ISSUES

Arctic ecosystems are largely experiencing alterations due to the continuing climate changes, and also due to the higher rates of resource exploitation in recent times. Factors such as the shrinkage of ice covers in the Arctic seas, atmospheric as well as oceanic transfer of heat are strongly linked to the community structure and composition of zooplankton. Continuing temperature hike can cause serious negative impacts on the local mesozooplankton community, which may also trigger certain species to flourish. The influx of warm saline Atlantic waters, which transfers sensible heat to the Arctic Ocean, can cause the extension of the biogeographical range of certain warm water species to the high Arctic seas. Furthermore, the increasing temperatures can also result in a decline in the primary production rates of Arctic marine systems which can affect the body size as well as the lifecycle of the Arctic zooplankton species. The subsequent shift from larger zooplanktons with ample lipid reserves too much smaller boreal zooplanktons with lesser reserves can cause serious implications in the energy transfer in the pelagic food web.

Arctic marine ecosystems are known to support a significant fishery stock, having excellent economic and nutritional value. The pelagic food web mainly supports the fishery stock in this ecosystem, and hence, disruption of the food web due to less sea ice and stronger stratification in the Arctic can also affect the fishery stock. Even though many studies focusing on the physical changes in the Arctic environment are taking place in the present, studies addressing the biological alterations are meagre. Furthermore, whether the pelagic food web of the fjord ecosystem is resilient or not with the climatic fluctuations of the Arctic environment also demands detailed research in the future. The other gap areas include insufficient quantitative data on production, key prey species abundance and the influence of advection on zooplankton communities in the Arctic fjord. Critical study and analysis of these areas can bring out multitudes of potential information on the zooplankton species of the Arctic region.

10.4 CONCLUSIONS

Polar zooplankton communities serve as a critical indicator system of climate change because of the temperature-driven changes in Arctic ice coverings. On analysing the previously published literature on Arctic zooplankton biomass and the biomass studies carried out from our collections made during the 2018 and 2019 Indian Arctic Expeditions, we could estimate that there is an uncertainity in the biomass of the Arctic zooplanktons in the past several years which might be associated with the higher global temperature rates. From the various gap areas discussed in this chapter, it is evident that despite the several well-recorded physical studies addressing the climate-driven alterations in the zooplankton community structure, thorough biological researches need to be undertaken which will be of efficient usage to the scientific world.

ACKNOWLEDGEMENTS

The authors would like to express their sincere gratitude to the Director, Zoological Survey of India, for all the encouragements to undertake the Arctic studies in ZSI.

We are also grateful to the Director, MOES, NCPOR and Arctic Studies coordinator of NCPOR, Dr Krishnan, Sci. E, and Dr Archana Singh, Scientist C, for all the help and support. We are grateful to the team members of the Indian Arctic Expedition (2018 and 2019) for the support and help provided during the expedition and the assistance provided to carry out this work.

REFERENCES

Aagaard, K., & Carmack, E.C. (1989). The role of sea ice and other fresh water in the Arctic circulation. *Journal of Geophysical Research: Oceans* 94(C10), 14485–14498.

Abe, Y., Matsuno, K., Fujiwara, A., & Yamaguchi, A. (2020). Review of spatial and inter-annual changes in the zooplankton community structure in the western Arctic Ocean during summers of 2008–2017. *Progress in Oceanography* 102391.

Ashjian, C.J., Campbell, R.G., Welch, H.E., Butler, M., & Van Keuren, D. (2003). The annual cycle in abundance, distribution, and size about hydrography of important copepod species in the western Arctic Ocean. *Deep-Sea Research Part I* 50, 1235–1261.

Astthorsson, O.S., & Gislason, A. (1995). Long-term changes in zooplankton biomass in Icelandic waters in spring. *ICES Journal of Marine Science* 52(3–4), 657–668.

Auel, H., & Hagen, W. (2002). Mesozooplankton community structure, abundance and biomass in the central Arctic Ocean. *Marine Biology* 140, 1013–1021.

Basedow, S.L., Eiane, K., Tverberg, V., & Spindler, M. (2004). Advection of zooplankton in an Arctic fjord (Kongsfjorden, Svalbard). *Estuarine, Coastal and Shelf Science* 60(1), 113–124.

Basedow, S.L., Sundfjord, A., von Appen, W.J., Halvorsen, E., Kwasniewski, S., & Reigstad, M. (2018). Seasonal variation in the transport of zooplankton into the Arctic basin through the Atlantic gateway, Fram Strait. *Frontiers in Marine Science*, 5, 194.

Berge, J., Cottier, F., Varpe, Ø., Renaud, P.E., Falk-Petersen, S., Kwasniewski, S., ... & Bjærke, O. (2014). Arctic complexity: a case study on diel vertical migration of zooplankton. *Journal of Plankton Research* 36(5), 1279–1297.

Blachowiak-Samolyk, K., Kwasniewski, S., Richardson, K., Dmoch, K., Hansen, E., Hop, H., ... & Mouritsen, L.T. (2006). Arctic zooplankton does not perform diel vertical migration (DVM) during periods of the midnight sun. *Marine Ecology Progress Series* 308, 101–116.

Borga° K, Gabrielsen GW, Skaare JU. (2002). Differences in contamination load between pelagic and sympagic invertebrates in the Arctic marginal ice zone: Influence of habitat, diet, and geography. *Marine Ecology Progress Series* 235, 157–169.

Brodsky, K.A., & Nikitin, M.N. (1955). Observational data of the scientific research drifting station of 1950–1951. *Hydrobiological Work. Morskoy Transport* 1, 404–410.

Brodsky, K.A., & Pavshtiks, E.A. (1976). Plankton of the central part of the Arctic Basin. *Polar Geography 1*, 143–161.

Buesseler, K.O., Lamborg, C.H., Boyd, P.W., Lam, P.J., Trull, T.W., Bidigare, R.R., ... & Honda, M. (2007). Revisiting carbon flux through the ocean's twilight zone. *Science* 316(5824), 567–570.

Carmack, E., Barber, D.G., Christensen, J.R., Macdonald, R.W., Rudels, B., & Sakshaug, E. (2005). Climate variability and physical forcing of the food webs and the carbon budget on panarctic shelves. *Progress in Oceanography* 71, 145–181.

Carmack, E.C., Aagaard, K., Swift, J.H., MacDonald, R.W., McLaughlin, F.A., Jones, E.P., Perkin, R.G., Smith, J.N., Elis, K.M., & Killius, L.R. (1997). Changes in temperature and tracer hdistributions within the Arctic Ocean results from the 1994 Arctic Ocean section. *Deep-Sea Research Part II. Topical Studies in Oceanography* 44, 1487–1502.

Comiso, J.C., Parkinson, C.L., Gersten, R., & Stock, L. (2008). Accelerated decline in the Arctic sea ice cover. *Geophysical Research Letters III* 35, L01703. doi: 10.1029/2007GL031972.

Conover, R.J., & Siferd, T.D. (1993). Dark-season survival strategies of coastal zone zooplankton in the Canadian Arctic. *The Arctic* 303–311.

Cottier, F.R., Tarling, G.A., Wold, A., & Falk-Petersen, S. (2006). Unsynchronised and synchronised vertical migration of zooplankton in a high Arctic fjord. *Limnology and Oceanography* 51(6), 2586–2599.

Dalpadado, P., Ingvaldsen, R., & Hassel, A. (2003). Zooplankton biomass variation in relation to climatic conditions in the Barents Sea. *Polar Biology* 26(4), 233–241.

Dalpadado, P., Ingvaldsen, R.B., Stige, L.C., Bogstad, B., Knutsen, T., Ottersen, G., & Ellertsen, B. (2012). Climate effects on Barents Sea ecosystem dynamics. *ICES Journal of Marine Science* 69, 1303–1316.

Dam, H.G., & Baumann, H. (2017). Climate change, zooplankton and fisheries. *Climate Change Impacts on Fisheries and Aquaculture: A Global Analysis* 2, 851–874.

Darnis, G., & Fortier, L. (2012). Zooplankton respiration and the export of carbon at depth in the Amundsen Gulf (Arctic Ocean). *Journal of Geophysical Research: Oceans* 117(C4).

Darnis, G., Hobbs, L., Geoffroy, M., Grenvald, J.C., Renaud, P.E., Berge, J., ... & Wold, A. (2017). From polar night to midnight sun: diel vertical migration, metabolism and biogeochemical role of zooplankton in a high Arctic fjord (Kongsfjorden, Svalbard). *Limnology and Oceanography* 62(4), 1586–1605.

Dawson, J.K. (1978). Vertical distribution of *Calanus hyperboreus* in the central Arctic Ocean. *Limnology and Oceanography* 23, 950–957.

Ellingsen, I.H., Dalpadado, P., Slagstad, D., & Loeng, H. (2008). Impact of climatic change on the biological production in the Barents Sea. *Climatic Change* 87(1–2), 155–175.

Falk-Petersen, S., Leu, E., Berge, J., Kwasniewski, S., Nygård, H., Røstad, A., ... & Gulliksen, B. (2008). Vertical migration in high Arctic waters during autumn 2004. *Deep-Sea Research Part II: Topical Studies in Oceanography* 55(20–21), 2275–2284.

Feng, M., Zhang, W., & Xiao, T. (2014). Spatial and temporal distribution of tintinnid (Ciliophora: Tintinnida) communities in Kongsfjorden, Svalbard (Arctic), during summer. *Polar Biology* 37(2), 291–296.

Fischer, J., & Visbeck, M. (1993). Seasonal variation of the daily zooplankton migration in the Greenland Sea. *Deep-Sea Research Part I: Oceanographic Research Papers* 40(8), 1547–1557.

Fisk, A.T., Stern, G.A., Hobson, K.A., Strachan, W.J., Loewen, M.D., & Norstrom, R.J. (2001). Persistent organic pollutants (POPs) in a small, herbivorous, Arctic marine zooplankton (Calanus hyperboreus): trends from April to July and the influence of lipids and trophic transfer. *Marine Pollution Bulletin* 43(1–6), 93–101.

Garzke, J., Ismar, S.M.H., & Sommer, U. (2016). Climate change affects low trophic level marine consumers: warming decreases copepod size and abundance. *Oecologia* 177, 849–860.

Gislason, A., & Astthorsson, O.S. (1998). Seasonal variations in biomass, abundance and composition of zooplankton in the subarctic waters north of Iceland. *Polar Biology* 20(2), 85–94.

Gluchowska, M., Trudnowska, E., Goszczko, I., Kubiszyn, A. M., Blachowiak-Samolyk, K., Walczowski, W., & Kwasniewski, S. (2017). Variations in the structural and functional diversity of zooplankton over vertical and horizontal environmental gradients en route to the Arctic Ocean through the Fram Strait. *PloS One* 12(2), e0171715.

Grainger, E.H. (1965). Zooplankton forms the Arctic Ocean and adjacent Canadian waters. *Journal of the Fisheries Board of Canada* 22(2), 543–564.

Grainger, E.H. (1989). Vertical distribution of zooplankton in the central Arctic Ocean. In *Proceedings of the sixth conference of the ComitéArctique International*. pp. 48–60.

Grebmeier, J.M., Overland, J., Moore, S.E., Farley, E.V., Carmack, E., Cooper, L.W., Frey, K., Helle, J.H., McLaughlin, F.A., & McNutt, S.L. (2006). A major ecosystem shift in the Northern Bering Sea. *Science* 311, 1461–1464. doi: 10.1126/science.1121365.

Grenvald, J.C., Callesen, T.A., Daase, M., Hobbs, L., Darnis, G., Renaud, P.E., ... & Berge, J. (2016). Plankton community composition and vertical migration during polar night in Kongsfjorden. *Polar Biology* 39(10), 1879–1895.

Haakstad, M., Kogeler, J.W., & Dahle, S. (1994). Studies of sea surface temperatures in selected northern Norwegian fjords using Landsat TM data. *Polar Research* 13(1), 95–10.

Hagen, W., & Auel, H. (2001). Seasonal adaptations and the role of lipids in oceanic zooplankton. *Zoology* 104(3–4), 313–326.

Haug, T., Bogstad, B., Chierici, M., Gjøsæter, H., Hallfredsson, E. H., Høines, Å.S., Hoel, A.H., Ingvaldsen, R.B., Jørgensen, L.L., Knutsen, T., Loeng, H., Naustvoll, L.J., Røttinger, I., & Sunnanå, K. (2017). Future harvest of living resources in the Arctic Ocean north of the Nordic and Barents seas: a review of possibilities and constraints. *Fisheries Research* 188, 38–57.

Hays, G.C., Kennedy, H., & Frost, B.W. (2001). Individual variability in diel vertical migration of a marine copepod: why some individuals remain at depth when others migrate. *Limnology and Oceanography* 46(8), 2050–2054.

Hirche, H.J., & Kosobokova, K.N. (2003). Early reproduction and development of dominant calanoid copepods in the ice zone of the Barents Sea – need for a change of paradigms? *Marine Biology* 143, 769–781.

Hirche, H.J., & Kosobokova, K.N. (2007). Distribution of Calanus finmarchicus in the northern North Atlantic and Arctic Ocean – expatriation and potential colonization. *Deep-Sea Research Part II* 54, 2729–2747.

Hirche, H.J., & Kosobokova, K.N. (2011). Winter studies on zooplankton in Arctic seas: the Storfjord (Svalbard) and adjacent ice-covered the Barents Sea. *Marine Biology* 158(10), 2359–2376.

Hirche, H.J., & Mumm, N. (1992). Distribution of dominant copepods in the Nansen Basin, Arctic Ocean, in summer. *Deep-Sea Research* 39(Suppl. 2), S485–S505.

Hop, H., Borga, K., Gabrielsen, G.W., Kleivane, L., & Skaare, J.U. (2002). Food web magnification of persistent organic pollutants in poikilotherms and homeotherms from the Barents Sea. *Environmental Science and Technology* 36, 2589–2597.

Hop, H., Pearson, T., Hegseth, E.N., Kovacs, K.M., Wiencke, C., Kwasniewski, S., ... & Lydersen, C. (2002). The marine ecosystem of Kongsfjorden, Svalbard. *Polar Research* 21(1), 167–208.

Hop, H., Wold, A., Vihtakari, M., Daase, M., Kwasniewski, S., Gluchowska, M., ... & Falk-Petersen, S. (2019). Zooplankton in Kongsfjorden (1996–2016) in Relation to Climate Change. In *The Ecosystem of Kongsfjorden, Svalbard*. pp. 229–300. Springer, Cham.

Hopcroft, R.R., Clarke, C., Nelson, R.J., & Raskoff, K.A. (2005). Zooplankton communities of the Arctic's Canada Basin: the contribution by smaller taxa. *Polar Biology* 28(3), 198–206.

Hopcroft, R.R., Kosobokova, K.N., & Pinchuk, A.I. (2010). Zooplankton community patterns in the Chukchi Sea during summer 2004. *Deep-Sea Research Part II: Topical Studies in Oceanography* 57(1–2), 27–39.

Hopkins, T.L. (1969a). Zooplankton standing crop in the Arctic Basin. *Limnology and Oceanography* 14(1), 80–85.

Hopkins, T.L. (1969b). Zooplankton biomass related to hydrography along the drift track of Arlis II in the Arctic Basin and the East Greenland Current. *Journal of Fisheries Research Board of Canada* 26, 305–310.

Hughes, K.H. (1968). Seasonal Vertical Distribution of Copepods in the Arctic Water in the Canadian Basin of the North Polar Sea. M.Sc. thesis, University of Washington.

Johnson, M.W. (1963). Zooplankton collections from the high polar basins with special reference to the Copepoda. *Limnology and Oceanography* 8, 89–102.

Kasyan, V.V. (2020). Variability and spatial distribution of zooplankton communities in the western Chukchi Sea during early fall. *Polar Science* 100508.

Kielhorn, W.V. (1952). The biology of the surface zone zooplankton of a boreo-arctic Atlantic Ocean area. *Journal of the Fisheries Board of Canada* 9(5), 223–264.

Klein Breteler, W.C.M., & Schogt, N. (1994). Development of Acartiaclausi (Copepoda, Calanoida) cultured at different conditions of temperature and food. *Hydrobiologia* 292–293, 469–479.

Koplin, J. (2020). Community composition of epipelagic zooplankton in the Eurasian Basin 2017 determined by ZooScan image analysis (Doctoral dissertation, University of Hamburg).

Kordas, R.L., Harley, C.D.G., & O'Connor, M.I. (2011.) Community ecology in a warming world: the influence of temperature on interspecific interactions in marine systems. *Journal of Experimental Marine Biology and Ecology* 400, 218–226.

Kosobokova, K., & Hirche, H.J. (2000). Zooplankton distribution across the Lomonosov Ridge, Arctic Ocean: species inventory, biomass and vertical structure. *Deep-Sea Research Part I: Oceanographic Research Papers* 47(11), 2029–2060.

Kosobokova, K., & Hirche, H.J. (2001). Reproduction of Calanus glacialis in the Laptev Sea, Arctic Ocean. *Polar Biology* 24, 33–43.

Kosobokova, K., & Hirche, H.J. (2009). Biomass of zooplankton in the eastern Arctic Ocean–a baseline study. *Progress in Oceanography* 82(4), 265–280.

Kosobokova, K.N. (1978). Diurnal vertical distribution of *Calanus hyperboreus*Kroyer and *Calanus glacialis*Jaschnov in the Central Polar Basin. *Oceanology* 18, 476–480.

Kosobokova, K.N. (1981). Zooplankton of the Arctic Ocean. PhD. Thesis, Institute of Oceanology, Academy of Sciences, Moscow, pp. 1–150.

Kosobokova, K.N. (1982). Composition and distribution of the biomass of zooplankton in the central Arctic Basin. *Oceanology* 22, 744–750.

Kosobokova, K.N. (1989). Vertical distribution of plankton animals in the eastern part of the central Arctic Basin. *Explorations of the Fauna of the Seas, Marine Plankton, Leningrad* 41(49), 24–31.

Kosobokova, K.N., Hopcroft, R.R., & Hirche, H.J. (2011). Patterns of zooplankton diversity through the depths of the Arctic's central basins. *Marine Biodiversity* 41(1), 29–50.

Kwasniewski, S., Gluchowska, M., Walkusz, W., Karnovsky, N.J., Jakubas, D., Wojczulanis-Jakubas, K., Harding, A.M.A., Goszczko, I., Cisek, M., Beszczynska-Möller, A., Walczowski, W., Weslawski, J.M., & Stempniewicz, L. (2012). Interannual changes in zooplankton on the West Spitsbergen Shelf in relation to hydrography and their consequences for the diet of planktivorous seabirds. *Journal of Marine Science* 69, 890–901.

Kwasniewski, S., Hop, H., Falk-Petersen, S., & Pedersen, G. (2003). Distribution of Calanus species in Kongsfjorden, a glacial fjord in Svalbard. *Journal of Plankton Research* 25(1), 1–20.

Kwasniewski, S., Walkusz, W., Cottier, F.R., & Leu, E. (2013). Mesozooplankton dynamics in relation to food availability during spring and early summer in a high latitude glaciated fjord (Kongsfjorden), with a focus on Calanus. *Journal of Marine Systems* 111, 83–96.

Lampert, W. (1989). The adaptive significance of diel vertical migration of zooplankton. *Functional Ecology* 3(1), 21–27.

Lane, P.V.Z., Llinás, L., Smith, S.L., & Pilz, D. (2008). Zooplankton distribution in the western Arctic during summer 2002: hydrographic habitats and implications for food chain dynamics. *Journal of Marine Systems* 70, 97–133.

Last, K.S., Hobbs, L., Berge, J., Brierley, A.S., & Cottier, F. (2016). Moonlight drives ocean-scale mass vertical migration of zooplankton during the Arctic winter. *Current Biology* 26(2), 244–251.

Lee, R.F. (1975). Lipids of Arctic zooplankton. *Comparative Biochemistry and Physiology Part B: Comparative Biochemistry* 51(3), 263–266.

Maslanik, J.A., Fowler, C., Stroeve, J., Drobot, S., Zwally, J., Emery, D., & Yi, W. (2007). A younger, thinner Arctic ice cover: Increased potential for rapid, extensive sea-ice loss. *Geophysical Research Letters* 34, L24501. doi: 10.1029/2007GL032043.

Melnikov, I.A., & Kolosova, E.G. (2001). The Canada Basin zooplankton in recent environmental changes in the Arctic. In Semiletov, I.P. (Ed.), *Proceedings of the Arctic Regional Center*, vol. 3. Pacific Institute of Oceanology, Vladivostok, pp. 165–176.

Minoda, T. (1967). Seasonal distribution of Copepoda in the Arctic Ocean from June to December 1964. *Records of Oceanographic Works in Japan* 9, 61– 168.

Miquel, J.C., Gasser, B., Martín de Nascimento, J., Marec, M., Babin, M., Fortier, L., & Forest, A. (2015). Downward particle flux and carbon export in the Beaufort Sea, Arctic Ocean; the role of zooplankton.

Møller, E.F., & Nielsen, T.G. (2020). Borealization of Arctic zooplankton—smaller and less fat zooplankton species in Disko Bay, Western Greenland. *Limnology and Oceanography* 65(6), 1175–1188.

Mumm, N. (1993). Composition and distribution of mesozooplankton in the Nansen Basin, Arctic Ocean, during summer. *Polar Biology* 13, 451–461.

Mumm, N., Auel, H., Hanssen, H., Hagen, W., Richter, C., & Hirche, H.-J. (1998). Breaking the ice, large-scale distribution of mesozooplankton after a decade of Arctic and transpolar cruises. *Polar Biology* 20, 189–197.

Nelson, R.J., Carmack, E.C., McLaughlin, F.A., & Cooper, G.A. (2009). Penetration of Pacific zooplankton into the western Arctic Ocean tracked with molecular population genetics. *Marine Ecology Progress Series* 381, 129–138.

NOAA National Centers for Environmental information, Climate at a Glance: Global Time Series, published November 2021, retrieved on November 25, 2021 from https://www.ncdc.noaa.gov/cag/

Ohman, M.D. (1990). The demographic benefits of diel vertical migration by zooplankton. *Ecological Monographs* 60(3), 257–281.

Olli, K., Wassmann, P., Reigstad, M., Ratkova, T., Arashkevich, E., Pasternak, A., Matrai, P.A., Knulst, J., Tranvik, L., Klais, R., & Jacobbsen, A. (2007). The fate of production in the central Arctic Ocean – top–downregulation by zooplankton expatriates? *Progress in Oceanography* 72, 84–113.

Ormańczyk, M.R., Głuchowska, M., Olszewska, A., & Kwasniewski, S. (2017). Zooplankton structure in high latitude fjords with contrasting oceanography (Hornsund and Kongsfjorden, Spitsbergen). *Oceanologia* 59(4), 508–524.

Pautzke, C.G. (1979). Phytoplankton primary production below the Arctic Ocean pack ice, an ecosystems analysis. PhD. Thesis, University of Washington, p. 181.

Perovich, D.K., Richter-Menge, J.A., Jones, K.F., & Light, B. (2008). Sunlight, water, and ice: Extreme Arctic sea ice melt during the summer of 2007. *Geophysical Research Letters* 35, L11501. doi: 10.1029/2008GL034007.

Raskoff, K.A., Purcell, J. E., & Hopcroft, R. R. (2005). Gelatinous zooplankton of the Arctic Ocean: in situ observations under the ice. *Polar Biology* 28(3), 207–217.

Richter, C. (1994). Regional and seasonal variability in the vertical distribution of mesozooplankton in the Greenland Sea. *BerichtezurPolarforschung (Reports on Polar Research)* 154.

Riser, C.W., Wassmann, P., Reigstad, M., & Seuthe, L. (2008). Vertical flux regulation by zooplankton in the northern Barents Sea during Arctic spring. *Deep-Sea Research Part II: Topical Studies in Oceanography* 55(20–21), 2320–2329.

Rudels, B., Muench, R.D., Gunn, J., Schauer, U., & Friedrich, H.J. (2000). Evolution of the Arctic boundary current north of the Siberian Shelves. *Journal of Marine Systems* 25, 77–99.

Saiz, E., Calbet, A., Isari, S., Anto, M., Velasco, E. M., Almeda, R., ... & Alcaraz, M. (2013). Zooplankton distribution and feeding in the Arctic Ocean during a Phaeocystispouchetii bloom. *Deep-Sea Research Part I: Oceanographic Research Papers* 72, 17–33.

Schauer, U., Muench, R.D., Rudels, B., & Timokhov, L. (1997). Impact of eastern shelf waters on the Nansen Basin intermediate layers. *Journal of Geophysical Research* 102(C2), 3371–3382.

Schmid, M.S., Aubry, C., Grigor, J., & Fortier, L. (2016). The LOKI underwater imaging system and an automatic identification model for the detection of zooplankton taxa in the Arctic Ocean. *Methods in Oceanography* 15, 129–160.

Sherr, E.B., Sherr, B.F., Wheeler, P.A., & Thompson, K. (2003). Temporal and spatial variation in stocks of autotrophic and heterotrophic microbes in the upper water column of the central Arctic Ocean. *Deep-Sea Research Part I* 50, 557– 571.

Slagstad, D., Ellingsen, I.H., & Wassmann, P. (2011). Evaluating primary and secondary production in an Arctic Ocean void of summer sea ice: an experimental simulation approach. *Progress in Oceanography* 90, 117–131.

Sobek, A., Reigstad, M., & Gustafsson, Ö. (2006). Partitioning of polychlorinated biphenyls between Arctic seawater and size-fractionated zooplankton. *Environmental Toxicology and Chemistry: An International Journal* 25(7), 1720–1728.

Taylor, A.H., Allen, I., & Clark, P.A. (2002). Extraction of a weak climatic signal by an ecosystem. *Nature* 416, 629– 632.

Thibault, D., Head, E.J.H., & Wheeler, P.A. (1999). Mesozooplankton in the Arctic Ocean in summer. *Deep-Sea Research Part I* 46, 1391–1415.

Trudnowska, E., Basedow, S., & Blachowiak-Samolyk, K. (2014). Mid-summer mesozooplankton biomass, its size distribution, and estimated production within glacial Arctic fjord (Hornsund, Svalbard). *Journal of Marine Systems* 137, 55–66.

Trudnowska, E., Dąbrowska, A. M., Boehnke, R., Zajączkowski, M., & Blachowiak-Samolyk, K. (2020). Particles, protists, and zooplankton in glacier-influenced coastal svalbard waters. *Estuarine, Coastal and Shelf Science* 106842.

Trudnowska, E., Stemmann, L., Błachowiak-Samołyk, K., & Kwasniewski, S. (2020). Taxonomic and size structures of zooplankton communities in the fjords along the Atlantic water passage to the Arctic. *Journal of Marine Systems* 204, 103306.

Turner, J.T. (2015). Zooplankton fecal pellets, marine snow, phytodetritus and the ocean's biological pump. *Progress in Oceanography* 130, 205–248.

Vihtakari, M., Welcker, J., Moe, B., Chastel, O., Tartu, S., Hop, H., Bech, C., Descamps, S., & Gabrielsen, G.W. (2018). Black-legged kittiwakes as messengers of Atlantification in the Arctic. *Scientific Reports* 8, 1178.

Vinogradov, M.E., & Melnikov, I.A. (1980). Studies of the pelagic ecosystems of the central Arctic basin. In Vinogradov, M.E. (Ed.), *BiologiyaTsentral'nogoArkicheskogobasse yna (Biology of the central Arctic Basin)*. Nauka, Moscow, pp. 5–14.

Virketis, M.A. (1959). Materials on the Zooplankton of the Central Part of the Arctic Ocean (in Russian). Results of Scientific Observations on the Drift Stations "North Pole 4" and North Pole 5", 1955–1956, Moscow, 132–206.

Walkusz, W., Kwasniewski, S., Petersen, S.F., Hop, H., Tverberg, V., Wieczorek, P., & Weslawski, J.M. (2009). Seasonal and spatial changes in the zooplankton community of Kongsfjorden, Svalbard. *Polar Research* 28(2), 254–281.

Wallace, M.I., Cottier, F.R., Berge, J., Tarling, G.A., Griffiths, C., & Brierley, A.S. (2010). Comparison of zooplankton vertical migration in an ice-free and a seasonally ice-covered Arctic fjord: an insight into the influence of sea ice cover on zooplankton behavior. *Limnology and Oceanography* 55(2), 831–845.

Wassmann, P., & others. (2015). The contiguous domains of Arctic Ocean advection: trails of life and death. *Progress in Oceanography* 139, 42–65.

Weydmann, A., Carstensen, J., Goszczko, I., Dmoch, K., Olszewska, A., & Kwasniewski, S. (2014). Shift towards the dominance of boreal species in the Arctic: inter-annual and spatial zooplankton variability in the West Spitsbergen Current. *Marine Ecology Progress Series* 501, 41–52.

Weydmann, A., Søreide, J.E., Kwaśniewski, S., Leu, E., Falk-Petersen, S., & Berge, J. (2013). Ice-related seasonality in zooplankton community composition in a high Arctic fjord. *Journal of Plankton Research* 35(4), 831–842.

Weydmann, A., Walczowski, W., Carstensen, J., & Kwaśniewski, S. (2018) Warming of Subarctic waters accelerates the development of a key marine zooplankton Calanus finmarchicus. *Global Change Biology* 24, 172–183.

Wheeler, P.A., Gosselin, M., Sherr, E., Thibault, D., Kirchman, D.L., Benner, R., & Whitledge, T.E. (1996). Active cycling of organic carbon in the central Arctic Ocean. *Nature* 380, 697–699.

Willis, K., Cottier, F., Kwasniewski, S., Wold, A., & Falk-Petersen, S. (2006). The influence of advection on zooplankton community composition in an Arctic fjord (Kongsfjorden, Svalbard). *Journal of Marine Systems* 61(1–2), 39–54.

11 Spectroscopic Characterizations of Humic Acids Isolated from Diverse Arctic Environments

Aswathy Shaji and Anu Gopinath
Kerala University of Fisheries and Ocean Studies

CONTENTS

11.1 INTRODUCTION

The sedimentary organic matter undergoes a combination of physical, chemical and biological processes resulting in an early diagenetic state. In this stage, the sedimentary system transforms both in terms of quantity and composition of organic matter. The pathway for this transformation from a starting material to the final products through the intermediates is always a challenge to organic geochemists and biogeo-chemists. Various studies on the diagenetic pathways reported providing information regarding the diagenetic alteration and transformation of precursor organic mole-cules (Henrichs 1992).

Sedimentary organic matter can be classified into humic substances and non-humic substances. Humic substances are the product of diagenesis, and non-humic substances are the survivors of diagenesis. The product (humic acid) is produced due to the biochemical transformation of sedimentary organic residues. They are a gen-eral class of biogenic refractory yellow, black organic substances that occur in both terrestrial and aquatic environments. The geological significance, nutritive proper-ties, involvement in biological processes and their chemical and structural complexi-ties make them an interesting material for researchers (Anuradha et al. 2011).

Humic substances do not correspond to a unique chemical entity, hence cannot be represented by a general structural formula. The humic materials, including humic acid, fulvic acid and humin, are formed through the humification process which pre-serves the organic carbon fixed during photosynthesis. This process also involves the condensation of various degradation products resulting from the decomposition of the organic debris (MacCarthy 2001).

The molecular weights of humic substances are reported within a range of a few hundred to several million (Rashid and King 1969, 1971). It contains carbon, hydro-gen, nitrogen, sulphur and oxygen as major elements with the predominance of car-bon and oxygen (Schnitzer 1977). The major oxygen-containing functional groups in humic substances are carboxyl, hydroxyl and carbonyl groups. The fundamental fea-tures of humic materials are their non-stoichiometric composition, irregular struc-ture, heterogeneity of structural elements and their polydispersity nature (Stevenson 1994). The structure of humic acids is described with the help of numerical parame-ters, representing atomic ratios of the constituent elements, their distribution between basic moieties and also the structural characteristics of molecular mass (Perminova 1999). The shape and sizes of humic and fulvic acid particles are known to vary with pH. They tend to aggregate into long fibres or bundles of fibres at low pH, but at high pH, they disperse, and the molecular arrangement becomes smaller but better oriented. The ability of a humic substance to form stable complexes has been well established. They form complexes with metal ions attributing to their high oxygen content containing functional groups (Stevenson 1982).

This study attempts to find out the characteristic nature of humic acids iso-lated from the sedimentary organic matter of diverse environments of Svalbard Archipelago, a Norwegian Arctic region. The study starts with the preliminary understanding of the state of the organic matter present in the sediments by analys-ing its biochemical composition, elemental composition and total organic carbon. This study's major focus is a qualitative study of the humic acid contents present in these sediments. They are isolated and then characterised spectroscopically. Apart from that, the presence of various metals bound within the macromolecular structure of the humic acid is also studied.

11.2 MATERIALS AND METHODS

11.2.1 STUDY AREA

The study area of this work is within the Svalbard Archipelago which comes within the Arctic circle of Norway. Diverse environments are selected in and around the area of Ny-Ålesund, a research base on the island. The environment under study includes a fjord system named Kongsfjorden, three lakes, viz., Tvillingvatnet, Storvatnet and Knudsenheia, and a major river system in the area named Bayelva.

The Kongsfjorden can be divided into two parts: the outer region of the fjord influenced by the Atlantic water masses and the inner region of the fjord influenced by the glacier meltwater (Cottier et al. 2014). Two sites were fixed in the outer basin and two sites were fixed within the inner basin of the fjord system. For the river Bayelva, three sampling sites were fixed in the order of origin (river mouth), middle and end. For lake environments, single sampling sites were considered for the study.

11.2.2 SAMPLE COLLECTION

Sediment samples were collected from all the stations within the fjord, lake and river systems. For the fjord sampling, a Van Veen grab is used, and for the lakes and river, a sampling scoop is used due to the shallow nature of the environment. The sediment samples collected were immediately transferred to zip-lock polythene bags and packed airtight. It is then stored in a deep freezer with a temperature of not more than $-20°C$. The samples are then cold shipped back to the lab, maintaining the samples' required temperature and were kept in a $-80°C$ deep freezer until analysed.

TABLE 11.1
Location Details of the Study Area

Stations	Latitude	Longitude
Fjord Stations (Kongsfjorden)		
K1	79°00′683	11°26′269
K2	78°57′534	11°49′529
K3	78°55′351	12°05′861
K4	78°53′716	12°19′239
Lake Stations		
L1 (Tvillingvatnet)	78°54′949	11°52′507
L2 (Storvatnet)	78°55′464	11°52′620
L3 (Knudsenheia)	78°56′443	11°49′459
River Stations		
R1 (River Mouth)	78°55′001	11°44′766
R2 (River Middle)	78°65′627	11°50′408
R3 (River End)	78°56′016	11°50′348

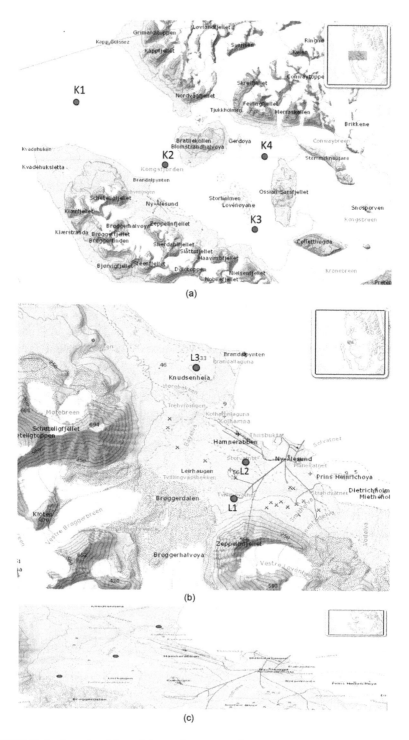

FIGURE 11.1   Location Map of the study area: (a) fjord stations, (b) lake stations and (c) river stations.

11.2.3   Assessment of Biochemical Composition

A small portion of each sample kept in the deep freezer is taken out and is freeze-dried for further analysis. The homogenisation of freeze-dried samples was done with an agate mortar and pestle. The biochemical composition was analysed based on UV-Visible spectrophotometric methods (Thermofischer Model No. 117). The major biochemical components analysed are proteins (PRT), carbohydrates (CHO) and lipids (LPD). The analysis of protein was performed with the modified Lowry procedure involving extraction with 1M NaOH for 2hrs and Albumin as standard (Lowry and Rosebrough 1951). Gerachov and Hatcher's (1972) procedure was used to determine sedimentary carbohydrates by expressing in terms of glucose equivalents (Gerachov and Hatcher 1972). The lipid extraction was done with direct elution of chloroform and methanol and assayed using the Barnes and Blackstock (1973) method. The standard used for lipid analysis is cholesterol. Along with the proteins, carbohydrates and lipids, the presence of a component named tannin and lignin was also done with a NaOH extraction method and estimated spectrophotometrically by sodium tungstate phosphomolybdic acid method (Nair et al. 1989; APHA 1995).

The calculation of Biopolymeric Carbon (BPC) was done by summing up the values of proteins, carbohydrates and lipids (Fabiano et al. 1995). The equivalents of carbon were obtained by multiplying each fraction with 0.49, 0.4 and 0.75 g of C/g, respectively (Pusceddu et al. 2000). The labile or easily assimilated organic fraction (LOM) is defined as the sum of all PRT, CHO and LPD (Danovaro et al. 1993; Cividanes et al. 2002).

11.2.4   Isolation of Humic Acid

The isolation of humic acid was performed with a modified method concerning the IHSS method (2010). About 1 kg of air-dried sediment samples was treated with 1 N HCl to eliminate weakly bound carbonates, sulphates and hydroxides. A few minutes later, the HCl solution is completely removed, and 1 N NaOH is added for extracting the humic acid from the sediments. The shaking of the sediment–NaOH mixture is done for 24 hours in a rotary shaker and allowed to settle. The supernatant solution was decanted, and the pH adjusted to 1 with 6 N HCl and kept overnight for complete precipitation. The precipitate was separated and washed with the help of a centrifuge and then freeze-dried for further spectroscopic characterisation (IHSS 2010a, b).

The characterisation of the isolated humic acid was done with UV-Vis and FTIR spectroscopy. The UV-Visible spectroscopic measurement was recorded with a Thermo Fisher UV-Visible spectrophotometer (Model no. 117). The FTIR spectrometer used for the FTIR spectral measurements is an IR prestige 21 models of Schimadzu.

Furthermore, the elemental composition and the total organic carbon content of the isolated humic acid, as well as the sedimentary contents, were analysed. PRIMACS[MCS] TOC analyser of SKALAR was used to analyse total organic carbon content, and the investigation of elemental composition was done using a CHNS elemental analyser of ElementarVario EL III. Relative quantities of C, H, N and S were measured directly from the analyser on an ash-free basis. The oxygen distribution was obtained by the difference between 100% and the sum of C, H, N and S percentages.

The study also includes the determination of metal content present in the humic acid and the sedimentary system with an acid digestion method involving nitric acid and perchloric acid on a 5:1 basis followed by instrumentation ICP-OES (Optima 8000) (Loring and Rantala 1992).

11.3 RESULTS AND DISCUSSIONS

11.3.1 BIOCHEMICAL COMPOSITION OF SEDIMENTARY ORGANIC MATTER

The biochemical composition of sedimentary organic matter provides information regarding nature and parameters that control the diagenetic process of organic matter. The major components of the biochemical composition are proteins, carbohydrates, lipids and chlorophyll pigment fractions. They are considered the best tool that provides a clear insight into the biogeochemical characterisation of sedimentary systems. The biochemical composition also evaluates the nutritional quality of organic matters which are the available food source for benthic consumers (Colombo et al. 1996; Dell'Anno et al. 2000; Joseph et al. 2008, Cividanes et al. 2002).

The labile fractions of sedimentary organic matter which are more readily available to the benthic consumers are evaluated by the estimation of main biochemical classes of organic compounds (Danovaro et al. 1993; Fabiano et al. 1995; Dell'Anno et al. 2002). The composition of the sedimentary organic matter is the major factor affecting the metabolism, distribution and dynamics of benthic organisms and is employed for the evaluation of the trophic status of marine ecosystems (Dell'Anno et al. 2002; Renjith et al. 2012). Apart from these, the protein-to-carbohydrate ratio and the lipid-to-carbohydrate ratio has been used as a valuable indicator for the investigation of the state of biochemical degradation processes (Galois et al. 2000).

The presence of Tannins and Lignins was also analysed along with other biochemical parameters. Tannins and Lignins are high-molecular-weight polycyclic aromatic compounds distributed throughout the plant kingdom. They have high resistance to the biological degradation process and hence can cause potential damage to the aquatic environments. The study of these compounds provides an idea regarding the land derived organic detritus in marine environments (Schnitzer and Khan 1972).

The biochemical composition of the sedimentary organic matter in the study area involving the fjord, lakes and rivers are given in Table 11.2.

Among the biochemical composition obtained from the sediments of the diverse Arctic environments, the carbohydrates (0.4959–3.0845 mg/g) were dominating in the study area except for L2 and R2 with lipid (0.0243–2.9100 mg/g) dominance. The protein content is found to be lower in all stations in comparison with the rest of the biochemical components. It was found within a range of 0.0005–0.8231 mg/g.

The protein concentration in sediments is a tool for understanding the productivity of an aquatic system. A classification of hypertrophic, eutrophic and meso-oligotrophic was done based on protein and carbohydrate concentrations. The hypertrophic condition was characterised by a protein concentration >4 mg/g and carbohydrate concentration >7 mg/g. For eutrophic status, the protein range is within 1.5–4 mg/g, whereas 5–7 mg/g is the range for carbohydrates. The protein and carbohydrate

TABLE 11.2

Biochemical Composition of Sedimentary Organic Matter Obtained from the Fjord, Lakes and River

Stations	Proteins (PRT) (mg/g)	Carbohydrates (CHO) (mg/g)	Lipids (LPD) (mg/g)	Tannins and Lignins (T&L) (mg/g)	PRT/ CHO	LPD/ CHO	Labile Organic Matter (LOM) (mg/g)	BPC (mgC/g)
K1	0.0452	1.5572	0.0910	0.1170	0.0290	0.0584	1.6934	0.7132
K2	0.0223	1.4486	0.0712	0.0742	0.0153	0.0491	1.5421	0.6437
K3	0.0230	1.2710	0.0404	0.0602	0.0180	0.0317	1.3344	0.5499
K4	0.0061	0.8441	0.0742	0.0207	0.0072	0.0879	0.9244	0.3962
L1	0.0038	0.9605	0.2132	0.0163	0.0039	0.2219	1.1775	0.5459
L2	0.8231	2.7777	2.9100	0.0693	0.2963	1.0476	6.5108	3.6968
L3	0.1723	3.0845	0.5987	0.0391	0.0558	0.1940	3.8555	1.7672
R1	0.0005	0.6059	0.0499	0.0072	0.0008	0.0823	0.6563	0.2800
R2	0.0325	1.3372	1.4720	0.0255	0.0243	1.1008	2.8417	1.6548
R3	0.0051	0.4959	0.0243	0.0103	0.0102	0.0490	0.5253	0.2190

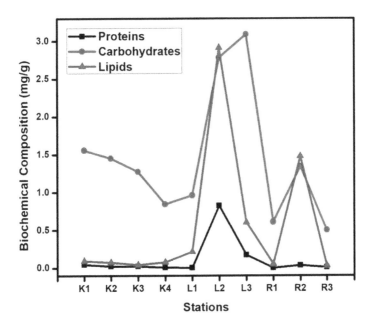

FIGURE 11.2 Biochemical composition of sedimentary organic matter obtained from the fjord, lakes and river.

concentrations of <1.5 and <5 mg/g, respectively, are the required level for meso-oligotrophic status of the sediment system (Dell'Anno et al. 2000, 2002).

The status of diverse Arctic sediments in this study was understood with the above-mentioned ranges and classifications. The protein concentration obtained

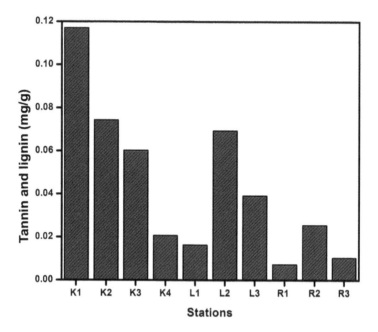

FIGURE 11.3   Tannin and lignin components in the sediments of fjord, lakes and river.

from all the sampling sites provided a value of <1.5 mg/g and a carbohydrate concentration of <5 mg/g. Hence, the sedimentary organic matter under study taken
from the Arctic fjord, lakes and river environment has a meso-oligotrophic status
with intermediate productivity and a medium level of nutritional content (Dell'Anno
et al. 2000, 2002).

The study of tannin and lignin components in the sedimentary system indicates
terrestrial inputs to the aquatic environments (Schnitzer and Khan 1972). The presence of tannin and lignin in the fjord, lake and river system was found in this study. It
ranges between 0.0072 and 0.1170 mg/g and provides information regarding a lower
degree of terrestrial input towards the different environments under investigation.
The lakes and fjord region has the highest input of terrestrial organic debris. In contrast, the river environment showed the lowest terrestrial input in terms of tannin and
lignin value with a minimum in the R1 region, i.e., the river origin.

The biochemical indices such as PRT/CHO ratio and LPD/CHO ratio are also used
for understanding the nature of the sedimentary organic matter from the diverse environments in this study. The PRT/CHO is an index used for the assessment of the
material origin and also used to differentiate between the fresh and aged organic matter in the sediments (Danovaro et al. 1993; Cividanes et al. 2002). If the PRT/CHO
ratio is >1, it indicates the presence of fresh organic matter, whereas a value <1 gives
an idea regarding the predominance of aged organic matter in the sediment system
(Dell'Anno et al. 2000). Along with the lipid concentration, the LPD/CHO ratio is
also used as an index for understanding the food nutritional quality of sedimentary
organic matter. Similar to PRT/CHO ratio, a value of LPD/CHO >1 is an indication of the good nutritional quality of organic matter and the ratio <1 corresponds to

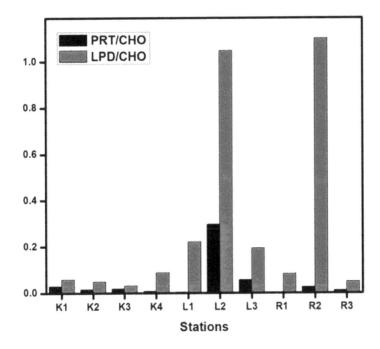

FIGURE 11.4 PRT/CHO and LPD/CHO ratios of the sediment organic matter from fjord, lakes and river.

the poor nutritional quality of organic matter (Fabiano and Danovaro 1998; Gremare et al. 2002).

The presence of aged organic matter is supported by the PRT/CHO ratio with values <1 in the entire study area. The stations L2 and R2 have a higher value for the LPD/CHO ratio which is >1 and supports the good nutritional quality of the sedimentary organic matter in these stations. Apart from these two, the rest of the stations have a <1 value for the LPD/CHO ratio and it points towards the poor nutritional quality of the organic matter present in the sediment system.

The sum of proteins, carbohydrates and lipids is known as the Labile Organic Matter (LOM) which is the easily available organic fraction of the sedimentary organic matter (Danovaro et al. 1993; Cividanes et al. 2002). This organic fraction is found to be higher in lake environments, especially L2 and L3. A river station R2 is also having an appreciably higher amount of this organic matter. The rest of the river station (R1 & R3) and an inner fjord station (K4) have the lowest content of LOM indicating less availability of organic matter.

The concentration of proteins, carbohydrates and lipids can be converted into carbon equivalents by using the conversion factors, viz., 0.49, 0.40 and 0.75 g of C/g, respectively, providing values for the BPC content in the sediments. According to the BPC data, the status of the sediment can be classified as eutrophic (BPC > 3 mgC/g), mesotrophic (BPC = 1–3 mgC/g) and oligotrophic (BPC < 1 mgC/g) (Pusceddu et al. 2009, 2011). With the BPC content present in the selected fjord, lake and river

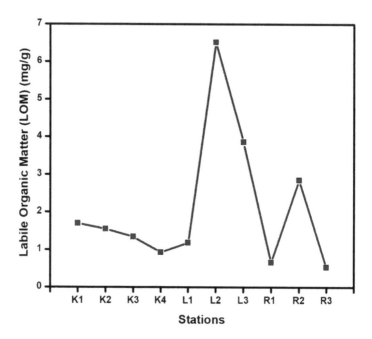

FIGURE 11.5   LOM in the sedimentary organic matter of fjord, lakes and river.

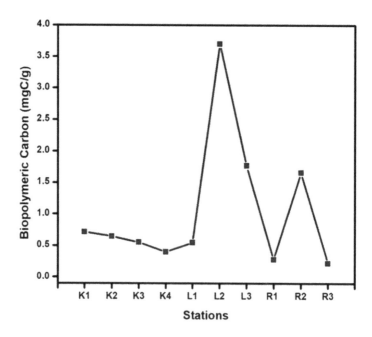

FIGURE 11.6   BPC content in the sedimentary organic matter of fjord, lakes and river.

stations of this study, the Lake L2 exhibits a value of >3 corresponding to the lake environment's eutrophic condition. The mesotrophic condition with a BPC value equalling 1–3 mgC/g is found to be persisting in Lake L3 and the river station R2 which is the middle portion of the river Bayelva having a higher terrestrial influence. Rest of the stations with <1 mgC/g value are having an oligotrophic condition with lower primary productivity and nutritional content according to the classification by Pusceddu (2009).

11.3.2 CHARACTERISATION OF HUMIC ACIDS ISOLATED FROM THE SEDIMENTARY ORGANIC MATTER

11.3.2.1 Elemental Analysis

The elemental composition of humic acid is a useful tool for the identification of humic acid origin, its aliphatic nature and the degree of carboxylation (Rashid 1985; Davies and Ghabbor 1998; Giovanela et al. 2004). The data obtained from the elemental analysis of humic acids isolated from diverse Arctic environments are given in Table 11.3. The lakes L2 and L3 exhibit higher elemental composition in comparison with the rest of the environments under study. The sulphur content was not detected in most of the stations. The presence of sulphur content is indirect evidence for the nature and type of sulphur functional groups present in the humic acid species. The presence of sulphides, disulphides, thiols of thiophenes, immediately oxidised sulphoxides or sulphonates and highly oxidised sulphates form attribute to the higher concentration of sulphur in humic acids (Solomon et al. 2003). The presence of an elevated sulphur concentration also reflects sulphate-reducing conditions (Christensen 1989). In the present study, sulphur was found to be present in the fjord region and in one of the lake region (L2). A higher concentration of sulphur was found in the K1 station of the fjord system where the oceanic water influence is more. While moving from the outer basin of the fjord system to the inner basin, a

TABLE 11.3

Elemental Composition of Humic Acids Isolated from the Fjord, Lakes and River Stations Understudy

Stations	C%	H%	N%	S%	O%	H/C	O/C	N/C	S/C
K1	0.63	0.02	0.14	0.86	98.35	0.0317	156.11	0.2222	1.3650
K2	0.34	0.01	0.08	0.33	99.24	0.0294	291.88	0.2352	0.9705
K3	0.43	0.05	0.11	0.13	99.28	0.1162	230.88	0.2558	0.3023
K4	0.13	0.03	0.06	-	99.78	0.2307	767.53	0.4615	-
L1	0.32	0.01	0.05	-	99.62	0.0312	311.31	0.1562	-
L2	5.51	1.18	0.70	0.31	92.3	0.2141	16.75	0.1270	0.0562
L3	4.04	0.53	0.44	-	94.99	0.1311	23.51	0.1089	-
R1	0.08	0.05	0.06	-	99.81	0.6250	1247.62	0.7500	-
R2	0.91	0.02	0.10	-	98.97	0.0219	108.75	0.1098	-
R3	0.09	0.01	0.06	-	99.84	0.1111	1109.33	0.6666	-

gradual decrease in the sulphur content can be observed. At the innermost region of the fjord, i.e. in the K4 station, the sulphur content is not detected or was practically nil. The sulphur content present in Lake L2 (0.33%) is almost comparable with that of the fjord K2 station (0.31%).

The aliphatic nature and the degree of carboxylation of humic acids can be understood with the CHNS data. A higher H/C ratio indicates the presence of aliphatic rich humic acid moieties (Rashid 1985; Giovanela et al. 2004). Among the various environments understudy, a higher H/C ratio was exhibited by the river station R1 with a value of 0.6250 and was immediately followed by the fjord (K4) and lake (L2) stations with a ratio of 0.2307 and 0.2141, respectively. In the fjord system, the humic acids isolated from the inner basin exhibit a more aliphatic nature than that of the outer basin. This may be due to the prevailing aromatic nature of humic acid species present in the outer region of the fjord (Rashid 1985; Giovanela et al. 2004; Rice and MacCarthy 1991).

During the humification process, the degree of carboxylation can vary significantly with spatial variability. The O/C ratio is an indicator of the degree of carboxylation in the humic acid moieties. The higher the O/C ratio, the higher will be the carboxylation degree. In this study, it is found that the highest degree of carboxylation was exhibited by the river stations R1 and R3, whereas the lowest degree was by the lake stations L2 and L3. The process of carboxylation occurring in the fjord region is in between that of the river and lake environments. Throughout the fjord system, the degree of carboxylation happening in the outermost part of the fjord is lower than that of the innermost part of the fjord (Rice and MacCarthy 1991; Sierra et al. 2004; Klavins et al. 2013).

The N/C ratio is another elemental ratio used for understanding the humic acid species in terms of lignin contribution (Saito and Hayano 1981). Humic acids with a lower N/C ratio are generally rich in lignin contents (Stuermer and Harvey 1978; Stuermer et al. 1978; Sierra et al. 2004). According to the N/C ratio, humic acids with higher lignin concentration are found in all the three lakes under study and in the middle region of the river (R2) which has a higher terrestrial input. The rest of the river regions R1 and R3 have lower lignin content with a higher N/C ratio when compared with that of R2. In the fjord system the innermost station with glacier input, i.e. K4 exhibits the lowest lignin contribution. In contrast, the rest of the three stations have a higher influence of lignin on humic acid species formation.

11.3.2.2 UV-Visible Spectroscopy

The UV-Visible spectrum was analysed by taking the absorbance of 250, 270, 280, 365, 400, 465, 472, 600, 664 and 665 nm wavelengths (Ghosh and Schnitzer 1979; Kumada 1987). The ratios such as E250/365 (E2/E3), E465/665 (E4/E6), E270/400, E472/664, E280/472 and E280/664 were used for the characterisation of humic acids under study. Among these ratios, E2/E3 and E4/E6 are the most commonly used ones for understanding the aromatic nature and molecular size of humic acids. Apart from the ratios, a term ΔlogK was also applied in studying the humic acid for providing information regarding their optical properties. This term is defined as the difference between the logarithms of the absorbance at 400 nm (logE400) and 600 nm (logE600) (Kumada 1987; Kachari et al. 2015).

11.3.2.2.1   E250/E365

It was reported that the ratio E250/E365 (E2/E3) ratio is a property of aquatic humic substances which increases with a decrease in the aromaticity and molecular size (Uyguner and Bekbolet 2004; Peuravuori and Pihlaja 1997). The E2/E3 ratio exhibited by the humic acids isolated from diverse environments is given in Figure 11.7a. All humic acids provided a value less than one except the R3 station of Bayelva River, even though it was not much greater than 1. Hence, a higher degree of aromaticity and molecular size of humic acids can be interpreted from these results, which are detailed and clarified with the rest of the ratios.

11.3.2.2.2   E465/E665

The E465/E665 (E4/E6) ratio is an indicator of the aromatic nature and degree of condensation of humic acids (Ghosh and Schnitzer 1979). It can also be used as a humification index for understanding the extent of the humification process happening in each environment. A decrease in the E4/E6 ratio indicates a progressive condensation of humic acids (Kumada 1987; Kachari et al. 2015; Ghosh and Schnitzer 1979; Stevenson and Schnitzer 1982). A higher value for the E4/E6 ratio reveals the presence of a dominating aliphatic nature of the humic acid in comparison with that of the condensed aromatic rings (Ching et al. 1997). The E4/E6 ratio is inversely proportional to the degree of aromaticity, particle size, molecular weight and acidity (Kachari et al. 2015). The E4/E6 values for the humic acids isolated from diverse Arctic environments are given in Figure 11.7b.

The humic acids isolated from the lake environment (L1 and L3) show the highest values for the E4/E6 ratio ranging between 4 and 6 which is almost closer to that of a soil humic acid (5.0) (Kononova 1966). The humic acid obtained from river station R2 provides an E4/E6 ratio closer to 3, revealing a condensation similar to that of a lignite humic acid (3.8) (Baruah 1983). The rest of the humic acids isolated from the fjord, lake and river environment has a value ranging between 1 and 2. This indicates a slightly higher percentage of aromaticity, particle size and molecular weight compared to that of the rest of the stations.

11.3.2.2.3   E270/E400

The phenolic/quinoid core of humic acid degrades to simple carboxylic aromatic compounds. Understanding this degradation can be done with the E270/E400 ratio (Uyguner and Bekbolet 2004). The degradation of the phenolic/quinoid core is found to be higher for the humic acids isolated from the river end station (R3) and lower for L2, L3 and R1 humic acids. The rest of the stations exhibits a similar range of ratio between 0.8 and 1.5. The variation of these ratios within these environments understudy is given in Figure 11.7c.

11.3.2.2.4   E472/E664

The ratio E472/E664 is inversely proportional to the degree of condensation and polymerisation of aromatic components (Kachari et al. 2015). It is also inversely correlated with organic matter humification (Albrecht et al. 2011). The trend of the E472/E664 ratio of humic acids isolated from the sedimentary organic matter of

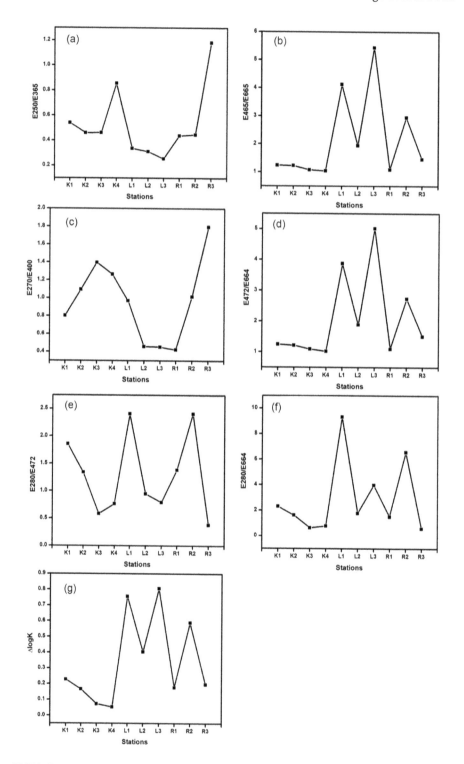

FIGURE 11.7 UV-Visible spectroscopic ratios of various humic acids isolated from the fjord, lakes and rivers.

various Arctic environments is similar to that of the E4/E6 ratio (Figure 11.7d). The higher value for the ratio is exhibited by the lake humic acids (L1 & L3), and the lower value is by the fjord humic acid. One of the river stations, R1 also provided a lower value for the ratio similar to that of the fjord. Rest of the stations, viz., L2, R2 and R3 have values within a mid-range between the higher and lower exhibited by other stations.

11.3.2.2.5   E280/E472

The proportion between lignins and other materials at the beginning of humification is reflected with the help of the E280/E472 ratio. The variation of this ratio obtained from diverse environments understudy is given in Figure 11.7e. A higher ratio indicates a higher lignin concentration in the isolated humic acids from various environments under study. The fjord station K1, lake station L1 and river station R2 exhibit a higher E280/E472 ratio. This may be due to the presence of lignin content in higher concentration as mentioned above. The lower ratios were observed in stations of K3, L3 and R3 with a lower lignin content and higher amount of other materials involved in the formation of humic acid during its initial stages.

11.3.2.2.6   E280/E664

The E280/E664 ratio reflects the relation between non-humified and strongly humified material. In a site-specific comparison, the lower the ratio, the higher the degree of aromatic condensation and the level of organic matter humification (Albrecht et al. 2011). The fjord region, lake L2, river stations R1 and R2 are having values in the lower range, which indicates a higher degree of aromatic condensation and humification process. The highest ratio for E280/E664 is observed in the L1 station of lake environment, followed by the R2 river station and L3 Lake. Here the degree of aromatic condensation and organic matter humification is lower due to its inverse relationship with the ratio. The variation in the E280/E664 values concerning each station understudy is given in Figure 11.7f.

11.3.2.2.7 $\Delta logK$

The humification degree of organic matter can be described with the help of $\Delta logK$ coefficient. The humic acids with a higher degree of humification have values of $\Delta logK$ coefficient up to 0.6. A value for the coefficient higher than 0.6 suggests that the organic matters have not undergone a higher humification process (Fong et al. 2006). Figure 11.7g represents the variability in the $\Delta logK$ coefficient values exhibited by each station under study. All the stations except L1 and L3 have a coefficient value less than 0.6, which indicates a higher degree of humification occurring in these environments.

11.3.2.3   FTIR Spectroscopy

The presence of band within the region of 3350–3450 cm^{-1} is attributed to the O-H stretching vibrations. The OH groups absorb near 3600 cm^{-1} when they are in a free state or in an undissociated state. The frequency of O-H stretching vibration reduces when there is hydrogen bonding in the association of molecules. The presence of broad bands in the spectrum in this range is associated with the O-H stretching vibration of the polymeric association of the molecules (Silverstain et al. 1981). In the FTIR spectrums of humic acids isolated from various Arctic environments, the presence of this spectral band is found to be absent in some stations. In the fjord region, the spectral band is present in K2 and K4 stations with a value closer to $3400\,cm^{-1}$. Among the lake environment lake L2 exhibits, this property in its spectrum and for the river environment stations R1 and R3 has the presence of this broadband indicating O-H stretching vibrations.

Within the FTIR spectral region of $2800–3100\,cm^{-1}$, the bands occurring at 2926 and $2853\,cm^{-1}$ are due to the asymmetrical and symmetrical stretching vibrations of C-H of methylene groups, respectively (Silverstain et al. 1981). At frequencies slightly higher than $3000\,cm^{-1}$ the presence of aromatic C-H stretching vibrations can be assigned. Apart from these, the occurrence of aliphatic content in humic acids is studied with the help of two weak absorption bands at near 2926 and $2853\,cm^{-1}$. These bands are assigned for the asymmetric and symmetric stretching vibrations of C-H aliphatic, respectively. The presence of these bands in the FTIR spectrums of humic acids isolated from the study area is observed with an exception in the spectrum of lake L1 and river R3 stations.

The spectral range of $1600–1800\,cm^{-1}$ has many bands to understand various functional groups' presence. One among those bands is around $1700\,cm^{-1}$, commonly seen at $1720\,cm^{-1}$ which is formed due to the stretching vibration of C=O of carboxylic acids, aldehydes and ketones. In the spectra understudy, the presence of these bands is observed in humic acids isolated from all the stations except lake L1. These bands are found to be broad and have some shifts in the spectrums with each station. Redshift was observed in stations K1, K2, L2, L3 and R2, whereas the rest of the station excepting L1 shows a blue shift in the spectrum for this functional groups moiety. In some stations, the bands are found to be weak. This weakness of the band is due to the H-bonding and resonance. The internal hydrogen bonding reduces the frequency of C=O stretching to a greater extent than the intermolecular hydrogen bonding. When the humic substances have undergone methylation, the bands near $1720\,cm^{-1}$ can shift to a higher frequency range. This may be due to the absorbance by C=O of esters than C=O of acids. The lower frequency of both acids and esters can be explained by the occurrence of conjugated C=O groups and/or aromatic acids (Silverstain et al. 1981; Naidja et al. 2002).

The presence of C=C stretching mode of unconjugated olefins shows weak absorptions at the $1667–1640\,cm^{-1}$ range. Monosubstituted olefins, i.e. vinyl groups, absorb near $1640\,cm^{-1}$ with moderate intensity. The di-substituted trans-olefins, tri- and tetra-alkyl substituted olefins absorb near or around $1670\,cm^{-1}$, whereas the di-substituted cis-olefins and vinylidene olefins absorb near the region of $1650\,cm^{-1}$. Hence, the presence of a moderate band in the region of $1650\,cm^{-1}$ can be attributed to C=C stretching vibrations of substituted olefins. Among the humic acids under study, the presence of substituted olefins is found in K1 and K2 stations of the fjord, L2 and L3 stations of lakes and all the three river stations (Silverstain et al. 1981).

The delocalisation of pi electrons of unsaturated groups occurs due to the conjugation with the C=C bond. This delocalisation of pi electrons of the C=O group reduces the C to O double-bond character and causes absorption at longer wavelengths. The absorption in the region of 1685–1666 cm^{-1} is caused by the conjugation with an olefinic or phenyl group. Additional conjugation reduces the frequency range from the normal values (Silverstain et al. 1981).

The region of 1300–1500 cm^{-1} provides information regarding the aromatic C-C ring stretching with a weak band at 1497 cm^{-1} and higher absorption at 1460 cm^{-1}. Apart from these, a band at 1430 cm^{-1} occurs due to the asymmetrical vibration of C-H bonds of bending methyl and methylene groups. Near the 1380 cm^{-1} region, the presence of symmetrical C-H bending vibrations of methyl groups is understood. The presence of aromatic C-C ring stretching is observed in all the stations of the fjord, one lake, L3 and R2, R3 of the river environment. Among these humic acids, an additional peak for the presence of symmetrical C-H bending vibrations of methyl groups is observed in the spectra of humic acids isolated from K3, L3 and R2 stations.

In the spectral region of 900–1100 cm^{-1}, a band appears within the range of 800–950 cm^{-1} due to asymmetric ring stretching, i.e. the C-C bond stretches during the contraction of C-O bonds. The band in the region of 1000–1100 cm^{-1} attributes to either the C-O stretching vibrations of polysaccharides (Stevenson and Goh 1971; Hatcher et al. 1980) or due to Si-O stretching vibration (Bailly 1976; Schnitzer 1971). It has been reported that the band near 1050 cm^{-1} is due to the C-O stretch of polysaccharides (Baruah 1986). The presence of a C-O stretch of polysaccharides was observed in the spectra of all the humic acids isolated from various environments under study. Also, the stations K1, K2, K3, L3 and R2 provide additional bands around the range of 900 cm^{-1} can be caused by the asymmetric ring stretching (Stevenson and Goh 1971; Hatcher et al. 1980).

The presence of aliphatic C-N stretching vibration occurs within the range of 1020–1250 cm^{-1}. The positional variation of the band can happen depending on the substitution. The aromatic C-N stretching vibrations can be observed in the range of 1180–1380 cm^{-1}. Most of the humic acids isolated from various Arctic environments have peaks within the mentioned ranges of aliphatic and aromatic C-N stretching. Hence, the presence of these groups in the macromolecular structure of humic acids can be interpreted from these results.

The spectral range of 400–700 cm^{-1} gives an idea regarding the substituted benzene groups. Monosubstituted benzene exhibits two strong bands between 690–710 and 730–770 cm^{-1} (Bailly 1976). The band corresponding to ortho-substituted benzene is found within the range of 735–770 cm^{-1}. For meta-substituted benzene, there are two bands located at two different positions. The intensity of one band (680–725 cm^{-1}) will be stronger than the other (750–810 cm^{-1}). In the case of para-benzene, a single strong band is observed at a higher frequency than the ortho-substituted benzene (790–850 cm^{-1}) (Sarma 2011). The presence of benzene in one form or other is found in all the spectrums associated with the humic acids isolated from various environments in the Arctic region.

Since the study is about the humic acids isolated from various environments, the uniqueness of each molecular species can be understood with the characteristic nature of the fingerprint region within the range of 600–1400 cm^{-1}.

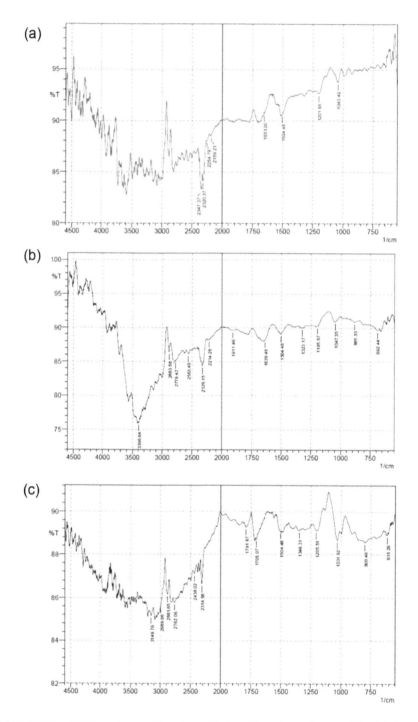

FIGURE 11.8 FTIR spectrums of humic acid isolated from the sediments of fjord, lakes and river: (a) K1, (b) K2, (c) K3, (d) K4, (e) L1, (f) L2, (g) L3, (h) R1, (i) R2 and (j) R3.

(*Continued*)

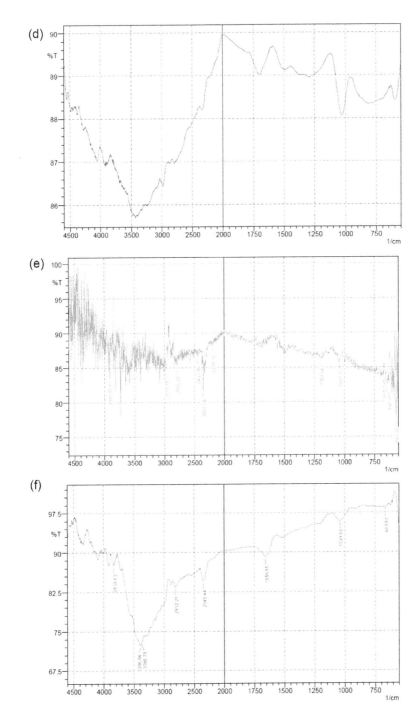

FIGURE 11.8 (*CONTINUED*) FTIR spectrums of humic acid isolated from the sediments of fjord, lakes and river: (a) K1, (b) K2, (c) K3, (d) K4, (e) L1, (f) L2, (g) L3, (h) R1, (i) R2 and (j) R3.

(*Continued*)

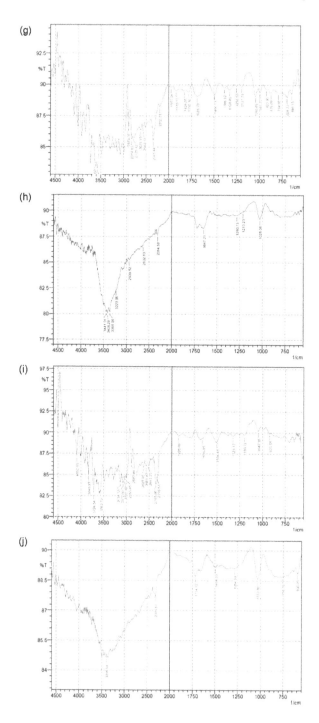

FIGURE 11.8 (*CONTINUED*)   FTIR spectrums of humic acid isolated from the sediments of fjord, lakes and river: (a) K1, (b) K2, (c) K3, (d) K4, (e) L1, (f) L2, (g) L3, (h) R1, (i) R2 and (j) R3.

11.3.3 METAL COMPLEXATION OF HUMIC ACIDS ISOLATED FROM SEDIMENTARY ORGANIC MATTER

Humic substances play a major role in the geochemistry of metals in the sediment system. The early diagenesis of the organic fractions has effects on the metal binding capacity (Fengler et al. 1994). About ten metals are analysed in this study which includes Pb, Ni, Cu, Cr, Co, Fe, Mn, Zn, Al and Mg. The predominance of Fe over Cr, Cu and Mg was observed in this study within the range of 100–10,000 ppm. The rest of the metals has a concentration <100 ppm, among which Pb and Co are found in very trace quantity with values <10 ppm.

The most common metals present in the sediments are Al, Fe, Ca and Mg. Hence, the presence of these compounds can be found in higher quantity in the humic acids (Nissenbaum and Swaine 1976). The excess concentration of Mg over Ca is found in the humic acids under study. The Ca was almost absent in the humic acids isolated from various environments under study. Also, the absence of Mg in the humic acid isolated from the fjord environment except K4 is visible in the results obtained. Apart from the fjord, one of the lake environments (L1) is also having the absence of Mg in the isolated humic acids. The ability of humic acids to complex with Al can be seen with its presence in the humic acids within a range of 1–33 ppm. All humic acids have the presence of Al in its complexed state, but lake L1 is an exception from the rest with its absence. Fe is found to be the most dominating metal species complexed with humic acid and is present in all stations within the range of 400–9000 ppm.

Naturally occurring organic materials are known to complex a wide variety of elements. Certain humic acid species can solubilise Co, Cu, Ni and Zn (Rashid 1972). In this study, Cu, Cr, Ni and Zn are present in all the humic acids isolated whereas, the Co is found to be absent in the humic acids obtained from the stations K1, K2, L3 and R2. The Co content present in the rest of the stations is not more than 10 ppm. The highest amount of Co is found within the humic acids from K4 and R1 stations which are closer to the glacier environment. The solubilising nature of humic acids

TABLE 11.4
Metals Complexed with the Humic Acids Isolated from the Sediments of Fjord, Lakes and River Stations Under Study

Stations	Pb	Ni	Cu	Cr	Co	Fe	Mn	Zn	Al	Mg
K1	8.75	20.25	667	294	-	6582.5	3.25	22.37	30.75	-
K2	-	9.37	859	313.62	-	8927.5	2.25	6.5	33.75	-
K3	-	47.12	236.62	197.75	6.62	2680	21.62	35	21.75	-
K4	1.375	55.75	356.62	163.25	9.25	3523.75	19.25	49.62	13.87	1862.5
L1	-	8.75	103.25	109.125	4.5	1194.375	8.625	63.75	-	-
L2	-	43.37	92.875	124.875	1.875	973.625	11.875	55.125	3.375	463.375
L3	-	17.75	655.5	262.375	-	6282.5	4.375	63.75	21.5	224
R1	-	12.62	116.25	62.875	9	1208.875	9	82	1.5	598.125
R2	-	15.75	716.25	263	-	7217.5	4.375	58.25	27.5	504.25
R3	2	31.12	33.125	83	2.125	466.5	9.5	44.125	3.5	407.25

for Mn is found to be lower than Cu, Ni and Zn. Hence, the amount of Mn obtained from the humic acids is lower than that of these trace elements. Apart from these metals, the presence of Pb complexed to humic acids was in trace quantities from stations K1, K4 and R3. The concentration of trace elements will be higher in the humic bounded state than that in the free sediments. The remobilisation of trace elements followed by its complexation with organic compounds make these metals available in the humic fractions of the sedimentary organic matter. The concentration of these metals can be considered as a minimum value due to the loss of these in trace quantities during the extraction procedure from the organic material (Fengler et al. 1994).

11.4   CONCLUSIONS

The preliminary understanding of the sedimentary organic matter present in the diverse environments in the Arctic region under study is done with the biochemical composition. The presence of meso-oligotrophic condition prevailing in these environments is understood from the biochemical indices. Along with these the state of an aged organic matter having poor nutritional quality is also understood from the study. The characterisation of humic acids isolated from these environments is done with elemental analysis, UV-Visible spectroscopy and FTIR spectroscopy. The degree of carboxylation and the lignin contribution towards the formation of humic acid is supported with O/C and N/C ratios, respectively. Various ratios obtained from the UV-Visible spectroscopy provide an insight into the general structural characteristics of humic acids. The degree of aromaticity and condensation of humic acid was studied with the well-known E4/E6 ratio. Other ratios also support the same and provide information on many other facts. A detailed study on the structural characteristics is understood with the FTIR spectroscopy by finding out various functional groups present in it. The uniqueness of each humic acid species isolated can be studied with the functional group specificities occurring in the fingerprint region of the spectrum. Metals present in the sediment system can complex with humic components. Among the ten metals under study, the presence of Pb and Co is found in trace quantity, whereas the metals present in bulk quantities are Fe, Cr, Cu and Mg. The property of humic acid to complex trace elements makes it more available in the humic components than from the sediments.

ACKNOWLEDGEMENTS

The authors would like to express their sincere gratitude to the MOES, NCPOR and the Indian Arctic Expedition (Batch 1–2018) team members for all the assistance provided to carry out this work.

REFERENCES

Anuradha, V., Nair, S.M., Kumar, N.C., 2011. Humic acids from the sediments of three ecologically different estuarine systems: a comparison. *Int. J. Environ.* 2, 174–184. doi: 10.6088/ijes.00202010018.

APHA, 1995. Standard methods for the examination of water and wastewater. *Am. Public Health Assoc.* 5, 47–48.

Aslbrecht, R., Petit, J.L., Terrom, G., Perissol, C., 2011. Comparison between UV spectroscopy and nirs to assess the humification process during sewage sludge and green wastes co-composting. *Bioresour. Technol.* 102, 4495–4500. doi: 10.1016/j.biortech.2010.12.053.

Bailly, V.R., 1976. Spectroscopie infra-rouge de quelques fractions d'acideshumiquesobtenues. *Plant Soil.* 45, 95–111. doi: org/10.1007/BF00011132.

Barnes, H., Blackstock, J., 1973. Estimation of lipids in marine animals and tissues: detailed investigation of the sulphophosphovanilun method for 'total' lipids. *J. Exp. Mar. Biol. Ecol.* 12, 103–118. doi: 10.1016/0022-0981(73)90040-3.

Baruah, M.K., 1983. Organo-geochemical aspects of the formation of pyrite and organic sulphur in coal. PhD Thesis. Gauhati University.

Baruah, M.K., 1986. Assignment of the I.R. absorption band at $1050 cm^{-1}$ in lignite humic acid. *Fuel* 65, 1756–1759. doi: 10.1016/0016-2361(86)90281-4.

Ching, Y., Aiken, G.R., Danielsen, K.M., 1997. Binding of pyrene to aquatic and commercial humic substances: the role of molecular weight and aromaticity. *Environ. Sci. Technol.* 31, 1630–1635. doi: org/10.1021/es960404k.

Christensen, J.P., 1989. Sulphate reduction and carbon oxidation rates in continental shelf sediments, an examination of off-shelf carbon transport. *Cont. Shelf Res.* 9, 223–246. doi: 10.1016/0278-4343(89)90025-3.

Cividanes, S., Incera, M., Lopez, J., 2002. Temporal variability in the biochemical composition of sedimentary organic matter in an intertidal flat of the Galician coast (NW Spain). *Oceanol. Acta.* 25, 1–12. doi: org/10.1016/S0399-1784(01)01178-1.

Colombo, J.C., Silverberg, N., Gearing, J.N., 1996. Biogeochemistry of organic matter in the Laurentian Trough, II. Bulk composition of the sediments and relative reactivity of major components during early diagenesis. *Mar. Chem.* 51, 295–314. doi: org/10.1016/0304-4203(95)00060-7.

Cottier, F.R., Nilsen, F., Skogseth, R., Tverberg, V., Skardhamar, J., Svendsen, H., 2014. Arctic fjords: a review of the oceanographic environment and dominant physical processes. *Geol. Soc. Spec. Publ.* 344, 35–50. doi: org/10.1144/SP344.4.

Danovaro, R., Fabiano, M., Dell Croce, N., 1993. Labile organic matter and microbial biomass in deep-sea sediments (Eastern Mediterranean Sea). *Deep-Sea Res. Pt. I.* 40, 953–965. doi: 10.1016/0967-0637(93)90083-F.

Davies, G., Ghabbor, E.A., 1998. *Humic Substances: Structures, Properties and Uses.* Royal Society of Chemistry, Cambridge.

Dell'Anno, A., Fabiano, M., Mei, M.L., Danovaro, R., 2000. Enzymatically hydrolyzed protein and carbohydrate pools in deep-sea sediments: estimates of the bioavailable fraction and methodological considerations. *Mar. Ecol. Prog. Ser.* 196, 15–23. doi: 10.3354/meps196015.

Dell'Anno, A., Mei, M.L., Pusceddu, A., Danovaro, R., 2002. Assessing the trophic state and eutrophication of coastal marine systems. A new approach based on the biochemical composition of sediment organic matter. *Mar. Pollut.* 44, 611–622. doi : 10.1016/s0025-326x(01)00302-2.

Dick, D.P., Burba, P., Herzog, H., 1999. Influence of extractant and soil type on molecular characteristics of humic substances from two Brazilian soils. *J. Braz. Chem. Soc.* 10, 140–145. doi: 10.1590/S0103-50531999000200012.

Fabiano, M., Danovaro, R., 1998. Enzymatic activity, bacterial distribution and organic matter composition in sediments of the Ross Sea (Antarctica). *App. Environ. Microbiol.* 64, 3838–3845. doi: 10.1128/AEM.64.10.3838-3845.1998.

Fabiano, M., Danovaro, R., Fraschetti, S., 1995. A three-year time-series of the elemental and biochemical composition of organic matter in sub-tidal sandy sediments of the Ligurian Sea (Northwestern Mediterranean). *Cont. Shelf Res.* 15, 1453–1469. doi: 10.1016/0278-4343(94)00088-5.

Fengler, G., Grossman, D., Kersten, M., Liebezeit, G., 1994. Trace metals in humic acids from recent Skagerrak sediments. *Mar. Pollut. Bull.* 28, 143–147. doi: org/10.1016/0025-326X(94)90389-1.

Field, J.A., Lettinga, G., 1987. The methanogenic toxicity and anaerobic degradability of a hydrolysable tannin. *Wat. Res.* 21, 367–374. doi: 10.1016/0043-1354(87)90217-X.

Finar, I.L., 1976. *Organic Chemistry*, Volume I, Longman, Singapore.

Fong, S.S., Lau, I.L., Chong, W.N., Asing, J., Faizal, M., Nor, M., Satirawaty, A., Pauzan, M., 2006. Characterization of coal-derived humic acids from Mukah, Sarawak as a soil conditioner. *J. Braz. Chem. Soc.* 17, 582–587. doi: 10.1590/S0103-50532006000300023.

Galois, R., Blanchard, G., Seguignes, M., Huet, V., Joassard, L., 2000. Spatial distribution of sediment particulate organic matter on two estuarine intertidal mudflats: a comparison between Marennes-Oleron Bay (France) and the Humber Estuary (UK). *Cont. Shelf Res.* 20, 1199–1217. doi: 10.1016/S0278-4343(00)00019-4.

Gerachov, S.M., Hatcher, P.G., 1972. Improved technique for analysis of carbohydrate in the sediment. *Limnol. Oceanogr.* 17, 938–943. doi: 10.4319/lo.1972.17.6.0938.

Ghosh, K., Schnitzer, M., 1979. UV and visible absorption spectroscopic investigations in relation to macromolecular characteristics of humic substances. *J. Soil Sci.* 30, 735–745. doi: 10.1111/j.1365-2389.1979.tb01023.x.

Giovanela, M., Parlanti, E., Soriano-Sierra, E.J., Soldi, M.S., Sierra, M.M.D., 2004. Elemental compositions, FTIR spectra and thermal behaviour of sedimentary fulvic and humic acids from aquatic and terrestrial environments. *Geochem. J.* 38, 255–264. doi: 10.2343/geochemj.38.255.

Gremare, A., Medernach, L., deBovee, F., Amouroux, J.M., Vetion, G., Albert, P., 2002. Relationships between sedimentary organics and benthic meiofauna on the continental shelf and the upper slope of the Gulf of Lions (NW Mediterranean). *Mar. Ecol. Prog. Ser.* 234, 85–94. doi: 10.3354/meps234085.

Hatcher, P.G., Breger, I.A., Mattingly, M., 1980. A structural characterization of fulvic acids from continental shelf sediments. *Nature* 285, 560–562. doi: 10.1038/285560a0.

Henrichs, S.M., 1992. Early diagenesis of organic matter in marine sediments: progress and perplexity. *Mar. Chem.* 39, 119–149. doi: org/10.1016/0304-4203(92)90098-U.

IHSS, 2010a. Natural organic matter research, Isolation of IHSS samples. http://www.humic-substances.org/ isolation.html.

IHSS, 2010b. Natural organic matter research, Elemental composition and stable isotopic ratios of IHSS samples. http://www.humicsubstances.org/elements.html.

Joseph, M.M., Kumar, C.R.S., Kumar, T.G.R., Renjith K.R., Chandramohanakumar, N., 2008. Biogeochemistry of surficial sediments in the intertidal systems of a tropical environment. *Chem. Ecol.* 24, 247–258. doi: 10.1080/02757540802119871.

Kachari, M., Belwar, P., Dutta, K., Sarmah, A., Saikia, P.P., Baruah, M.K., 2015. Ultraviolet-visible and Infrared spectroscopic studies of soil humic acids. *Int. J. Plant Soil Sci.* 6, 194–202.

Kalesh, N.S., Sujatha, C.H., Nair, S.M., 2001. Dissolved Folin phenol active substances (Tannin and Lignin) in the seawater along the west coast of India. *J. Oceanogr.* 57, 29–36. doi: 10.9734/IJPSS/2015/16032.

Klavins, M., Purmalis, O., Rodinov, V., 2013. Peat humic acid properties and factors influencing their variability in a temperate bog ecosystem. *Est. J. Ecol.* 62, 35–52. doi: 10.3176/eco.2013.1.04.

Kononova, M.M., 1966. *Soil Organic Matter, Its Nature, Its Role in Soil Formation and Soil Fertility.* Second English edition, Pergamon, Oxford.

Kumada, K., 1987. *Chemistry of Soil Organic Matter.* Japan Scientific Societies Press, Tokyo.

Loring, D.H., Rantala, R.T.T., 1992. Manual for the geochemical analyses of marine sediments and suspended particulate matter. *Earth Sci. Rev.* 32, 235–283. doi: 10.1016/0012-8252(92)90001-A.

Lowry, O.H., Rosebrough, N.J., 1951. Protein measurement with the Folin phenol reagent. *J. Biol. Chem.* 193, 265–275.

Maccarthy, P., 2001. The principles of humic substances. *Soil Sci.* 166, 738–751. doi: 10.1097/00010694-200111000-00003.

Naidja, A., Huang, P.M., Anderson, D.W., Kessel, C.V., 2002. Fourier Transform Infrared, UV-Visible, and X-ray Diffraction analyses of organic matter in humin, humic acid and fulvic acid fractions in soil exposed to elevated CO_2 and N fertilization. *Appl. Spectrosc.* 56, 318–324. doi: 10.1366/0003702021954908.

Nair, S.M., Balchand, A.N., Nambisan, P.N.K., 1989. On the determination and distribution of hydroxylated aromatic compounds in estuarine waters. *Toxicol. Environ. Chem.* 23, 203–213. doi: 10.1080/02772248909357439.

Nissenbaum, A., Swaine, D.J., 1976. Organic matter-metal interactions in recent sediments: the role of humic substances. *Geochim. Cosmochim. Acta* 40, 809–816. doi: 10.1016/0016-7037(76)90033-8.

Perminova, I.V., 1999. Size exclusion chromatography of humic substances: Complexities of data interpretation attributable to non-size exclusion effects. *Soil Sci.* 164, 834–840. doi: 10.1097/00010694-199911000-00008.

Peuravuori, J., Pihlaja, K., 1997. Molecular size distribution and spectroscopic properties of aquatic humic substances. *Anal. Chim. Acta.* 337, 133–149. doi: 10.1016/ S0003-2670(96)00412-6.

Pusceddu, A., Bianchelli, S., Gambi, C., Danovaro, R., 2011. Assessment of the benthic trophic status of marine coastal ecosystems: Significance of nneiofaunal rare taxa. Estuar. *Coast. Shelf Sci.* 93, 420–430. doi: 10.1016/j.ecss.2011.05.012.

Pusceddu, A., Dell Anno, A., Fabiano, M., 2000. Organic matter composition in coastal sediments at Terra Nova Bay (Ross Sea) during summer 1995. *Pol. Biol.* 23, 288–293. doi: 10.1007/s003000050446.

Pusceddu, A., Dell'Anno, A., Fabiano, M., Danovaro, R., 2009. Quantity and bioavailability of sediment organic matter as signatures of benthic trophic status. *Mar. Ecol. Prog. Ser.* 375, 41–52. doi: 10.3354/meps07735.

Rashid, M.A., 1972. Amino acids associated with marine sediments and humic compounds and their role in the solubility and complexing of metals. *Proc. 24th Int. Geol. Congr., Montreal 1972, Section 10 (Geochemistry).* pp. 346–353.

Rashid, M.A., 1985. *Geochemistry of Marine Humic Compounds.* Springer Verlag, New York.

Rashid, M.A., King, L.H., 1969. Molecular weight measurements on humic and fulvic acid fractions from marine clays on the Scotian Shelf. *Geochim. Cosmochim. Acta* 33, 147–151. doi: 10.1016/0016-7037(69)90099-4.

Rashid, M.A., King, L.H., 1971. Chemical characteristics of fractionated humic acids associated with marine sediments. *Chem. Geol.* 7, 37–43. doi: 10.1016/0009-2541(71)90030-1.

Renjith, K.R., Joseph, M.M., Ghosh, P., Rahman, K.H., Kumar, C.S.R, Chandramohanakumar, N., 2012. Biogeochemical facsimile of the organic matter quality and trophic status of a micro-tidal tropical estuary. *Environ. Earth Sci.* 70, 729–742. doi: 10.1007/ s12665-012-2159-0.

Rice, J.A., MacCarthy, P., 1991. Statistical evaluation of the elemental composition of humic substances. *Org. Geochem.* 17, 635–648. doi: 10.1016/0146-6380(91)90006-6.

Saito, Y., Hayano, S., 1981. Characterization of humic and fulvic acids isolated from marine sediments of Sagami and Suruga Bays with C-13 and proton nuclear magnetic resonance. *J. Oceanogr. Soc. Jpn.* 36, 286–292. doi: 10.1007/BF02070157.

Sarma, B.K., 2011. *Spectroscopy.* Goel Publishing House, Meerut.

Schnitzer, M., 1971. Characterization of humic constituents by spectroscopy. In: McLaren, A.D., Skujins, J., (Eds.), *Soil Biochemistry,* Marcel Dekker, New York.

Schnitzer, M., 1977. Recent findings on the characterization of humic substances extracted from soils from widely differing climatic zones. *Soil Org. Matter Stud.* 117–132.

Schnitzer, M., Khan, S.U., 1972. *Humic Substances in the Environment*. Marcel Dekker, New York.

Sierra, M.M.D., Giovanel, M., Parlanti, E., Esteves, V.I., Duarte, A.C., Fransozo, A., Soriano, S.E.J., 2004. Structural description of humic substances from subtropical coastal environments using elemental analysis, FT-IR and 13Csolid-state NMR data. *J. Coast. Res.* 42, 219–231.

Silverstain, R.M., Bassler, G.C., Morrill, T.C., 1981. *Spectroscopic Identification of Organic Compounds*, John Wiley & Sons, New York.

Solomon, D., Lehmann, J., Martinez, C.E., 2003. Sulfur K-edge XANES spectroscopy as a tool for understanding sulfur dynamics in soil organic matter. *Soil Sci. Soc. Am. J.* 67, 1721–1731. doi: 10.2136/sssaj2003.1721.

Stevenson, F.J., 1982. Extraction, Fraction and General Chemical Composition of Soil Organic Matter, In Stevenson, F.J. (Ed.), *Humus Chemistry, Genesis, Composition, Reactions*. John Wiley and Sons, New York.

Stevenson, F.J., 1994. *Humus Chemistry- Genesis, Composition, Reactions*. 2nd edition, Wiley, New York.

Stevenson, F.J., Goh, K.M., 1971. Infra-red spectra of humic acids and related substances. *Geochim. Cosmochim. Acta* 35, 471–483. doi: 10.1016/0016-7037(71)90044-5.

Stevenson, I.L., Schnitzer, M., 1982. Transmission electron microscopy of extracted fulvic and humic acids. *Soil Sci.* 133, 179–185.

Stuermer, D.H., Harvey, G.R., 1978. Structural studies on marine humus: a new reduction sequence for carbon skeleton determination. *Mar. Chem.* 6, 55–70. doi: 10.1016/0304-4203(78)90006-3.

Stuermer, D.H., Peters, K.E., Kaplan, I.R., 1978. Source indicators of humic substances and protokerogen: stable isotope ratios, elemental compositions and electron-spin resonance spectra. *Gechim. Cosmochim. Acta* 42, 989–997. doi: 10.1016/0016-7037(78)90288-0.

Uyguner, C.S., Bekbolet, M., 2004. Implementation of spectroscopic parameters for practical monitoring of natural organic matter. *Desalination* 176, 47–55. doi: 10.1016/j. desal.2004.10.027.

12 Arctic Phyto-Technology

Rajesh Kumar Dubey
JNV University Jodhpur

Priyanka Babel
Mohanlal Sukhadia University

CONTENTS

12.1   INTRODUCTION

Sir Gottlieb Haberlandt is a German botanist who discovered artificial cell culture of a plant in 1898 and based on his work (1902) on 'Toti-potentiality', he is recognised as the father of plant tissue culture (PTC). PTC is an economically viable and climate action-driven asexual technique very useful in the 21st century to promote, maintain and grow plant cells, tissues or organs under controlled sterile conditions. PTC is widely used to produce clones of a plant in a method known as micropropagation.

DOI: 10.1201/9781003265177-12

In this regard, the micropropagation holds significant promise for true-to-type, rapid and mass multiplication under disease-free conditions. White (1937) first time established successfully root culture of tomato. In 1941, Johannes van Overbeek used coconut milk for growth and development of young *Datura* embryos on a nutrient culture medium of known composition. This became a major breakthrough in the advancement of phyto-culture science and practice. In the Polar Regions, due to extreme variations in physical and geological conditions, the vegetation phenology is different and dominated by bryophytes like mosses, grasses, and sedges, symbionts like lichens, and small woody shrubs in most Arctic tundra. In polar deserts, dwarf variety of angiosperm is commonly present with fruiting and flowering characteristics. In PTC, both fundamental investigations of science and applied aspects of industrial research can be pursued with equal intensity. Gottlieb Haberlandt's (1902) concept of culturing plant cells *in vitro* and expecting vegetative cells to grow 'into embryos', which developed into the science of PTC, forms the basis of both fundamental and applied studies. The remarkable insight Haberlandt had and the prophecy he made can be appreciated by his statement: 'Without permitting myself to pose more questions, or to prophesy too boldly, I believe, in conclusion, that it will be possible to grow, in this manner, artificial embryos from vegetative cells. In any case, the method of growing isolated plant cells in nutrient solutions could be a new experimental approach to various important problems'. In fact, it has been the concept of 'totipotency', which was later demonstrated by Steward with carrot cells in 1958.

A fallacy that the PTC is empirical is largely based on ignorance of the fact that we have hardly understood morphogenesis so much so that the genetic make-up and the state of development of an individual have to decide the response that could be evoked by the morphogenetic stimuli, the suitability of which has to vary greatly in a biological system. For example, when the regeneration phenomenon is under genetic control and there are specific regeneration genes deciding the extent of ease with which a genotype can be or cannot be regenerated with particular morphogenetic stimuli, how can all plant species or genotypes behave similarly? For example, Oelack and Schieder (1983) have demonstrated that regenerant differentiation in legume calli is under direct genetic control, implying that selection of species and cultivars can be critical and more important than the choice of culture media. It has been demonstrated by Lazar and colleagues studying with seven addition lines, in which chromosomes from rye (*Secale cereal*) were incorporated into Chinese Spring wheat (*Triticum saestivum*) that rye chromosome 4 contains genes that control organogenesis in anther cultures and chromosomes 6 and 7 contain genes that control organogenesis in immature embryo cultures and there has been no correlation between organogenesis in anther culture and that in embryo culture (Lazar et al. 1987). A particular genotype of tobacco may be induced to regenerate simply by submerging its somatic callus in liquid nutrient medium with no supplement of growth substances, i.e. perhaps by creating O_2 tension or certain others by adenine or kinetin, whereas that of other plant species may require 6-benzylaminopurine (BAP) or of still others may require isopentenyladenosine (2iP) or zatin or thidiazuron (TDZ) and so on. To make the point more explicit, Gautheret (1948) did not succeed in regenerating shoots from cambial callus tissue cultures of *Ulmus* and *Salix*, since cytokinins were not discovered by that time. Likewise, certain human beings may be treated with sulpha drugs or penicillin, while others may develop deadly allergic reactions. Similarly, some persons especially of 'O' blood group may have an enzyme of disintegrate sulphone drug, depsone (DDS), making it ineffective for *Mycobacterium leprae*, while in others it has been found very effective in controlling leprosy. There could be many such examples, but they do not go to prove that the science of medicine is empirical.

The discovery of kinetin by Miller et al. (1955) led to the convincing first demonstration by Skoog and Miller (1957), using tobacco callus, or regulation of organ formation, shoots or roots, by changing kinetin-auxin ratio, where more kinetin than auxin favours shoot formation, while the reverse favours root formation, which lays the foundation of study of morphogenesis *in vitro*. Similarly, the astonishing rate of multiplication of an orchid *Cymbidium* by meristem culture demonstrated by Morel (1960) forms the basis of micropropagation and precisely clonal multiplication. It cannot be over-emphasised that right from the very beginning PTC has been envisaged to solve applied problems. For example, White (1934) initiated tomato excised root culture actually to study multiplication of viruses in it and their elimination. Similarly, Morel and Martin (1952) undertook shoot meristem culture for virus elimination in *Dahlia* and later in *Cymbidium* (Morel 1960). The former investigations by White led to the fundamental discovery that vitamins B1 (thiamine) and B6 (pyridoxine) are the root growth factors (Robbins and Bartley 1937, 1939), while the latter investigations by Morel led to the most applied discovery of mericloning. Along another dimension, i.e. production of useful secondary metabolites by PTC, first significant studies were conducted through large-scale cultivation of cells or mainly tobacco from the late 1950s to the early 1960s at Pfizer, Inc. by Tulecke and Nickell (1959), but it took about 20 years for its industrial utilisation. Now with the technological advancements, increasing interest and being a futuristic technology in the challenging areas like Arctic and Antarctic tissue and cell culture led production of several high-value compounds through cell suspension cultures using fermentation technology is being investigated, while a few have already been produced.

It is now also known that in many cases, active biochemicals and secondary metabolites are produced in optimum and even higher quantities in cell cultures than by the plants, e.g. shikonin, ginsenosides, anthraquinones, ajmalicine and berberine, which are produced at the levels (% dry weight) of 20, 27, 18, 1 and 10 from cell cultures as compared to their concentrations in the respective plants, i.e. 1.5 (*Lithospermum erythrorhizon*), 4.5 (*Panax ginseng*), 0.3 (*Morinda citrifolia*), 0.3 (*Catharanthus roseus*) and 0.01 (*Thalictrum minor*).

PTC is an important facet of biotechnology, and in many cases, it becomes a limiting factor for the fruition of the goals of biotechnology. A general connotation of biotechnology is industrial utilisation of biological processes. As such, the examples that first come to mind are things like the use of yeast for fermentation or production of antibiotics, but if looked closely, there are several important ramifications of this powerful branch of biology. Broadly speaking, biotechnology can be differentiated into aspects of biomedicine and those of agriculture, including horticulture and forestry. There are some glaring examples of useful products in the medicinal biotechnology, like production of new drugs, vaccines and diagnostic procedures for detecting genetic disorders and, of course, a single example of production of human insulin from transformed bacteria cells (Johnson 1983) (employing recombinant DNA technology in *Escherichia coli*, achieved by Eli Lilly and Co., Indiana, USA, in 1978 and its production started by 1982) can outweigh the achievements of the agricultural biotechnology so far as the improvement of quality of life is concerned. But with regard to human welfare and survival, the agricultural biotechnology is more important, particularly in view of population explosion in the third-world countries. Biotechnology may further be distinguished between the area dealing with genetic modifications precisely molecular biology and the area where the principles of morphogenesis are utilised, i.e. tissue culture. However, tissue culture becomes the nerve centre of molecular biology as well in certain situations.

On the global scene and studying science for achieving novel things, the molecular biology approach is very attractive, but in the Indian context, the priorities are to be drawn keeping in view the financial resources, time period involved and the extent of dependence on import that can be permitted or sustained. The priorities of developed countries, like the USA, can hardly be matched with that of a developing country like ours. Transgenosis can render wonderful and very useful results, like those of transgenic cotton recently produced by introducing in it the insect control protein gene of *Bacillus thuringiensis* by Monsanto Company, USA, which is resistant to lepidopteran insect pests, which take a heavy toll of cotton produce in field. Likewise, if maize and some other field crops, like *Cajanus cajan* and *Cicer arietinum*, can be made resistant to such insects; their yields may tremendously be improved. And if, like the transgenic tomato with enhanced firmness and shelf life, produced by Calgene, Inc., USA, by incorporating a bacterial 1-aminocyclopropane-1-carboxylate (ACC) deaminase gene, which limits the ethylene production, the mango fruit is also transformed, it will be of immense commercial value. However up to 1991, though 395 transgenic plants have been field released, the list does not include any from India (Dale et al. 1993). But, with this impressive list, a parallel disappointing phenomenon of inactivation of transgenes following propagation and field cultivation is also at work, leading to the loss of engineered traits (Finnegan and Mc Elory 1994). Thus, the situation is not so rosy as once thought, but on the contrary is alarming in view of the News recently published in *Nature* (Mc Cilwain 1996): 'A large question mark is hanging over the effectiveness and safety of one of the first genetically-engineered crops to be extensively planted, after cotton bollworms were found to have infested thousands of acres planted with the new breed of cotton in Texas', i.e. *Bt cotton*. Besides, production of transgenic plants, despite spending enormous amount of money and time and employing sophisticated infrastructure, may not be possible in certain other cases because of several intricacies involved in the process. Leaving insect resistance, most of the useful traits are polygenic. The frustrating experience on this account faced in respect of *nif*-gene transfer to non-leguminous plants is of near past only. Techniques are not available to effect integration of many genes in the plant genome. Besides poor acceptability of bacterial genes by higher plants, there may be several other problems, like if multi-copies of single genes are incorporated at random and at multiple sites, it will certainly adversely affect the activity of other vital genes. Furthermore, genes for a multi-genic trait must be integrated at specific loci in order to have position effect. While such problems are yet to be resolved, the very basic requirement of identification of genes for specific traits and their cloning in most of the polygenic traits are not yet accomplished and may take several years before being solved. Thus, in the national context, first the concerted efforts need to be made where the problems are practicable and objectives realisable immediately or in the near future for economic growth of the country, while the strides in long-term realisable problems of biotechnology requiring much basic research may be taken up later. The creation of blue rose, red tuberose and yellow petunias may be extremely fascinating, but commercial multiplication of ornamentals is definitely down to earth proposition, on which the economy of a commercially advanced country, like the Netherlands, is based. Certainly, the second approach has to be a priority in the Indian context too, if we have to differentiate between the facts and fantasies.

Though it is true that micropropagation is still the most commercially applicable aspect of tissue culture, there are some other facets too which are of immense practical value, like embryo rescue, *in vitro* pollination and intra-ovarian fertilisation, meristem culture for production of clean stocks, androgenesis and gynogenesis for haploid production, endosperm culture for triploid production, induced nucellar polyembryony, soma clonal variation, somatic hybridisation, synthetic seed production, germplasm preservation for conservation of phyto-diversity and, of course, also production of active principles. The viewpoint expressed above may be appreciated by knowing some instances of achievements in these various aspects.

12.2   MICROPROPAGATION

Application of plant cell, tissue and organ culture has proven its potentials for the practical application in the improvement of important and threatened medicinal plants. PTC represents the most promising areas of application at present time and giving an out look into the future. The areas range from micropropagation of ornamental and forest trees, production of pharmaceutically interesting compounds and plant breeding for improved nutritional value of staple crop plants, including trees to cryopreservation of valuable germplasm. All biotechnological approaches like genetic engineering, haploid induction or soma clonal variation to improve traits strongly depend on an efficient *in vitro* plant regeneration system.

Multiplication of plants under micropropagation can be affected by inducing callogenesis, i.e. shoot bud differentiation and somatic embryogenesis in explants. Callogenesis can be adventitious, without intervening callusing, at the cut end of an excised shoot tip as well as in the axil of nodal stem explants, or it can be in callus tissue; the former results into direct regeneration and clonal multiplication, whereas the latter results into indirect regeneration, which may be of non-clonal nature. The somatic embryogenesis is advantageous since the embryoids possess a ready-made root system, while in regenerated shoots, rooting is to be induced. The embryoids differentiated from somatic tissue directly develop into plants without intervening callus phase. Very young zygotic embryos of homozygous plants are expected to produce clonal plants.

12.2.1   SOMATIC EMBRYOGENESIS VS. CALLOGENESIS

Regeneration via somatic embryogenesis or callogenesis is generally species and/or explants specific (Chaturvedi and Mitra 1975). For example, the vegetative explants of carrot produce somatic embryos, while those of tobacco differentiate shoot buds under the same morphogenetic stimuli. Similarly, tissues of vegetative parts of *Citrus* differentiate through callogenesis, while those from reproductive parts differentiate via somatic embryogenesis under the influence of similar growth substances (Chaturvedi and Sharma 1988). The aspect of somatic embryogenesis, particularly to understand the factors governing this phenomenon, is of current interest, so that somatic embryogenesis could be induced at will in plant species which do not regenerate through this pathway (Thorpe 1995). The role of arabinogalactan proteins in somatic embryogenesis (Kreuger and Van Holst 1993) and that of polar transport of auxin in establishing bilateral symmetry at an early stage of embryo development (Liu et al. 1993) are yet to be established.

12.2.2 SOME EXAMPLES

Orchid micropropagation is the best example of tissue culture applications in the applied field that, within a period of a decade of its initiation (Morel 1964), revolutionised the orchid multiplication and got stabilised as a multi-million-dollar trade in such small countries like Thailand and Singapore. Tremendous scope in this field exists in India, which is extremely rich in orchid flora. Besides orchids, there are a number of ornamentals valued for their flowers and foliage now being commercially multiplied throughout the world, where the Netherlands remains at the top. Tissue culture of ornamentals is more lucrative than other plants and that of foliage ornamentals still more, since their individual specimens are quite costly, even Rs. 1000 per plant in case of *Anthurium scherzerianum*. A number of ornamentals, like orchids (*Vanda hybrid, Dendrobium chrysotoxum, Rhynchostylis retusa*), *Chrysanthemum morifolium, Petunia hydrida, Rosa hybrid, Gladiolus, Amaryllis, Peperomia obtusifolia, Ficus elastic, Bignonia chamberlaynii, Monstera pinnatifida* and *Bougainvillea*, have been multiplied by tissue culture, albeit to different extents (Chaturvedi et al. 1988, 1995). Similarly, a number of medicinal plants, namely, *Dioscorea floribunda, D. deltoidea* (the richest source of raw material for steroidal drug production), *Rauvolfia serpentine, Rosmarinus officinalis, Atropa belladonna, Digitalis purpurea* and *Chrysanthemum cinerariaefolium*, have been multiplied at a very fast rate. PTC has been advantageously used for multiplication of a number of fruit plants. Citrus production in Spain has been remarkably increased by the use of tissue culture, precisely through micrografting (Navarro and Juarez 1977).

Recently, success has been achieved also in micropropagating forest trees, albeit they pose several difficulties as compared to herbaceous plants. Nevertheless, some woody plants, like *Eucalyptus, Pinus radiate*, oil palm, rhododendrons, etc., are commercially multiplied. The success in proliferating the shoots of *Shorea robusta in vitro* also deserves special mention, since this tree is most intractable-to-multiply even in nature. Multiplication of forest trees at a rapid rate for afforestation is imminent in view of the mass deforestation obtained all over the world. Worldwide many plant species are threatened with extinction because of the gradual disappearance of the terrestrial natural ecosystem caused by various human activities; often this is due to the clearing of indigenous vegetation for agriculture and the resulting erosion. Apart from salinisation and invasion by alien species, more recently, climate change is looming as a significant new threat (Varshney and Anis 2012). Besides this demographic change, the growing size of population of the world and increasing urbanisation have had and will continue to have a major impact on forest cover and condition (Kataria et al. 2013). By a rough estimate, the forest cover has been reduced by 55% over the years. In the USA alone, there is a need of planting 1.5 million trees per day (Durzan 1985). Such a high demand cannot be fulfilled without using PTC.

12.2.3 COMMERCIAL MICROPROPAGATION

In recognition of the great potential of micropropagation, it is indeed heartening that in our country also, the wave of Tissue Culture Revolution is spreading rapidly. Commercialisation of tissue culture in India started in the year 1987 by M/s A.V. Thomas and Company, Cochin. Its success with tissue culture of cardamom and orchids and their export was an encouragement for other firms, like Indo-American Hybrid Seeds, Bangalore, which followed the suit to create the capacity to produce 10 million plants annually. Commercial tissue culture is taken as synonymous to micropropagation of ornamentals, which is valid for earning profits, particularly in global market, but may be erroneous if considered in the interest of the survival of mankind on the planet. In the latter context, micropropagation of plants necessary for restoring the green cover on earth should be given top priority. But, how to make it commercial is a big question? There is going to be a global demand of over 2 billion plants by the turn of the century against around 500 million plants produced presently by micropropagation and the market for it is expected to be U.S. $ 50billion/ year. There are about 100 commercial companies of the capacity for producing more than a million plants, of which in India there are hardly 6. So far, the commercial micropropagation is confined to ornamentals, but a trend towards fruit plants has already set in. Covered area under floriculture in the world is about 20,000 ha. There are some startling figures which at one hand expose our miserable status in the world market of about U.S. $40 billion of hardly having 0.1% contribution, but on the other hand assure of a great scope for expanding micropropagation of ornamentals. The leading countries in commercial tissue culture trade are the Netherlands, the USA, Japan, Israel, France, Germany, Columbia, etc. However, per capita consumption of cut-flower and other floral products is maximum in Norway followed by Switzerland, Denmark and Austria, where it runs into U.S. $100–130. The cost of production/ ft^2 in Europe is around Rs. 350 as compared to Rs. 75 in India. Unfortunately, with all the seasonal and agro-climatic and area-wise blessings obtained in India, where though about 35,000 ha land is under floriculture, its world share remains only 0.1% in contrast with the Netherlands, where the area under floriculture is only 3600 ha, but its share is as much as 67%. The Indian government has announced certain incentives and concessions for promoting the export of floriculture with the target of about Rs. 100 crores in 1995–1996. The main reason of India lagging even behind the small countries is that so far there is no self-reliance worth the name in respect of exploitation in indigenous technology. Whatever trade in micropropagation exists in the country is mostly based on tie-ups of Indian companies with the multinationals in the Netherlands, Israel, France, etc. How to change the scenario is again a major question, for which new strategies are to be drawn without losing any more time and ground. With the renewed interest in herbal drugs, its market in the USA alone is expected to be of about US \$47 billion per year by the turn of this century; however, the role played by biotechnology in this area so far is negligible. This makes rapid clonal multiplication of medicinal plants imperative, in which respect PTC is indispensable. Besides, it also ensures production of high-yielding plants all the year round unaffected by seasonal or ecological barriers. The plants so produced yield standard raw materials in respect of both quality and quantity of their active principles.

12.3   EMBRYO RESCUE AND *IN VITRO* POLLINATION AND FERTILISATION

A very practical and useful approach of immediate application is the embryo rescue, by which new hybrids even through wide crosses have been created, which otherwise are unexpected to be produced in nature. In fact, embryo culture is amongst the oldest applications of tissue culture when Laibach (1925) successfully created hybrid seedlings from a cross between *Linum perenne* and *L. austriacum*. A new crop, rich in protein triticale, owes its existence to embryo rescue of the cross between *Triticum durum* and *Secale cereal* (Hulse and Spurgeon 1974). Likewise, there are a number of examples where wide crosses in cotton, legumes, medicinal plants, etc., have been successfully utilised to create useful hybrids and variation. In our laboratory, seed sterility in *Rauvolfia serpentine* due to degeneration of endosperm has been overcome by embryo culture from young seeds. In a more or less similar approach, the *in vitro* pollination and fertilisation have resulted in intergeneric hybrids, which have been unknown in nature, involving plant species of Solanaceae, Caryophyllaceae, Cruciferae, etc. (Zenkteler 1984).

12.4   SOMATIC HYBRIDISATION

Contrary to embryo rescue and *in vitro* fertilisation, the most expensive research of somatic hybridisation pursued for nearly 30 years could not yield any worthwhile parasexual hybrid, which could not be produced in nature or by other methods. Such investigations must have consumed millions of dollars before being given up few years ago. The anomalous situation can be assessed by the fact that somatic hybridisation involving *Nicotiana glauca* $(2n=24) \times N.$ *longsdorfii* $(2n=18)$, worked by three groups (Carlson et al. 1972; Smith et al. 1976; Chupeau et al. 1978), has been reported to yield parasexual hybrids having variable chromosome numbers (28–183) in each case and only occasionally they showed the sum total of chromosome numbers of the two species involved. That is why, the chapter of parasexual hybridisation has now been closed for creation of novel hybrids amongst the plant species or genera, which are genetically incompatible. However, its role in creating cybrids, transfer of cytoplasmic male sterility and, to some extent, imparting disease resistance still appears valid. To give a few examples, a parasexual hybrid, *Brassica naponigra* produced from protoplast fusion of *B. napus* and *B. nigra* is resistant to infection of *Phoma lingam* and the parasexual hybrid of potato involving *Solanum tuberosum* and *S. brevidens* is resistant to several viral diseases. Similarly, glaring examples of producing disease-resistant plants are in the cases of maize and tobacco. The cells were screened against resistance to methionine sulphoximine, added to the medium, which is similar in effect to the toxin produced by *Helminthosporium* in case of maize and *Pseudomonas tabaci* in the case of tobacco. Such procedures are definitely simpler, realisable within a short time and far less expensive than transgenosis.

12.5 SHOOT MERISTEM CULTURE FOR PATHOGEN ELIMINATION

Elimination of viruses and other pathogens through shoot meristem culture is one of the most established and trusted methods. The clean stocks are more high-yielding than their diseased counterparts. The tremendous increase in *Citrus* production through *in vitro* micrografting in Spain is an event of the near past (Navarto and Juarez 1977). And to cite the earliest instance, the potato varieties King Edward and Arran Victory have been freed from viruses through shoot meristem culture and thus saved from extinction (Kassanis 1957). Remarkable success has been achieved in our laboratory in regenerating excised shoot meristems, measuring 1 mm in length, of field-grown trees of *C. aurantifolia*, for the first time, and in raising plantlets of *C. aurantifolia* and *C. sinensis* through micrografting of meristems measuring less than 1 mm in length.

12.6 ANDROGENESIS AND GYNOGENESIS FOR HAPLOID PRODUCTION

Production of haploids through androgenesis or in certain cases through gynogenesis cannot be over-emphasised for their role in creation of new improved varieties. Recently, a treatise devoted to *in vitro* haploid production has been published in three volumes, of which one is already out of press (Jain et al. 1996). Its significance is immense in the case of perennial crops, like trees having long reproductive cycles, and in the case of outbreeding plants that too with seedless character for making homozygous plants in respect of desired genetic characters. Haploids produced through anther culture in rubber, horse chestnut, *Populus*, and litchi are some examples in trees besides some other woody plants like grape and tea. But the breakthrough in creating new improved varieties of rice, wheat and tobacco achieved in China is a real success story in this respect (Hu 1978). In our laboratory, androgenic plants of *Citrus aurantifolia* have been raised and grown in soil (Chaturvedi and Sharma 1985). Haploids can also be produced by the chromosome elimination method, also called as bulbosum method. By crossing *Hordeum vulgare* with *H. bulbosum*, the haploid *H. vulgare* is produced, while the chromosome complement of *H. bulbosum* is eliminated.

Likewise, crosses between *Nicotiana tabacum* and *N. Africana*, *Solanum tuberosum* and *S. phurjea*, *Medicago sativa* and *M. falcate*, and *Triticum aestivum* cv. Solamon and *Aegilops* result into haploids of the former species by chromosome elimination procedure. Although there are fewer examples of haploid production through gynogenesis than by androgenesis, these include such economic plants as *Triticum aestivum*, *Oryza sativa*, *Solanum tuberosum* and *Hevea brasiliensis* (Yang and Zhou 1990).

12.7   ENDOSPERM CULTURE FOR TRIPLOID PRODUCTION

Triploids have at least two outstanding traits compared to their diploid counterparts, viz. more vigour and seedlessness of fruit (Elliott 1958). Triploids of many fruit crops are already in commerce because of their seedless fruits, namely, grape, banana, apple, mulberry, watermelon, etc. Triploids of a few plant species have been raised through regenerant differentiation in endosperm culture. The first triploid has been formed in *Putranjiva roxburghii* (Srivastava 1973), while some other prominent examples are *Prunus persica* (Shu-quiong and Jia-qu 1980), *Pyrus malus* (Mu et al. 1977), *Citrus grandis* (Wang and Chang 1978) and *Citrus* hybrid (*C. reticulata* x *C. paradise*) (Gmitter and Ling 1991). In our laboratory, regeneration of shoots has been obtained from excised endosperm tissue of *Santalum album*, and efforts are being made to produce triploids of *Emblica officinalis* by endosperm culture with promising results.

12.8   INDUCED NUCELLAR POLYEMBRYONY

Induction of nucellar polyembryony is of immense practical value in case of a number of fruit trees where true-to-type disease-free plants are required, while cloning by other methods is far more difficult as even if somatic embryos are differentiated, they pose difficulty in producing high-frequency normal plants. And if somatic embryos are not produced, the adventitiously rooted cuttings are inferior in their performance. There are some examples in fruit trees, where nucellar polyembryony is either induced or augmented, such as *Vitis vinifera*, *Malus domesticum*, *Eriobotrya japonica* (loquat), *Eugenia jambos* and *Carica papaya* (Thorpe 1995). Enormous work has been done on nucellar polyembryony in Citrus in our laboratory and abroad. Nucellars of polyembryonic species, *C. aurantifolia* and *C. sinensis*, and of a mono-embryonic species, *C. grandis*, and of root-stocks, like *C. karna* and *C. jambhiri*, have been produced and grown in field conditions (Chaturvedi and Sharma 1988). Success has been achieved in obtaining proliferation and development of nucellar embryos of *Mangifera indica*, both scion and root-stock varieties (Chaturvedi et al. 1994).

12.9   SOMACLONAL VARIATION

Somaclonal variation for creating variation through somatic cell cultures has been introduced as a big hope (Larkin and Scoweroft 1981). There may be several causes of somaclonal variation besides involving chromosome and nuclear DNA; it may be of epigenetic nature as well. However, when variation is caused by transposable elements or demethylation of DNA, there are great chances of reversions, but still there are hopes of crop improvement by this phenomenon (Semal and Lepoivre 1990). In any case, so far a number of somaclonal variants have been produced in several important crop plants, like sugarcane, potato, tobacco, etc. In sugarcane, besides creation of new varieties in terms of yield, certain varieties resistant to diseases, like eye spot disease (by *Helminthosporium sacchari*), Fiji disease (by a virus transmitted by aphids) and downy mildew (by *Sclerospora sacchari*), have also been produced. Similarly, in potato, the variants were produced in respect of both characters of tuber and resistance to early blight (by *Alternaria solani*) and late blight (by *Phytophthora infestans*). Amongst the ornamentals, a new cultivar of geranium known as "Velvet Rose" has been created from callus cultures. The new cultivar differs from the original cultivar 'Rober's Lemon Rose' in having symmetrical flowers and seedlings. In our laboratory also, several stable variants of *Dioscorea floribunda* and one of *Solanum surattense* have been created and the one in *Citrus sinensis* is being evaluated.

12.10   SYNTHETIC SEED PRODUCTION

Synthetic seeds produced by encapsulating somatic embryos or other regenerants may prove to be of great practical value both in propagation and storage of germplasm (Redenbaugh et al. 1986; Redenbaugh 1993). Synthetic seeds if developed in crops, like sugarcane, may revolutionise its cultivation, as the stem pieces in storage are not only difficult to manage because of their big size but are also susceptible to several fungal diseases. In our laboratory, the basal pieces (3–5 mm in length) of *in vitro* regenerated and proliferating tillers when encapsulated in calcium alginate have been made to regenerate multiple shoots on morphogenetic medium. Microtubers in potato, microcorns in gladioli and protocorms in orchids have been likewise encapsulated to produce synthetic seeds, which on storage at 4°C for several months have shown normal germination.

12.11   GERMPLASM PRESERVATION AND CONSERVATION

Germplasm or the theory of inheritance was first described by a German scientist Weismann (1834–1940) for living organisms. Germplasm conservation is the most successful method to conserve the living genetic traits of endangered and commercially valuable species. This is known as germplasm bank, seed bank or gene bank where it can be preserved for centuries without losing the trait. Germplasm is a live information source for all the genes present in the respective plant, which can be conserved for long periods and regenerated whenever it is required in the future.

The application of micropropagation and tissue culture for germplasm preservation in conservation of phyto-diversity is one of the emerging areas of paramount significance for human welfare by establishing "Gene Banks" for climate action. Of late, there is global awakening on this score. The tissue culture strategy is essential in the case of hybrids, which must be propagated vegetatively and, in those cases, where either the seeds recalcitrant or not produced or the plant material is very limited.

Cryopreservation (freezing-drying) of plant propagules, like shoot meristems and somatic embryos, at very low temperatures using cryoprotectants like glycerol, mannitol, dimethyl-sulfoxide (DMSO) and proline is used to store plant tissues. These cryoprotectants prevent damage caused to cells by freezing or thawing. But this has not yielded satisfactory results commensurate to the efforts made, time spent and resources utilised. The main drawbacks of the process are being low recovery of plants and loss of regeneration potential within few weeks or months of storage.

There are only a few examples to cite where the plants have been stored for two-and-a-half years, viz. strawberry and pea (Kartha 1985).

Under laboratory conditions, two approaches have been successfully examined for long-term germplasm preservation. First, by limited growth cultures of shoots of a number of economic plants, such as *Dioscorea floribunda, D. deltoids, Solanum khasianum, Costus speciosus, Rauvolfia serpentine, Chrysanthemum cinerariaefolium, Atropa belladonna, Rosmarinus officinalis, Solanum tuberosum, Chrysanthemum morifolium*, orchids, *Rosa hybrid, Ficus elastic, Simmondsia chinensis*, bamboos, *Glycyrrhiza glabra*, etc. In the second approach which is novel, the plant species are preserved in terms of their root cultures, which may be induced to regenerate, after several years of their establishment, normal plants. The root-regenerated plants have been grown in soil under field conditions. The method has been successfully tested in *S. khasianum* (spiny and spineless), *S. torvum, A. belladonna, Kalankoe fedtschenkoi* and, to some extent, also *R. serpentine* (Chaturvedi et al. 1991). Also, a tree species, like *P. deltoides*, has been regenerated from more than 3-year-old excised root cultures and normal plantlets raised. Recently, success has been achieved in inducing somatic embryogenesis in root explants of *Elaeis guineensis*, which may prove to be a breakthrough for conservation of germplasm of this important tree, a major source of oil, which cannot be cloned by any other means. Generally, the period of preservation of germplasm of these plant species ranges from 4 to 16 years, tested so far.

12.12   PRODUCTION OF ACTIVE PRINCIPLES

This important commercial aspect of tissue culture for producing active principles or drugs from cultures of plant tissues in bio-reactors has, of late, taken strides mainly in Japan involving private companies (Morris et al. 1986). A vegetable dye shikonin used for making harmless lipsticks and other cosmetics besides being used as a natural antiseptic agent is commercially produced by Mitsui Petrochemical Industry Co. Ltd., Japan, from cultures of cells of *Lithospermum erythrorhizon*. Similarly, the elixir of life is produced from large-scale culture, in 20 kL fermenters, of roots and cells of ginseng (*Panax ginseng*) by Nitto Denko Co. Ltd., Japan. Berberine, digitoxin and codeine are some other drugs, which are near to commercial production through suspension culture of cells of *Coptis japonica*, foxglove and poppy, respectively.

Commercial production of vanilla flavour from cell cultures of *Vanilla fragrance* by Escagenetics and Sanguinarine from cell cultures of *Papaver somniferum* by Vipon Research Laboratories, at the behest of Colgate Company, have been undertaken. Several firms and academic institutions are engaged in production of taxol, a very promising antitumour compound from cell cultures of *Taxus brevifolia* and *T. baccata* spp. *wallichiana* (an Indian species).

Similarly, commercial production of vinblastine, one of the most potent anticancer compounds, costs 5 million U.S. $/kg, employing a process by combining the cell cultures of *Catharanthus roseus* and a chemical coupling reaction. In laboratory, a number of active principles, including alkaloids, steroids, a cardenolide and an essential oil, have been produced from *in vitro*-grown plant tissues and organs (Chaturvedi 1979; Jain et al. 1991). Diosgenin and solasodine, the base materials for producing steroidal hormones, have been biosynthesised from somatic calli of *Bioscorea deltoidea* and *Solanum khasianum*, respectively, whereas atropine and hyocyamine have been biosynthesised from excised root cultures of *Atropa belladonna*. A correlation between organogenesis and enhanced degitoxin biosynthesis has been demonstrated in seedling callus of *Digitalis purpurea*, while essential oil, rosemary oil, has been obtained from *in vitro*-proliferating shoots of *Rosmarinus officinalis*.

12.12.1 CONSTRAINTS IN MICROPROPAGATION

Micropropagation has been extensively used for the rapid multiplication of many plant species, but the ultimate success of *in vitro* propagation depends on the ability to transfer plants out of culture on a large scale at low cost and with high survival rates (Hazarika 2006; Xiao et al. 2011). Plantlets or shoots that are grown *in vitro* have been continuously exposed to a unique microenvironment that has been selected to provide minimal stress and nearly optimum conditions for plant multiplication (Hazarika 2006). The high cost of tissue production is a drawback for laboratories with limited resources. Moreover, higher production cost has been an impediment to tissue culture adoption in developing countries (Sahu and Sahu 2013). Besides the cost, the reproducibility and efficiency of protocol are also of much importance in commercial micropropagation. Some of the problems associated with micropropagation include the following.

12.12.2 BROWNING OF CULTURED TISSUE

This phenomenon results from physiological changes within the cultured tissues that lead to gradual browning and eventual death of tissue. The browning appears due to the oxidation of phenols within the tissues. The problem of phenolic browning can be minimised to a great extent by leaching of phenolic compounds due to agitation in an antioxidant solution and by proper drying of explants prior to inoculation (Meghwal et al. 2001).

12.12.3 CONTAMINATION

Contamination of cultures is a serious problem that not only reduces the frequency of shoot culture initiation from the source explants but also reduces the total number of shoots produced at various cycles due to loss of cultures. Bacteria and fungi are the frequently encountered contaminants in culture of many plant species particularly in large-scale commercial operations (Young et al. 1984; Skirvin et al. 1999). Such microorganisms are either present in the explant or arise as laboratory contaminants both natural and man-made (Shields et al. 1984). Contamination caused by some common microorganisms can result in large losses during micropropagation, and their control is the most serious problem encountered in many commercial laboratories.

Regardless of the application of PTC, it is essential that the culture should be established *in vitro* free of biological contamination and be maintained as aseptic cultures during manipulation growth and storage.

12.12.4   VARIABILITY IN CULTURE (SOMACLONAL VARIATION)

Tissue culture is recognised as one of the key areas of biotechnology for large-scale propagation and conservation, but it is severely hindered due to somaclonal variations (Larkin and Scowcraft 1981). *In vitro* culture has long been associated with the occurrence of somaclonal variation ranging from moderate to strikingly high rates.

12.12.5   VITRIFICATION OF SHOOTS

Hyperhydricity, which is characterised by a glassy or swollen appearance to the tissue, usually results in reduced multiplication rates, poor-quality shoots and tissue necrosis (Ziv 1991). High relative humidity, poor gaseous exchange between the internal atmosphere of the culture vessel and its surrounding environment, and the accumulation of ethylene may induce hyperhydricity during micropropagation. This physiological malformation is associated with chlorophyll deficiency, poor lignification and excessive hydration of tissues, which result in malformed plantlets that cannot withstand *ex vitro* conditions after transplanting (Hazarika 2006).

12.12.6   MORPHO-PHYSIOLOGICAL AND BIOCHEMICAL STATUS

The special conditions during *in vitro* culture result in the formation of plantlets of abnormal morphology, anatomy and physiology. After *ex vitro* transfer, these plantlets might easily be impaired by sudden changes in environmental conditions and so need a period of acclimatisation to correct the abnormalities (Pospisilova et al. 1999). The environment inside the culture vessels, normally used in micropropagation, is characterised by high relative humidity low photon flux density (PFD) and poor exchange of gases between the internal atmosphere and its surrounding environment which may cause morpho-physiological disorders. During transfer to *ex vitro* conditions, *in vitro* plants are exposed to higher PFD and lower air humidity, which can induce photoinhibition and water stress accompanied by production of reactive oxygen species.

12.12.7   FUTURISTIC APPLICATIONS

There are different strategies for development of better micropropagation technologies and for restoration of abnormalities developed during *in vitro* culture. They include the use of liquid culture system, replacing agar with other low-cost gelling agents, growing cultures under CO_2-enriched environment, improvement in culture vessel environment by ventilation, use of salicylic acid, endosperm and ovary culture. Such approaches allow production of microclones which are comparatively cheaper, better, vigorous and low temperature resistance and can easily be grown under the changing climate of the Polar Regions.

12.13   SUMMARY

From a biological perspective, there is no universally accepted definition of the Arctic, but Arctic plants are generally considered to be those living in Tundra and

Polar deserts beyond the northern climatic limits of forests, i.e. generally north of the boreal zone. The boundary between boreal forests and the Arctic is often broad and ambiguous. Arctic plants exist along a global continuum of decreasing floristic diversity with increasing latitude. This gradient starts well outside of the Arctic and continues within the Arctic to the northernmost reaches of land.

Arctic plants come in a wide variety of forms. Mosses, lichens and low-growing woody and herbaceous perennials characterise Arctic vegetation. Trees, succulents, ferns and annual plants are rare or absent from most Arctic plant communities. Combinations of mosses, lichens, sedges, grasses and dwarf woody shrubs domi-nate most Arctic tundra, and miniature flowering plants dominate the polar des-erts. Adaptations of Arctic plants to cold and short growing seasons as well as other aspects of their physical environment are evident in their morphologies, physiolo-gies and life histories. Arctic plants are also adapted to their biotic environment. Extremely low temperatures are less characteristic of the Arctic than they are of some other regions, but the Arctic is consistently cold, resulting in permafrost and direct and indirect environmental challenges to plants. During short growing sea-sons, Arctic plants utilise seasonally thawed soils above the permafrost and tolerate frozen soils in winter. Low temperatures affect the availability of mineral nutrients, frequently limiting the growth and productivity of Arctic plants. Usable soil is lim-ited by permafrost, and low temperatures retard soil genesis, microbial activity and uptake by roots. Birds and mammals play a key role in nutrient redistribution and the creation of local sites with high fertility.

Arctic vegetation patterns are closely correlated with moisture, and steep local moisture gradients are characteristic of the Arctic. Although the Arctic is climato-logically a desert, few Arctic plants experience water stress. Moisture affects thermal characteristics and oxygenation of soils, which in turn affects decomposition rates and the availability of mineral nutrients. Patterns of moisture are strongly influenced by topography due to the combined effects of low precipitation, low evaporation and water ponding due to permafrost. Mechanical stresses associated with freezing and thawing of soils and substrates shape the habitats of Arctic plants. Geomorphic pro-cesses unique to cold regions produce vegetational patterns and can lead to cyclic plant succession.

The Arctic zone climate is dynamic and different, that is, changing fast in the present context of climate change. It is very difficult, if not impossible, to monitor and document the effects of a changing climate on the Arctic due to the diversity of plants and habitats and due to non-linear interactions between environmental factors within Arctic ecosystems.

REFERENCES

Carlson PS, Smith HN, and Dearing RD. 1972. *Proc. Natl. Acad. Sci.* 69: 2292–2294.

Chaturvedi HC, and Mitra GC. 1975. *Ann. Bot.* 39: 683–687.

Chaturvedi HC, and Sharma AK. 1985. *Planta* 165: 142–144.

Chaturvedi HC, and Sharma AK. 1988. Plant Tissue Culture, In: *Proc. Natl. Seminar on Plant Tissue Culture*, ICAR, New Delhi, pp. 36–46.

Chaturvedi HC, and Sharma M. 1990. *Proc. VII Intl. Cong., Plant Tissue and Cell Culture*, Amsterdam, The Netherlands, p. 247.

Chaturvedi HC, and Sinha M. 1979a. *Indian J. Exp. Biol.* 17: 153–157.

Chaturvedi HC, and Sinha M. 1979b. *Mass Propagation of Dioscorea floribunda by Tissue Culture. Extn. Bull. No. 6*, EBIS, NBRI, Lucknow, p. 12.

Chaturvedi HC, Prasad RN, Sharma AK, Mishra P, Sharma M, Jain M, and Bhattacharya A. 1995. Advances in Horticulture, In: vol. 12, *Ornamental Plants.* Eds. K.L. Chadha and S.K. Bhattacharjee, Malhotra Publishing House, New Delhi, pp. 562–581.

Chaturvedi HC, Sharma AK, Prasad RN, and Sharma M. 1988. Plant Tissue Culture, In: *Proc. Natl. Seminar,* ICAR, New Delhi, pp. 76–90.

Chaturvedi HC, Sharma AK, Sharma M, and Sane PV. 1991. Conservation of Plant Genetic Resources through *In Vitro* Methods, In: *FRIM/MNCPGR,* Univ. Kebangsaan, Malaysia, pp. 29–41.

Chaturvedi HC, Sharma M, and Sharma AK. 1994. *Proc. 81st Ind. Sci. Cong. Pt. III (Bot.),* p. 183.

Chaturvedi HC. 1975. *Curr. Sci.* 44: 839–841.

Chaturvedi HC. 1979. Progress in Plant Research, In: vol. 1. Eds. T.N. Khoshoo and P.K.K. Nair, Today and tomorrow's Printers and Publishers, New Delhi, pp. 265–288.

Chupeau Y, Missonier C, Hommel MC, and Goujaud J. 1979. *Mol. Gen. Genet.* 165: 239–245.

Dale PJ, Trwin JA, and Scheffler JA. 1993. *Plant Breed.* 111: 1–22.

Dubey RK. 2013, *Indian J. Appl. Res.* 3(11): 18–21.

Durzan DJ. 1985. *Tissue Culture in Forestry and Agriculture.* Eds. R.R. Henke, K.W. Hughes, M.J. Constantin and A. Hollaender, Plenum Press, NY, pp. 233–256.

Elliott FC. 1958. *Plant Breeding and Cytogenetics.* McGraw Hill, NY.

Finnegan J, and McElory D. 1994. *Bio/Technology* 12: 883–888.

Gautheret RJ. 1948. *C.R. Soc. Biol.* 142: 807–808.

Gmitter FG, Jr. and Ling X. 1991. *J. Am. Soc. Hort. Sci.* 116: 317–321.

Haberlandt G. 1902. *Sitzber. Kais Akad. WIss, Math-natur wiss* I, 69–92.

Hazarika BN. 2006. *Sci. Hort.* 108: 105–120.

Hu H. 1978. *Proc. Symp. On Plant Tissue Culture.* Science Press, Peking, pp. 3–10.

Hulse JH, and Spurgeon D. 1974. *Sci. Am.* 231: 72–80.

Jain M, Banerji R, Nigam SK, Scheffer JJC, and Chaturvedi HC. 1991. *Planta Medica* 57: 122–124.

Jain, S.M., Sopory, S.K. and Veilleux, R.E. (Eds.) 1996. *In Vitro* Haploid Production in Higher Plants, In: vol. 1, *Fundamental Aspects and Methods,* Kluwer Academic Publishers, The Netherlands, p. 356.

Johnson, I.S. 1983. *Science* 219: 632–637.

Kartha KK. 1985. *Cryopreservation of Plant Cells and Organs,* CRC Press Inc., Boca Raton, pp. 116–134.

Kassanis B. 1957. *Ann. Appl. Biol.* 45: 422–427.

Kreuger M, and van Holst G. 1993. *Planta* 189: 243–248.

Kumar DR, and Kumar SA. 2015, *Eur. Acad. Res.* III(6): 6134–6157.

Kumar DR, and Kumar SA. 2015, *Int. J. Innov. Res. Sci. Eng. Technol.* 4(8): 7007–7007.

Laibach F. 1925. *Z. Bot.* 17: 417–559.

Larkin PJ, and Scowcroft WR. 1981. *Theor. Appl. Genet.* 60: 197–214.

Lazar MD, Chen THH, Scoles GJ, and Kartha KK. 1987. *Plant Sci.* 51: 77–81.

Liu C, Xu Z, and Chua N. 1993. *The Plant Cell* 5: 621–630.

Macilwain C. 1996. *Nature* 882: 289.

Meghwal PR, Sharma HC, Goswami AM, and Srivastava KN. 2001. *Indian J. Hort.* 58: 328–331.

Miller CO, Skoog F, Okumura FS, von Saltza MH, and Strong FM. 1955. *J. Am. Chem. Soc.* 77: 2662–2663.

Morel G, and Martin C. 1952. *C.R. Acad. Sci.* 235: 1324–1325.

Morel G. 1960. *Am. Orchid Soc. Bull.* 29: 495–497.

Morel G. 1964. *Am. Orchid Soc. Bull.* 31: 473–477.

Morris P, Scragg AM, Stafford A, and Fowler MW. (Eds.) 1986. *Secondary Metabolism in Plant Cell Cultures*, Cambridge Univ. Press, Cambridge.

Mu S, Liu S, Zhou Y, Quan N, Zhang P, Xie H, Zhang F, and Yen Z. 1977. *Sci. Sin.* 55: 370–376.

Navarro L, and Juarez J. 1977. *Acta Hortic.* 78: 425–435.

Oelck MM, and Schieder O. 1983. *Z. Pflanzenzucht.* 91: 312–321.

Pospisilova J, Ticha I, Kadlecek P, Haisel D, and Plazakova S. 1999. *Biol. Plant.* 42: 481–489.

Redenbaugh K, Paasch BD, Nichol JW, Kossler ME, Viss PR, and Walker KA. 1986. *Bio/Technology* 4: 797–801.

Redenbaugh K. 1993. *Synseeds, Applications of Synthetic Seeds to Crop Improvement*, CRC Press Inc., Boca Raton, p. 481.

Robbins WJ, and Bartley MA. 1937. *Science* 86: 290–291.

Robbins WJ, and Bartley SM. 1939. *Prof. Natl. Acad. Sci.* 25: 1–3.

Sahu J, and Sahu RK. 2013. *J. Pharm. Biosci.* 1: 38–41.

Semal J, and Lepoivre P. 1990. *Plant Tissue Culture: Applications and Limitations*, Ed. S.S. Bhojwani, Elsevier Science Publishing Co. Inc., The Netherlands, pp. 301–315.

Shields R, Robinson SJ, and Anslow PA. 1984. *Plant Cell Rep.* 3: 33–36.

Shu-quiong L, and Jai-qu L. 1980. *Acta Bot. Sin.* 22: 198–199.

Skirvin RM, Motoike S, Norton MA, Ozgur M, Al–Juboory K, and McMeans OM. 1999. *In Vitro Cell. Dev. Biol.-Plant* 35: 278–280.

Skoog F, and Miller CO. 1957. *Symp. Soc. Exp. Biol.* 11: 118–131.

Smith HH, Kao KN, and Combatti NC. 1976. *J. Hered.* 67: 123–128.

Srivastava PS. 1973. *Z. Pflanzenphysiol.* 69: 270–273.

Steward FC. 1958. *Am. J. Bot.* 45: 709–713.

Thorpe A. (Ed.) 1995. *In Vitro Embryogenesis in Plants*, Kluwer Academic Publishers, The Netehrlands, p. 558.

Tulecke W, and Nickell LG. 1959. *Science* 130: 863–864.

Wang T, and Chang C. 1978. *Proc. Symp. on Plant Tissue Cutlure*, Science Press, Peking, pp. 463–467.

White PR. 1934. *Phytopath* 24: 1003–1011.

Xiao Y, Niu G, and Kozai T. 2011. *Plant Cell Tiss. Org. Cult.* 105(2): 149–158.

Yang HY, and Zhou C. 1990. *Plant Tissue Culture: Applications and Limitations*, Ed. S.S. Bhojwani, Elsevier Sci. Publishing Co. Inc., The Netherlands.

Zenkteler M. 1984. Cell Culture and Somatic Cell Genetics of Plants, In: vol. 1, *Laboratory Procedures and their Applications*, Ed. I.K. Vasil, Acad. Press Inc., NY, pp. 269–275.

13 Bio-Optical Characteristics in Relation to Phytoplankton Composition and Productivity in a Twin Arctic Fjord Ecosystem during Summer

Sarat Chandra Tripathy
National Centre for Polar and Ocean Research

CONTENTS

DOI: 10.1201/9781003265177-13

13.1 INTRODUCTION

Due to their sensitiveness to temperature, the Arctic fjords are considered as indicators of continuously warming global climate (Cohen et al. 2014). The warming climate is changing the Arctic oceanic ecosystems progressively to a more temperate condition (Vihtakari et al. 2018) which is popularly known as 'Atlantification'. The Kongsfjorden (KG)-Krossfjorden (KR) ecosystem, positioned at the interface of Arctic and Atlantic oceanic regimes, experiences physicochemical and biological variations, and hence is an ideal fjord system for studying warming-induced changes in the Arctic (Bischof et al. 2019). These high Arctic fjords are influenced by the warm West Spitsbergen Current (WSC) carrying saline Atlantic water (AW) and the coastal current that brings in cold and less saline Arctic water (Vihtakari et al. 2018). An amplified inflow of warm AW displacing the cool Arctic water has resulted in accelerated melting of the tidewater glaciers that terminates straight into the sea (Walczowski et al. 2017). The glacial meltwater discharge normally begins in June and ends around September with a peak during (summer) July–August (Darlington 2015). The meltwater run-off ladened with sediments and large quantity of mineral particles forms a strong freshwater and turbidity gradient in the water column from inner to outer fjords (Svendsen et al. 2002). Concurrent intrusion of AW from the outer fjord and, fresh, turbid water run-off from the glacial end results in sharp horizontal gradients in the thermohaline as well as underwater light environment (reduced euphotic layer) along the fjords affecting the phytoplankton assemblages and growth conditions that serves as the basis of marine food-web dynamics (Piquet et al. 2014; Calleja et al. 2017; Hegseth et al. 2019).

Previous study has shown significant differences in phytoplankton community and production between the geographically closely placed fjords of Spitsbergen that are influenced by the same water masses (Eilertsen et al. 1989). The last couple of decades witnessed detailed studies involving hydrography, nutrients, phytoplankton assemblages and/or optical parameters in Kongsfjorden (Svendsen et al. 2002; Hop et al. 2002, 2006; Piwosz et al. 2009; Hodal et al. 2012; Pavlov et al. 2014; Kulk et al. 2018; van de Poll et al. 2018; Halbach et al. 2019), and it is acting as a hotspot for research activities in the Arctic region. However, this cannot be said about the adjacently located Krossfjorden, which has geological resemblance with the Kongsfjorden. There are few observations in relation to phytoplankton assemblages, light regime, nutrients in relation to meltwater and hydrography in Krossfjorden (Piquet et al. 2014). Moreover, a comparative account between Kongsfjorden and Krossfjorden (Svendsen et al. 2002; Piquet et al. 2010, 2014; Singh and Krishnan 2019) is rare.

The quantitative studies of underwater light-fields (bio-optical properties) in oceanic/fjord waters have vital applications in ecology and remote sensing. For example, the underwater light-field plays an important role in determining photosynthesis rates of marine autotrophs including the phytoplankton and therefore can limit the phytoplankton productivity in oceanic ecosystems (Kirk 1994). Further, it has implications on visual interactions between predators-prey species, and determination of volume reflectance of seawater that is an essential input for optical remote sensing. The underwater light-fields show strong diurnal and seasonal variability, and these variations depend on (i) the optical characteristics of seawater, which is controlled by the in-water constituents such as by suspended particulate materials (phytoplankton, organic detritus and sediments) and coloured dissolved organic matter (CDOM), and (ii) the conditions of solar illumination that include solar angle, the nature of cloud cover and the state of the sea surface (Kirk 1994). The optical properties of waters are basically divided into two categories, i.e., inherent optical properties (IOPs) and apparent optical properties (AOPs). Magnitude of the IOPs (e.g., absorption and scattering coefficients) depends only on the characteristics of the water but not the radiation field (solar angle), whereas magnitude of AOPs (e.g., diffused attenuation, remote sensing reflectance) is a function of both optical characteristics of water (the IOPs) and the geometric (directional) structure of the radiance distribution. In terms of difficulties, the AOPs can be relatively easily measured as compared to IOPs and give useful and consistent/bulk information about the optical properties of a water body.

This chapter discusses the comparison of phytoplankton pigments, inorganic nutrients limitation, light-absorption coefficients and hydrography in these twin fjords of West Spitsbergen that are significantly influenced by the warm AW and rapid glacial melting. The study was carried out during summer (August 2016) when the glacial run-off is expected to be maximum (Darlington 2015). Documentation of the probable alteration that might occur in the bio-optical properties and phytoplankton pigment signatures as a consequence of increased 'Atlantification' and/or warming is quite sparse till date. Concerted efforts are put in place to investigate the variability in underwater light characteristics and its ecological consequences in these high Arctic fjords. It is hypothesised that environmental settings influence phytoplankton assemblages (size-class, diversity), photosynthetic (light-absorption) efficiency and biomass along and between the fjords through reduced underwater light intensities caused by elevated suspended matter load near to the glacial end. This study would further improve our understandings of the factors that control variability in phytoplankton biomass, composition (diversity), light-absorption efficiency and productivity in these two closely located fjords during summer.

13.2 MATERIALS AND METHODS

13.2.1 STUDY AREA AND SAMPLING

The Kongsfjorden-Krossfjorden is a dynamic glacial fjord ecosystem in the high Arctic situated between 78° 40′–77° 30′N and 11° 3′–13° 6′E on the west coast of Spitsbergen in the Svalbard archipelago. The Kongsfjorden (KG) is oriented from south-east to north-west, while Krossfjorden (KR) is oriented from north to south. KG is 20 km in length and its width varies from 4 to 10 km, whereas KR is 30 km long and its width varies between 3 and 6 km. This double fjord ecosystem shares a common mouth located at 79°N and 11°E, and is influenced by the freshwater discharge from surrounding glaciers, which is minimal in winter and maximum during summer season (Svendsen et al. 2002). Intrusion of relatively warm water brought in by the WSC at irregular intervals changes the water mass within these fjords from Arctic predominance during winter to Atlantic predominance during summer season (Svendsen et al. 2002), thereby strongly influencing the physicochemical and biological properties of the ecosystem inside the fjords (Hop et al. 2002).

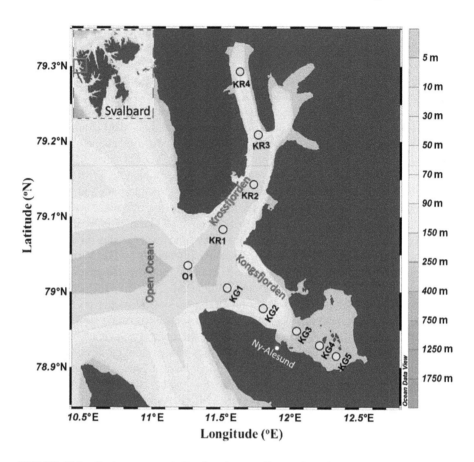

FIGURE 13.1 Study area map indicating the sampling stations (orange-coloured circles) in the Kongsfjorden (KG) and Krossfjorden (KR). Background colour indicates the bathymetry (m), and the inset picture shows a larger perspective of the study area in Svalbard.

For this study, ten stations were sampled along the fjords allowing a comparison of variables in the meltwater gradient. Five in Kongsfjorden (KG1, KG2, KG3, KG4, KG5) and four in Krossfjorden (KR1, KR2, KR3, KR4) and one in the open ocean (O1), which is located in the common mouth region (Figure 13.1). Water sampling and profiling of hyperspectral underwater radiometer were carried out on board research boat *MS-Teisten*. Water samples were collected from three depths (0, 30, 60 m) using 10 L Niskin's samplers. The overall depth along the sampling transects ranges from <50 m near glacial end to >300 m near the mouth. To study the temporal variation along these dynamic fjords, sampling was carried out in two phases, i.e., phase 1 (2–3 August 2016) and phase 2 (11–12 August 2016) (Table 13.1).

13.2.2 HYDROGRAPHY AND NUTRIENTS

A hyperspectral underwater radiometer (HyperPro II, Satlantic) was deployed in free-fall profiling mode to record hydrography in the euphotic zone (Z_{eu}; described as

TABLE 13.1
Sampling Details and Some of the Variables Measured in Both the Fjords

	Stations ID									
Variables	KG5	KG4	KG3	KG2	KG1	O1	KR4	KR3	KR2	KR1
Latitude (°N)	78.895	78.905	78.940	78.959	79.010	79.035	79.250	79.217	79.117	79.067
Longitude (°S)	12.320	12.228	12.141	11.822	11.608	11.284	11.700	11.867	11.750	11.500
Bottom Depth (m)	40	67	189	235	205	333	265	109	158	288
Phase 1 [2–3 August 2016]										
SST (°C)	5.3	5.5	6.1	4.4	-	6.4	3.3	2.3	6.0	6.2
Chl_{int} (mg m^{-2})	8.98	14.53	0.78	8.91	-	4.74	3.16	12.73	8.09	2.13
Avg. TSM (mg L^{-1})	14.41	12.91	10.83	12.22	-	15.13	15.06	14.65	14.3	14.35
K_d (m^{-1})	0.56	0.91	0.41	0.25	-	0.14	0.33	0.22	0.31	0.34
Z_{eu} (m)	12.31	7.52	16.68	26.86	-	49.13	20.79	30.15	21.631	19.84
Phase 2 [11–12 August 2016]										
SST (°C)	4.5	4.5	4.5	5.1	5.5	5.5	5.1	5.7	4.3	4.8
Chl_{int} (mg m^{-2})	5.21	2.97	1.91	2.65	17.18	9.62	38.05	6.29	5.98	3.55
Avg. TSM (mg L^{-1})	13.6	10.16	11.0	10.79	12.0	10.37	15.31	12.76	12.29	10.33
K_d (m^{-1})	0.71	0.50	0.42	0.12	0.15	0.11	0.21	0.21	0.33	0.25
Z_{eu} (m)	9.61	13.56	16.06	56.13	44.86	61.82	32.40	32.51	20.87	27.60

Avg. (average) indicates mean value of the three depths.

the depth interval down to 0.1% light). Vertical profiles of temperature (°C), salinity (psu), phytoplankton fluorescence (µg/L), photosynthetically active radiation (PAR, µmol/m^2/s^1), CDOM (in ppb) and turbidity concentrations (NTU at 700 nm) at different stations were measured by the profiling HyperPro II equipped with ECO triplet pack (WET Labs). Vertical diffuse attenuation coefficient of PAR (K_d, m^{-1}) was derived by linear regression fitting of log-transformed PAR data. The Z_{eu} was calculated as: $Z_{eu}(m) = -\ln(0.001)/K_d$. Sea surface temperature (SST) was measured using a bucket thermometer (Theodor Friedrichs & Co., Germany) with a measurement accuracy of +0.2°C. An aliquot (100 mL) of water samples were collected directly from the Niskin's sampler into acid-cleaned plastic bottles and were stored at −20°C until used for estimating concentrations (µM) of nitrate ($NO_3^- + NO_2^-$), nitrite (NO_2^-), phosphate (PO_4^{3-}) and silicate (SiO_4^{4-}) by a continuous-flow autoanalyser (Seal Analytical Ltd., Germany) following standard colorimetric methods (JGOFS 1994).

13.2.3 LIGHT-ABSORPTION COEFFICIENTS AND SUSPENDED MATTER

The water samples collected from discrete depths by the Niskin's sampler were analysed for measuring light-absorption coefficients of phytoplankton and detritus through QFT (quantitative filter technique) according to Mitchell (1990). Sample from each depth was filtered separately onto a GF/F filter (0.7 µm, 25 mm) under a maintained low suction pressure. Subsequently, the filters were measured within 400–700 nm wavelength at 1 nm interval using a double-beam UV-VIS spectrophotometer

(UV-2600, Shimadzu, Japan) equipped by an integrating sphere. A GF/F filter paper soaked with freshly prepared filtered seawater (FSW) was used as reference while carrying out these measurements. The path-length amplification effect of the glass fibre filter caused due to multiple scattering was corrected according to Cleveland and Weidemann (1993) as given below:

$$OD_s(\lambda) = 0.378 \, OD_f(\lambda) + 0.523 \, OD_f(\lambda)^2 \tag{13.1}$$

where $OD_s(\lambda)$ and $OD_f(\lambda)$ are optical densities (ODs) of the particulate matter in suspension and filter, respectively. The clearance area of each filter paper was measured thrice using a digital Vernier caliper, and average of the three was taken as the measure for clearance area. Further, light-absorption coefficient of the total particles ($a_p(\lambda)$) present in the suspension was calculated as follows:

$$a_p(\lambda) = 2.303 \, OD_s(\lambda) * S/V \tag{13.2}$$

where 2.303 = conversion factor for $\log_{10}$ to $\log_e$, S = clearance area measured for the filter paper (m²), V = filtered volume (m³) and S/V = approximate geometrical light pass length. Subsequently, the filter papers were soaked in 2% solution of calcium hypochlorite (Ca (ClO)₂) for 20 minutes to bleach the algal pigments (Woźniak et al. 1999). After rinsing with FSW, absorbance of the decolorised papers was re-measured to obtain OD of the non-phytoplankton particles or detritus ($a_d(\lambda)$) using equation (13.2). Phytoplankton light-absorption coefficient ($a_{ph}(\lambda)$) was calculated by substituting $a_d(\lambda)$ from $a_p(\lambda)$, and using the $a_{ph}(750)$ value as the null-point correction (Mitchell et al. 2002; Stramska et al. 2003) as follows:

$$a_{ph}(\lambda) = a_p(\lambda) - a_d(\lambda) \tag{13.3}$$

The mean values of duplicate spectra were analysed in this study. The chlorophyll-specific phytoplankton pigments absorption coefficient ($a^*_{ph}(\lambda)$) for a particular sample was calculated as the ratio of the $a_{ph}(\lambda)$ to Chl-a concentration.

Light-absorption coefficient for CDOM (a_{CDOM}) was measured by filtering 200 mL of water samples through 47 mm membrane filters (Millipore, 0.2 μm). Subsequently, the filtrates were equilibrated to room temperature by storing in dark condition. The absorption measurements were carried out spectrophotometrically (UV-2600, Shimadzu, Japan) using quartz cuvettes (path-length: 10 cm) and Milli-Q water as reference. Absorbance spectra were obtained between 350 and 750 nm at 1 nm interval. Spectral absorbances were normalised to zero at 600 nm due to temperature-dependent artefacts. The absorption coefficient of CDOM (m⁻¹) was calculated adopting the equation of Mitchell et al. (2002) as given below:

$$a_{CDOM}(\lambda) = \frac{2.303}{L} \left[\left\{ OD_s(\lambda) - OD_{fsw}(\lambda) \right\} - OD_{null} \right] \tag{13.4}$$

where 2.303, L, OD_s, OD_{fsw} and OD_{null} are the conversion factor from $\log_{10}$ to $\log_e$, optical (cuvette) path-length in metres, optical densities of sample, purified FSW and at null absorption wavelength, respectively. The spectral slope of CDOM absorption coefficient (S) was derived by non-linear exponential regression fitting of a_{CDOM} coefficient versus wavelength (350–650 nm) as per the approach of Stedmon et al. (2000). The S is the exponential slope coefficient, which indicates decrease in absorption with respect to wavelength. The S value varies with the source of the CDOM.

Water samples (2 L) were filtered through prewashed and pre-weighed (w_1 in mg) 47 mm Millipore filters (pore size 0.45 μm) under maintained low vacuum pressure (approx. 120 mm Hg), and then the filters were oven dried at 60°C for 4 hours and re-weighed to obtain the final weight (w_2 in mg). The concentration of total suspended matter (TSM in mg/L) was obtained by subtracting w_1 from w_2.

13.2.4 PHYTOPLANKTON PIGMENTS

Phytoplankton biomass or chlorophyll-a (Chl-a) was quantified by filtering 2–3 L water samples onto a 47 mm GF/F filters (Whatman®) under low suction pressure followed by overnight extraction of the filters with 10 mL of AR grade 90% acetone at 4°C. The extracts were then measured fluorometrically (10-AU, Turner® Designs) before and after the addition of 10% HCl to quantify Chl-a and phaeopigments (Pheo: degraded phytoplankton pigments) concentrations (JGOFS 1994). The entire assay was carried out in dim light or dark to avoid photodegradation of phytoplankton pigments. The Chl-a concentrations (mg/m^3) at different depths were trapezoidally integrated to estimate the water column Chl-a value (Chl$_{int}$, mg/m^2) at each sampling location.

For analysis of other phytoplankton pigments, 3 L of seawater was filtered onto GF/F filters (0.7 μm, 25 mm, Whatman®) in a dark and cold room, and stored at −80°C until further analysis. For logistical limitation, samples for high-performance liquid chromatography (HPLC) analysis were collected for phase 1 only. The preserved filters were soaked in 100% methanol for extraction of pigments and subsequently analysed by a HPLC (Agilent Technologies) equipped with an Eclipse XDB C8 column as detailed in Kurian et al. (2012). For phytoplankton taxa classification (Table 13.2), seven pigments such as fucoxanthin, peridinin, alloxanthin, 19′-hexanoyloxyfucoxanthin (19′HF), 19′-butanoyloxyfucoxanthin (19′BF), zeaxanthin and TChl-b (Chl-b+divinyl chlorophyll-b (DivChl-b)) were identified as diagnostic pigments (DP) according to Uitz et al. (2006). Using DP, size-fractioned (microphytoplankton (>20 μm), nanophytoplankton (>2 and <20 μm) and picoplankton (<2 μm)) contributions of phytoplankton (f_{micro}, f_{nano}, and f_{pico}) were derived as follows:

$$f_{micro}\% = 100 \times \left(1.41\,[\text{fucoxanthin}] + 1.41[\text{peridinin}]\right) / \sum DP\,(> 20\,\mu m) \quad (13.5)$$

$$f_{nano}\% = 100 \times \left(0.60\,[\text{alloxanthin}] + 0.35[19'\text{BF}]\right)$$
$$+ 1.27[19'\text{HF}] / \sum DP\,(2\ \text{to}\ 20\,\mu m) \quad (13.6)$$

$$f_{pico}\% = 100 \times \left(0.86\,[\text{zeaxanthin}] + 1.01[\text{TChlb}]\right) / \sum DP\,(< 2\,\mu m) \quad (13.7)$$

TABLE 13.2

(A) The Observed Phytoplankton Pigments, Their Abbreviations and Phytoplankton Groups to Which They Belong to; (B) Description of Pigment Sums; and (C) Pigment Index

(A)	Observed Pigments	Abbreviation	Phytoplankton Groups
	Chlorophyll-c2	chl-c2	
	Chlorophyll-c3	chl-c3	
	Chlorophyll-b	chl-b	Chlorophyceae
	Chlorophyll-a	chl-a	
	Peridinin	PER	Dinophyceae I
	Fucoxanthin	FUC	Bacillariophyceae
	19-Butanoyloxyfucoxanthin	19BUT	Haptophytes
	Neoxanthin	NEO	Euglenophyceae
	Prasinoxanthin	PRAS	Prasinophyceae
	Violaxanthin	VIO	Raphidophyceae
	19-Hexanoyloxyfucoxanthin	19HEXA	Prymnesiophyceae
	Diadinoxanthin	DIADINO	
	Alloxanthin	ALLO	Cryptophyceae
	Diatoxanthin	DIATO	
	Zeaxanthin	ZEA	Cyanobacteria
(B)	Pigment Sum	Abbreviation	Description
	Total DPs	DP	PSC + ALLO + ZEA + TChl-b
	Photoprotective carotenoids	PPC's	ALLO+DIAD+VIO + ZEA
	Photosynthetic carotenoids	PSC's	FUCO + BUT + HEX + PER
(C)	Pigment Index	Abbreviation	Description
	Microphytoplankton proportion factor	Mpf	(FUCO + PER)/DP
	Nanophytoplankton proportion factor	Npf	(HEX + BUT + ALLO)/DP
	Pico-phytoplankton proportion factor	Ppf	(Zea + chl b)/DP

The Shannon–Weaver index for phytoplankton group diversity (H') was calculated according to Shannon and Weaver (1949), whereas group evenness (J') was derived ($J' = H'/\log_2 G$) as suggested by Pielou (1966). Evenness index is the ratio of observed diversity to highest diversity, and it is attained when most groups in a sample are equally abundant.

13.3　RESULTS AND DISCUSSION

13.3.1　Hydrography and Nutrients Variability

Cooler and fresher water (avg. temp: 4.83°C ± 0.68°C; avg. salinity: 34.45 ± 0.93 psu) in the inner Kongsfjorden (phase 1) was observed with a clear thermal front-like feature in the mid-fjord. Low-salinity water spreads outward forming clear halocline.

Though temperature and salinity ranges were similar in both phases, intrusion of warmer water to inner fjord was discernible, and shoaling of salinity contours in mid-fjord was observed indicating wind- and/or tide-induced mixing. The presence of less freshwater in the upper 10 m at glacial end indicated probable lessened melting in phase 2 or the presence of intruded warmer water. Conversely, in Krossfjorden, more uniform distribution of temperature (avg. 4.50°C ± 0.58°C) and salinity (avg. 34.28 ± 1.08 psu) was observed in the upper 10 m with lesser gradients. Salinity difference of nearly 2 psu was observed between inner and outer fjord. In phase 2, a tongue-like structure from outer fjord was sandwiched between cooler water from sub-surface and meltwater discharge from *Julibreen* glacier (near to station KR2), which significantly influenced the fjord water and resulted in unconventional lower salinity at mouth than the glacial-end region. Relatively higher salinity at glacial end implied less melting and/or uplifting of more saline outer sub-surface water in inner fjord.

Irrespective of proximity to the glacial end, where high discharge of freshwater occurs, stratification was observed throughout the fjord with occasional shoaling of thermohaline contours (phase 2) in the mid-fjord indicating upward movement of the water masses probably caused by wind flow or bottom topography in Kongsfjorden. Difference in the intrusion of AW into the fjords and/or strength of glacial melting could be clearly discerned between the sampling phases indicating the dynamic oceanographic features of both the fjords. Both the fjords were observed to be influenced by Atlantic-origin warm and saline water (AW) masses superimposed by a slim layer of fresh and cool surface water (SW) run-off from the glaciers (SW: $T = 1-5°C$, $S < 34.7$ psu, 0–15 m) and AW ($T = 5.5-7°C$, $S > 34.9$ psu, 15–60 m). The observed stratification in this twin-fjord ecosystem signified a shift from Arctic to Atlantic predominance in summer (Svendsen et al. 2002; Piwosz et al. 2009; David and Krishnan 2017).

Nearly identical concentrations of the macronutrients, i.e., NO_3^-, NO_2^-, PO_4^{3-} and SiO_4^{4-}, were observed in Kongsfjorden ((avg. 1.89 ± 1.47 μM), (avg. 0.24 ± 0.13 μM), (avg. 0.35 ± 0.12 μM), (avg. 1.64 ± 0.40 μM)) and Krossfjorden ((avg. 1.24 ± 1.26 μM), (avg. 0.16 ± 0.12 μM), (avg. 0.32 ± 0.11 μM), (avg. 1.40 ± 0.52 μM)). Not much variation was recorded in PO_4^{3-} and SiO_4^{4-} along the fjord, but NO_3^- availability, which is generally considered as the principal limiting nutrient in coastal Arctic region (van de Poll et al. 2016), was more towards outer fjord compared to the mid- and inner fjord as was observed by Calleja et al. (2017). Low elemental (N:P < 5 and N:Si ≥ 1) ratio implied N- and Si-limited condition throughout the fjord was indicative of conducive environmental settings for growth of non-siliceous phytoplankton such as *Phaeocystis* spp. and heterotrophic/mixotrophic dinoflagellates (van de Poll et al. 2016; Bhaskar et al. 2016, 2020). Nitrogen and Si limitation was also ubiquitous in Krossfjorden, and it was even more severe. Very low Si:P ratio (<5) insinuated probable succession of Green Algae/Cyanobacteria over microplankton (Harrison et al. 1977; Levasseur and Therriault 1987) in Krossfjorden. It has been reported that in summer, increase in glacial meltwater input leads to build-up of ammonium and low concentrations of NO_3^- caused by reduced nitrification during non-bloom circumstances and the utilisation of existing NO_3^- (Calleja et al. 2017).

13.3.2 LIGHT ATTENUATION IN THE WATER COLUMN

In Kongsfjorden, the vertical-profile PAR in the upper 60 m varied between phase 1 and phase 2. The corresponding surface values and indicated lesser incident PAR during phase 2 (avg. $470 \pm 243 \,\mu mol/m^2/s^1$) measurements are compared to phase 1 (avg. $749 \pm 252 \,\mu mol/m^2/s^1$). Similarly, condition was also observed in Krossfjorden, where surface PAR was $771 \pm 348 \,\mu mol/m^2/s^1$ and $241 \pm 50 \,\mu mol \; m^{-2}s^{-1}$, respectively, indicating less light availability during phase 2, which could be ascribed to the overcast sky conditions. In Kongsfjorden, vertical light extinction coefficient (K_d) varied from 0.11 to 0.92 (avg. $0.39 \,m^{-1}$), whereas it varied between 0.21 and 0.35 (avg. $0.28 \,m^{-1}$) in the Krossfjorden, indicative of relatively clear water column and deeper euphotic depth ($Z_{eu,}$ 1% light depth) in Krossfjorden. Usually, shallower Z_{eu} observed at glacial end deepened towards the mouth region in Kongsfjorden as observed previously (Piquet et al. 2014); nevertheless, Krossfjorden witnessed shallower Z_{eu} towards the outer fjord stations, which could be ascribed to impact meltwater from the nearby (*Julibreen*) glacier that brings in optically active substances (OAS) like turbidity and CDOM.

Interrelationship between K_d and other environmental variables was dissimilar among and between the fjords. In Kongsfjorden (P-I+P-II), the K_d showed significant positive relationship with CDOM ($R^2 = 0.71$, $n = 12$, $p < 0.05$) and turbidity ($R^2 = 0.69$, $n = 12$, $p < 0.05$), and significant negative relationship with salinity ($R^2 = 0.60$, $n = 12$, $p < 0.05$) implying that the freshwater-generated CDOM and turbidity were primarily controlling the K_d in Kongsfjorden (Pavlov et al., 2014; Payne and Roesler 2019), whereas no significant association was observed for Krossfjorden where K_d showed moderate negative relationship with salinity ($R^2 = 0.47$, $n = 8$) and Chl_{int} ($R^2 = 0.32$, $n = 8$) and weak negative relationship with CDOM ($R^2 = 0.15$, $n = 8$) implying that unlike Kongsfjorden (where OAS were more dominant), phytoplankton biomass was the major controlling factor for K_d variation in this fjord.

The turbidity and CDOM concentrations were comparatively higher in phase 1 (avg. 1.36 ± 1.70 NTU; 1.39 ± 0.37 ppb) than phase 2 (avg. 0.83 ± 1.46 NTU; 1.33 ± 0.30 ppb) in Kongsfjorden. Elevated concentrations of turbidity and CDOM were observed at the inner and mid-fjord stations (KG5, KG4, KG3), which rapidly decreased towards outer fjord. High turbidity and CDOM in the inner fjord were basically originated from glacial inputs or mixing-induced resuspension of bottom sediments (Halbach et al. 2019). Conversely, in Krossfjorden, both the variables were uniformly distributed throughout in the upper 10–15 m during phase 1 (avg. 0.75 ± 0.76 NTU; 1.26 ± 0.16 ppb) and phase 2 (avg. 0.58 ± 0.63 NTU; 1.28 ± 0.12 ppb), and even higher concentrations were observed in outer fjord station (KR2), which could be linked to meltwater from the nearby (*Julibreen*) calving glacier. Turbidity and CDOM concentrations showed significant inverse linear relationships with salinity in both the fjords (Figure 13.2), whereas significant positive linear relationship between these two variables in Kongsfjorden ($R^2 = 0.95$, $p < 0.001$) and Krossfjorden ($R^2 = 0.82$, $p < 0.001$) indicated their similar distribution pattern, and probable origin from the same source (i.e., from glacial meltwater).

Occurrence of CDOM is ubiquitous in natural waters bodies and is generated from the degradation of plant materials both of terrestrial and aquatic origin (Kirk 1994). It

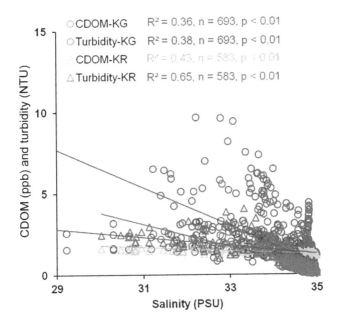

FIGURE 13.2 Scatter plots showing relationship of CDOM and turbidity with salinity in Kongsfjorden and Krossfjorden.

plays a significant role in underwater light attenuation. CDOM spectral slope (S) is a proxy for origin (marine vs. terrestrial) of CDOM, with usually lower slopes for fresh water and coastal waters as compared to the high-saline open ocean waters (Stedmon and Markager 2001), which contains marine humic substances and newly formed biological CDOM (Ferrari 2000), and photo-bleached CDOM (Para et al. 2010). The mean values of CDOM spectral slope ($S_{350-650}$) were 0.014 ± 0.003 ($0.008-0.25 \, nm^{-1}$) and 0.011 ± 0.001 ($0.008-0.15 \, nm^{-1}$) in Kongsfjorden and Krossfjorden, respectively, signifying higher S values in the Kongsfjorden. Thus, the lower average S value noticed in Krossfjorden could be ascribed to more of autochthonous in origin (*in situ* production of organic matter by phytoplankton), whereas the higher S values in Kongsfjorden could be due to input of terrestrial organic matter during the ice-melting (Calleja et al. 2017) advection of CDOM by AW (Pavlov et al. 2014).

13.3.3 FACTORS INFLUENCING PHYTOPLANKTON BIOMASS, COMMUNITY STRUCTURE AND DIVERSITY

13.3.3.1 Phytoplankton Biomass and Phaeopigments

In Kongsfjorden, the *in situ* Chl-*a* and Pheo varied from ND-1.02 (0.17 ± 0.27 mg/m³) and ND-0.64 (0.14 ± 0.19 mg/m³), respectively, while it varied from ND-2.29 (0.21 ± 0.48 mg/m³) and ND-0.72 (0.12 ± 0.18 mg/m³), indicative of relatively higher Chl-*a* and lower Pheo in Krossfjorden. Fluorescence profiles measured by HyperPro II corroborated the fluorometric measurements. Higher Chl-*a* fluorescence was observed in the outer fjords compared to the inner fjord (Hegseth and Tverberg 2013;

Piquet et al. 2014; Bhaskar et al. 2016; Payne and Roesler 2019) that reported higher phytoplankton biomass and productivity at the outer Kongsfjorden compared to inner region. In general, higher fluorescence was observed in the phase 2 in both the fjords, which was also reflected in column-integrated Chl-a (Chl$_{int}$, mg m^{-2}) values in the fjords (Table 13.1) during phase 2. Sigma-t (density, kg/m^3–1000) increased with depth indicating that both the fjords' water column was stable, and the chlorophyll fluorescence observed during this study existed in the most stable water layers in both the fjords.

Vertical distribution of *in situ*-measured Chl-a and Pheo coexisted in the water column and yielded significant positive relationship between them for Kongsfjorden ($R^2 = 0.72$, $n = 35$, $p < 0.01$) and Krossfjorden ($R^2 = 0.75$, $n = 31$, $p < 0.01$). The Chl-a:Pheo ratio usually decreased with increasing depths and varied from 0.04 to 2.56 (0.90 ± 0.72) and from 0.08 to 5.99 (1.67 ± 1.49) in Kongsfjorden and Krossfjorden, respectively. Higher average Chl-a:Pheo in Kongsfjorden was indicative of stronger grazing pressure and/or cell senescence under reduced growth environments (Welschmeyer and Lorenzen 1985) compared to Krossfjorden. The observed nutrient limitation in both the fjords corroborates the underlying fact for Pheo formation.

Chlorophyll and Pheo are known to be the direct products of phytoplankton growth and zooplankton grazing, respectively. The stoichiometry between Chl-a and Pheo (Chl-a:Pheo ratio) has been used to indicate herbivore grazing, and Pheo is usually considered as an indicator for herbivore grazing (Shuman and Lorenzen 1975). It is well known that photodegradation, which decreases exponentially with depth, is largely responsible for disappearance of Pheo in the presence of light; i.e. Pheo would be removed at an accelerated rate near the surface (well-lit zone), resulting in decreased Chl-a:Pheo ratio with depth (Welschmeyer 1994). This study showed typically higher Chl-a:Pheo ratio at the surface, where PAR was also highest and decreased with increasing depth showing the lowest Chl-a:Pheo ratio at deeper depths. Furthermore, the significant positive relationship between log-transformed Pheo and PAR ($R^2 = 0.69$, $n = 31$, $p < 0.05$) showed higher concentrations of Pheo in the higher PAR region (upper layers). From these observations, it can be inferred that microzooplankton grazing was prevalent in the study area. These results provide an assumption that Pheo may be used as a tag for microzooplankton grazing in these fjord ecosystems.

13.3.3.2 Phytoplankton Community Structure and Diversity Index

Phytoplankton pigments analysis of the SW samples by HPLC revealed the presence of 15 phytoplankton pigments during this study (Table 13.2). The marker pigments were selected in reference to the phytoplankton groups observed in previous studies, in the Kongsfjorden (Bhaskar et al. 2016; va de Poll et al. 2016) and Svalbard (Pettersen et al. 2011) regions. The DPs were taken as the sum of seven pigments such as alloxanthin, zeaxanthin, Chl b, fucoxanthin, 19-butanoyloxyfucoxanthin, 19-hexanoyloxyfucoxanthin and peridinin (Table 13.2). Difference in spatial distribution of pigments was observed inside the two fjords and between different sampling locations of the same fjord. The pigment concentrations were very less near glacial-end (KG5, KG4) and mid-fjord (KG2) stations in the Kongsfjorden and suddenly increased at the mouth and open ocean, whereas comparatively higher pigment concentrations were observed in the Krossfjorden. The very low pigments at glacial end indicated inhibited phytoplankton growth due to turbidity and CDOM-induced light limitation (Kowalczuk et al. 2019). Unlike in the Kongsfjorden, higher pigment concentrations were noticed at the glacial end (KR4) of the Krossfjorden, which decreased in the mid-fjord (KR3, KR2) stations and then increased towards mouth (KR1). The observed dominant pigment was fucoxanthin (FUCO), which was present at all the stations in both the fjords. The average contribution (%) of FUCO was 48.76% (Kongsfjorden) and 36.38% (Krossfjorden) followed by Chl c2, which contributed 21.16% and 31.90% in Kongsfjorden and Krossfjorden, respectively. Average contribution of Chl c3 was higher in the Kongsfjorden (12.14%) than that in Krossfjorden (9.76%). These three pigments (FUCO, Chl c2 + c3) are specific pigment markers for phytoplankton groups such as Bacillariophyceae, Chrysophyceae and Pavlovophyceae (Johnsen and Sakshaug 2007; Pettersen et al. 2011) in the North AWs. Their contribution accounted for 56%–95% of the total accessory pigments concentrations with nearly equal average (78%) values in both the fjords.

This study categorised the phytoplankton size-class based on the DP index and observed prevalence of microplankton (presumably flagellates) in the surface waters of both the fjords. Micro-sized assemblages contributed nearly 58% (KG1) to 100% (KG4) of the total phytoplankton community in Kongsfjorden. Though nanoplankton was absent in KG4 (cause not known), it was present in all other stations (Figure 13.3) and their contribution ranged from 9% (KG1) to 19% (KG2), signifying relatively higher abundance in the near glacial than the outer stations. The nanoplankton contribution further decreased to 4% at open ocean (O1) station. Picoplankton community was conspicuously absent in the mid- and near glacial (KG2, KG4 and KG5) stations, while their presence was noticed towards outer (33% at KG1) and open ocean (7% at O1) regions, respectively. Though picoplankton is often considered to be less abundant in polar waters, their increasing numbers were recorded while moving from Arctic to AW masses in the Barents Sea (Not et al. 2005). Though the phytoplankton community structure in Krossfjorden was different from Kongsfjorden, dominance of microplankton was also observed in this fjord with contribution varying from 49% (KR1) to 89% (KR4) and gradually increasing from outer to inner stations (Figure 13.3). Contrasting to Kongsfjorden, nano- and pico-sized individuals were recorded at all the stations of the Krossfjorden, where % contribution of nanoplankton and picoplankton prevailed over each other in the inner (KR4, KR3) and outer (KR2, KR1) stations, respectively, indicating increase in pico-sized assemblages towards outer stations as was the case in Kongsfjorden. Prevailing very low Si:P ratio (<5) insinuated probable succession of smaller plankton over microplankton (Harrison et al. 1977; Levasseur and Therriault 1987) in this fjord, which was possibly starting during the observation period.

Phytoplankton group diversity index (H') ranged from a maximum of 0.93 (KG1) to a minimum of 0.495 (KG2), implying relatively lesser diversity in the mid- and inner fjord compared to outer fjord, whereas Krossfjorden displayed overall more group diversity than Kongsfjorden with relatively higher diversity towards outer stations (KR1, KR2). The group evenness index (J') was found to be higher at outer stations in both the fjords. Surface waters of inner Krossfjorden are often relatively less turbid than inner Kongsfjorden (Piquet et al. 2010), which results in deeper euphotic

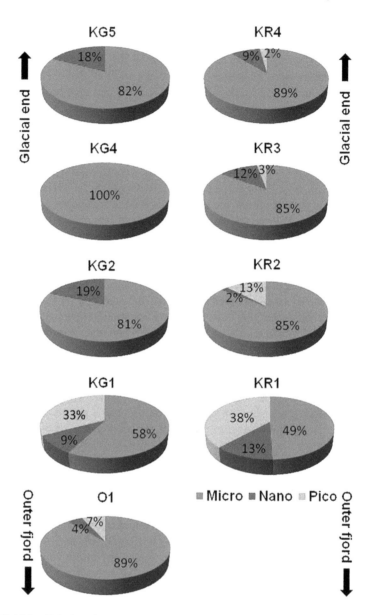

FIGURE 13.3 Relative abundance (%) of size-fractionated phytoplankton composition in the surface waters of Kongsfjorden (left column) and Krossfjorden (right column) during phase 1.

layer in Krossfjorden, thereby supporting higher phytoplankton growth, abundance and diversity in the clearer waters (Piwosz et al. 2009; Calleja et al. 2017; Halbach et al. 2019). The observed high K_d values associated with turbidity and CDOM in this study are concurrent with previous reports and imply unfavourable condition for the growth of phytoplankton at glacial end especially in Kongsfjorden.

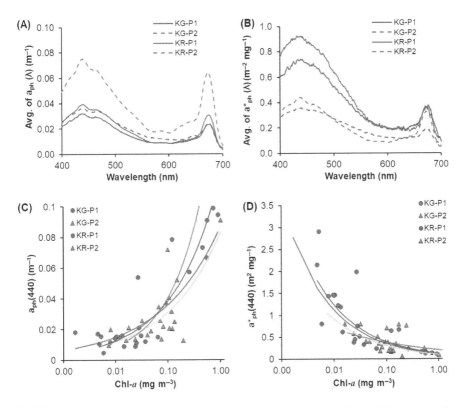

FIGURE 13.4 Spectral average of (a) phytoplankton absorption ($a_{ph}(\lambda)$), and (b) chlorophyll-specific phytoplankton absorption ($a^*_{ph}(\lambda)$). Power function relationships showing (c) $a_{ph}(\lambda)$ and (d) $a^*_{ph}(\lambda)$ at 440 nm as a function of Chl-a in the fjords. P-1: Phase 1, P-2: Phase 2.

13.3.4 PHYTOPLANKTON LIGHT ABSORPTION AND ASSOCIATED PACKAGE EFFECT

The average $a_{ph}(\lambda)$ was distinctly different in both the fjords showing comparatively lower values in Kongsfjorden than Krossfjorden during both the phases (Figure 13.4a). The interphase difference in average $a_{ph}(\lambda)$ in Kongsfjorden was much lower than that in Krossfjorden. Phase 2 in Krossfjorden showed the highest values among the spectra. The largest difference in magnitude of $a_{ph}(\lambda)$ was observed in blue (440 nm) wavelength compared to red (675 nm). Likewise, the average Chl-a-specific a_{ph} ($a^*_{ph}(\lambda)$) displayed dissimilarity in both magnitude and spectrum shape among the fjords and between the phases (Figure 13.4b). The highest average $a^*_{ph}(\lambda)$ in the blue spectral region was observed in Krossfjorden phase 1 and the lowest average $a^*_{ph}(\lambda)$ in Kongsfjorden phase 2. Moreover, the Krossfjorden showed higher average $a^*_{ph}(\lambda)$ values than Kongsfjorden during both the phases.

The relationship between $a_{ph}(440)$ with Chl-a was different between the phases and among the fjords, and showed significant positive power correlations with Chl-a (Figure 13.4c). The equations that describe the best-fit between Chl-a and $a_{ph}(440)$ are KG-P1: $a_{ph}(440) = 0.0862[\text{Chl-}a]^{0.3823}$ ($R^2 = 0.66$, $n = 14$, $p < 0.001$), KG-P2: $a_{ph}(440) = 0.0824$ $[\text{Chl-}a]^{0.4262}$ ($R^2 = 0.62$, $n = 15$, $p < 0.001$),

KR-P1: $a_{ph}(440) = 0.1189[\text{Chl-}a]^{0.4917}$ ($R^2 = 0.73$, $n = 12$, $p < 0.01$) and KR-P2: $a_{ph}(440) = 0.1809[\text{Chl-}a]^{0.6509}$ ($R^2 = 0.60$, $n = 11$, $p < 0.01$). Similar to $a_{ph}(\lambda)$, the relationship between $a^*_{ph}(\lambda)$ and Chl-a was different between the phases and among the fjords (Figure 13.4d). Significant negative power correlations were obtained between $a^*_{ph}(440)$ and Chl-a except at KR-P2 (weak correlation). The regression formula and the best-fit parameters between Chl-a and $a^*_{ph}(440)$ are KG-P1: $a^*_{ph}(440) = 0.107[\text{Chl-}a]^{-0.509}$ ($R^2 = 0.80$, $n = 14$, $p < 0.001$), KG-P2: $a^*_{ph}(440) = 0.0967[\text{Chl-}a]^{-0.485}$ ($R^2 = 0.59$, $n = 14$, $p < 0.001$), KR-P1: $a^*_{ph}(440) = 0.1189[\text{Chl-}a]^{-0.508}$ ($R^2 = 0.74$, $n = 12$, $p < 0.01$) and KR-P2: $a^*_{ph}(440) = 0.2036[\text{Chl-}a]^{-0.266}$ ($R^2 = 0.17$, $n = 11$).

Elevated Chl-a concentration is often connected to increase in intracellular pigment concentration or cell size, which leads to lack in correlation between light-harvesting efficiency and pigment packaging originating from intracellular overlapping of the chloroplasts is termed as 'package effect'. The a^*_{ph} is an effective proxy for measuring 'package effect' (Bricaud et al. 1995). The coefficient $a^*_{ph}(440)$ is generally inversely correlated to Chl-a concentration because of increasing package effect with Chl-a and an inverse co-variation between the relative abundance of accessory pigments to Chl-a. The observed negative power correlations of $a^*_{ph}(440)$ with Chl-a is indicative of prevailing package effect (Wang et al. 2014; Tripathy et al. 2014; Alcantara et al. 2016; Chakraborty et al. 2017; Kerkar et al. 2020, 2021) in this region dominated by microplankton (in the surface layer) during the study period. To confirm this, we also calculated the blue-to-red (B/R) ratio of a^*_{ph} (i.e., $a^*_{ph}(440/a^*_{ph}(675))$, which is used as an indicator of phytoplankton size-classes (Lohrenz et al. 2003). The B/R ratio of a^*_{ph} in the water column varied from 0.98 to 5.05 (avg. 2.18 ± 1.17) and 0.97 to 6.91 (avg. 2.66 ± 1.86) in KG-P1 and KR-P1, respectively. In principle, if the B/R is >3.0, it denotes dominance of picoplankton (<2 μm); if the ratio is <2.5, predominance of microphytoplankton (>20 μm) is suggested (Bricaud et al. 2004; Stramski et al. 2001), whereas ratios between 2.5 and 3.0 indicate nano-phytoplankton predominance (Aguilar-Maldonado et al. 2018). The observed average B/R ratio is parallel with our phytoplankton pigment analysis confirming that the surface waters of the study area were dominated by microplankton (presumably flagellates), which is prone to package effect, thereby resulting in reduced photosynthetic efficiency which will lead to low productivity (Bricaud et al. 1995; Ferreira et al. 2017). Slightly higher average B/R ratio observed in Krossfjorden can be linked to the presence of nano- and picoplankton throughout this fjord, which are less prone to package effect, thereby having more photosynthetic efficiency. The $a^*_{ph}(\lambda)$ values reported in this study are quite higher than the reported values in this region because $a^*_{ph}(\lambda)$ is derived based on the concentration of the main pigment (Chl-a) but since there are accessory pigments that also absorb, thereby influences the $a^*_{ph}(\lambda)$ estimates across the blue-to-red spectral range (Bricaud et al. 1995; Ferreira et al. 2017). This effect can also change within a species due to low-light acclimatisation in the Polar Regions (Palmisano et al. 1986; Wang et al. 2005; Matsuoka et al. 2009; Mendes et al. 2015; Kowalczuk et al. 2019).

13.4 SUMMARY

This chapter elucidates a comparative account of spatiotemporal variability in phytoplankton biomass and other OASs in the water column in relation to the phytoplankton community structure and photosynthetic efficiency in two high Arctic fjords during summer (Figure 13.5). Interphase variability of thermohaline features within the fjord implied AW inflow and mixing in these dynamic fjords, which influenced the nutrient availability. Krossfjorden was found to be less influenced by OASs (i.e.,

FIGURE 13.5 Sample collection and analysis activities in Ny-Ålesund, Arctic. (a) Water sample collection by Niskin's sampler, (b) deployment of hyperspectral radiometer, (c) filtration of water samples, (d) residual of water sample deposited on a filter, (e) UV-VIS spectrophotometer used for light-absorption measurements and (f) filtration area measurement by digital caliper.

turbidity, CDOM) than the Kongsfjorden, where the inner fjord is tremendously affected by glacial melting-induced turbidity and CDOM. The produced CDOM was more of autochthonous (*in situ*) nature in Krossfjorden, whereas in Kongsfjorden, it was probably a mixture of both autochthonous and advected CDOM by AW. Higher OASs in Kongsfjorden resulted in restricted euphotic zone and unfavourable condition for phytoplankton growth.

Nitrate and silicate limitations in both the fjords were conducive for growth of non-silicious phytoplankton. Dominance of microplankton (heterotrophic flagellates and/or flagellates) was observed in the surface layer of both the fjords followed by nano- and picoplankton, which marked their appearances in relatively clearer waters of the outer fjord stations. Phytoplankton light-absorption efficiency was higher in Krossfjorden than that in Kongsfjorden, signifying its higher productivity potential. Nevertheless, the presence of microplankton in the surface layers was found to be affected by pigment packaging which would reduce their photosynthetic efficiency/productivity. Furthermore, this study showed that chlorophyll to phaeopigments ratio can probably be used as an assumption for microzooplankton grazing in these fjord ecosystems. This study reveals that despite the close geographical proximity, environmental settings in both fjords were quite dissimilar, where less OASs create more favourable underwater light conditions for phytoplankton photosynthesis enabling Krossfjorden to be more productive and richer in phytoplankton diversity than the Kongsfjorden. Thus, increasing climate warming can have different levels of impacts on fjord ecosystems despite their close geographical proximity.

ACKNOWLEDGEMENTS

The author is thankful to the Ministry of Earth Sciences (MoES), Govt. of India, for funding to conduct this study. Sincere thanks to the Director, NCPOR, for his constant encouragement and support. Logistical support rendered by the Arctic Logistics Division of NCPOR is highly acknowledged. I thank the volume Editor for his efforts in effective handling of the review process of the manuscript. Thanks to the managerial staff of Kings Bay, Ny-Ålesund and Captain of *MS-Teisten* for their exemplary assistance. This work is "First published in [*International Journal of Environmental Science and Technology*, https://doi.org/10.1007/s13762-021-03767-4, 2021] by Springer Nature". Some part of the published article is "Reproduced with permission from Springer Nature". This is NCPOR contribution number B-5/2021-22.

REFERENCES

Aguilar-Maldonado J, Santamaría-del-Ángel E, González-Silvera A, Cervantes-Rosas O, López L, Gutiérrez-Magness A, Cerdeira-Estrada S, Sebastiá-Frasquet MT (2018) Identification of phytoplankton blooms under the index of Inherent Optical Properties (IOP index) in optically complex waters. *Water* 10(2), 129.

Alcântara E, Watanabe F, Rodrigues T, Bernardo N (2016) An investigation into the phytoplankton package effect on the chlorophyll-a specific absorption coefficient in Barra Bonita reservoir, Brazil. *Remote Sensing Letters* 7(8), 761–70.

Bhaskar JT, Parli BV, Tripathy SC (2020) Spatial and seasonal variations of dinoflagellates and ciliates in the Kongsfjorden, Svalbard. *Marine Ecology* 41(3), 1–12.

Bhaskar JT, Tripathy SC, Sabu P, Laluraj CM, Rajan S (2016) Variation of phytoplankton assemblages of Kongsfjorden in early autumn 2012: a microscopic and pigment ratio-based assessment. *Environmental Monitoring Assessment* 188(4), 1–13.

Bischof K. et al. (2019) Kongsfjorden as Harbinger of the Future Arctic: Knowns, Unknowns and Research Priorities. In: Hop H., Wiencke C. (eds.), *The Ecosystem of Kongsfjorden, Svalbard. Advances in Polar Ecology*, Vol 2. Springer, Cham.

Bricaud A, Babin M, Morel A, Claustre H (1995) Variability in the chlorophyll-specific absorption coefficients of natural phytoplankton: analysis and parameterization. *Journal of Geophysical Research: Oceans* 100(C7), 13321–32.

Bricaud A, Claustre H, Ras J, Oubelkheir K (2004) Natural variability of phytoplanktonic absorption in oceanic waters: influence of the size structure of algal populations. *Journal of Geophysical Research: Oceans* 109(C11).

Calleja ML, Kerhervé P, Bourgeois S, Kędra M, Leynaert A, Devred E, Babin M, Morata N (2017) Effects of increase glacier discharge on phytoplankton bloom dynamics and pelagic geochemistry in a high Arctic fjord, *Progress in Oceanography* doi.: 10.1016/j.pocean.2017.07.005.

Chakraborty S, Lohrenz SE, Gundersen K (2017) Photophysiological and lightabsorption properties of phytoplankton communities in the river-dominated margin of the northern Gulf of Mexico. *Journal of Geophysical Research: Oceans* 122, 4922–4938.

Cleveland JS, Weidemann AD (1993) Quantifying absorption by aquatic particles: A multiple scattering correction for glass-fiber filters. *Limnology and Oceanography* 38(6), 1321–7.

Cohen JH, Screen JA, Furtado JC, Barlow M, Whittleston D, Coumou D, et al. (2014) Recent arctic amplification and extreme mid-latitude weather. *Nature Geoscience* 7, 627–637.

Darlington E (2015) Meltwater Delivery from the Tidewater Glacier Kronebreen to Kongsfjorden, Svalbard; Insights from in situ and Remote-Sensing Analyses of sediment plumes (PhD Thesis). Loughborough University.

David DT, Krishnan, KP (2017) Recent variability in the Atlantic water intrusion and water masses in Kongsfjorden, an Arctic fjord. *Polar Science* 11, 30–41.

Eilertsen HC, Taasen JP, Weslawski JM (1989) Phytoplankton studies in the fjords of West Spitzbergen: physical environment and production in spring and summer. *Journal of Plankton Research* 11, 1245–1260.

Ferrari GM (2000) The relationship between chromophoric dissolved organic matter and dissolved organic carbon in the European Atlantic coastal area and in the West Mediterranean Sea (Gulf of Lions). *Marine Chemistry* 70(4), 339–357.

Ferreira A, Ciotti AM, Mendes CRB, Uitz J, Bricaud A (2017) Phytoplankton light absorption and the package effect in relation to photosynthetic and Photoprotective pigments in the northern tip of Antarctic Peninsula *Journal of Geophysical Research: Oceans* 122, 7344–7363. doi: 10.1002/2017JC012964.

Halbach L, Vihtakari M, Duarte P, Everett A, Granskog MA, Hop H, Kauko HM, Kristiansen S, Myhre PI, Pavlov AK, Pramanik A, Tatarek A, Torsvik T, Wiktor JM, Wold A, Wulff A, Steen H, Assmy P (2019) Tidewater glaciers and bedrock characteristics control the phytoplankton growth environment in a fjord in the arctic. *Frontiers in Marine Science* 6, 254.

Harrison PJ, Conway HL, Holmes RW, Davis CO (1977) Marine diatoms in chemostats under silicate or ammonium limitation.III. Cellular chemical composition and morphology of three diatoms. *Marine Biology* 43, 19–31.

Hegseth EN, Assmy P, Wiktor J, Kristiansen S, Leu E, Piquet AM-T, et al. (2019) Phytoplankton seasonal dynamics in Kongsfjorden, Svalbard and the adjacent shelf, In *The Ecosystem of Kongsfjorden, Svalbard, Advances in Polar Ecology 2*, H. Hop and C. Wiencke (eds.), Cambridge, Springer, pp. 173–228.

Hegseth EN, Tverberg V (2013) Effect of Atlantic water inflow on timing of the phytoplankton spring bloom in a high Arctic fjord (Kongsfjorden, Svalbard). *Journal of Marine Systems* 113–114, 94–105.

Hodal H, Falk-Petersen S, Hop H, Kristiansen S, Reigstad M (2012) Spring bloom dynamics in Kongsfjorden, Svalbard: nutrients, phytoplankton, protozoans and primary production. *Polar Biology* 35, 191–203.

Hop H, Falk–Petersen S, Svendsen H, Kwasniewski S, Pavlov V, Pavlova O, Søreide JE (2006) Physical and biological characteristics of the pelagic system across Fram Strait to Kongsfjorden. *Progress in Oceanography* 71, 182–231.

Hop H, Pearson TH, Hegset EN, Kovacs KM, Wiencke C, Kwasniewski S, et al. (2002) The marine ecosystemof Kongsfjorden, Svalbard. *Polar Research* 21, 167–208.

JGOFS (1994) Protocols for the joint global ocean flux study (JGOFS) core measurements. *IOC Manual and Guides* 29, 181.

Johnsen G, Sakshaug E (2007) Bio-optical characteristics of PSII and psi in 33 species (13 pigment groups) of marine phytoplankton, and the relevance for pulse amplitude-modulated and fast-repetition-rate fluorometry. *Journal of Phycology* 43, 1236–1251.

Kerkar AU, Tripathy SC, David JH, Pandi SR, Sabu P, Tiwari M (2021) Characterization of phytoplankton productivity and bio-optical variability of a polar marine ecosystem. *Progress in Oceanography* 195, 102573.

Kerkar AU, Tripathy SC, Minu P, Baranval N, Sabu P, Patra S, Mishra RK, Sarkar A (2020) Variability in primary productivity and bio-optical properties in the Indian sector of the Southern Ocean during an austral summer. *Polar Biology* 43(10), 1469–1492.

Kirk JTO (1994) *Light and Photosynthesis in Aquatic Ecosystems.* Cambridge University Press, p. 509.

Kowalczuk P, Sagan S, Makarewicz A, Meler J, Borzycka K, Zabłocka M, et al. (2019) Bio-optical properties of surface waters in the Atlantic Water inflow region off Spitsbergen (Arctic Ocean). *Journal of Geophysical Research: Oceans* 124, 1964–1987.

Kulk G, van de Poll WH, Buma AGJ (2018) Photophysiology of nitrate limited phytoplankton communities in Kongsfjorden, *Spitsbergen Limnology and Oceanography* 63, 2606–2617.

Kurian S, Roy R, Repeta DJ, Gauns M, Shenoy DM, Suresh T. et al. (2012) Seasonal occurrence of anoxygenic photosynthesis in Tillari and Selaulim reservoirs, western India. *Biogeosciences* 9(7), 2485–2495. doi: 10.5194/bg-9-2485-2012.

Levasseur ME, Therriault JC (1987) Phytoplankton biomass and nutrient dynamics in a tidally induced upwelling: the role of $NO_3:SiO_4$ ratio. *Marine Ecology Progress Series* 39, 87–97.

Lohrenz SD, Weidemann AD, Tuel M (2003) Phytoplankton spectral absorption as influenced by community size structure and pigment composition. *Journal of Plankton Research* 25(1), 35–61.

Matsuoka A, Larouche P, Poulin M, Vincent W, Hattori H (2009) Phytoplankton community adaptation to changing light levels in the southern Beaufort Sea, Canadian Arctic. *Estuarine, Coastal and Shelf Science* 82, 537–546.

Mendes CRB, Kerr R, Tavano VM. et al. (2015) Cross-front phytoplankton pigments and chemotaxonomic groups in the Indian sector of the Southern Ocean. *Deep-Sea Research II* 118, 221–232.

Mitchell BG (1990) Algorithms for determining the absorption coefficient for aquatic particulates using the quantitative filter technique. In *Ocean Optics X*, Vol 1302, International Society for Optics and Photonics, pp. 137–148.

Mitchell BG, Kahru M, Wieland J, Stramska M (2002) Determination of spectral absorption coefficients of particles, dissolved material and phytoplankton for discrete water samples. In GS Fargion, JL Mueller (eds.), *Ocean Optics Protocols For Satellite Ocean Color Sensor Validation, Revision 3, NASA Tech. Memo.*, Vol 2, Greenbelt, MD, National Aeronautics and Space Administration, Goddard Space Flight Center, pp. 231–257.

Not F, Massana R, Latasa M. et al. (2005) Late summer community composition and abundance of photosynthetic picoeukaryotes in Norwegian and Barents Sea. *Limnology Oceanography* 50, 1677–1686.

Palmisano AC, SooHoo JB, SooHoo SL, Kottmeier ST, Craft LL, Sullivan CW (1986): Photoadaptation in *Phaeocystispouchetii* advected beneath annual sea ice in McMurdo Sound, Antarctica. *Journal of Plankton Research* 8(5), 891–906.

Para J, Coble PG, Charriere B, Tedetti M, Fontana C, Sempere R (2010) Fluorescence and absorption properties of chromophoric dissolved organic matter (CDOM) in coastal surface waters of the northwestern Mediterranean Sea, influence of the Rhone River. *Biogeosciences* 7, 4083–4103.

Pavlov AK, Silyakova A, Granskog MA, Bellerby RGJ, Engel A, Schulz KG, Brussaard, CPD (2014) Marine CDOM accumulation during a coastal Arctic mesocosm experiment: no reposne to elevated pCO$_2$ levels. *Journal of Geophysical Research* 119, 1216–1230.

Payne CM, Roesler CS (2019) Characterizing the influence of Atlantic water intrusion on water mass formation and phytoplankton distribution in Kongsfjorden, Svalbard. *Continental Shelf Research* 191, 104005. doi: 10.1016/j.csr.2019.104005.

Pettersen R, Johnsen G, Berge J, Hovland EK (2011) Phytoplankton chemotaxonomy in waters around the Svalbard archipelago reveals high amounts of Chl *b* and presence of gyroxanthin-diester. *Polar Biology* 34:627–635.

Pielou EC (1966) The measurement of diversity in different types of biological collections. *Journal of Theoretical Biology* 13, 131–144.

Piquet AMT, Scheepens JF, Bolhuis H, Wiencke C, Buma AGJ (2010) Variability of protistan and bacterial communities in two Arctic fjords (Spitsbergen). *Polar Biology* 33, 1521–1536. doi: 10.1007/s00300-010-0841-9.

Piquet AMT, van de Poll WH, Visser RJW, Wiencke C, Bolhuis H, Buma AGJ (2014) Springtime phytoplankton dynamics in Arctic Krossfjorden and Kongsfjorden (Spitsbergen) as a function of glacier proximity. *Biogeosciences* 11, 2263–2279.

Piwosz K, Walkusz W, Hapter R, Wieczorek P, Hop H, Wiktor J (2009) Comparison of productivity and phytoplankton in a warm (Kongsfjorden) and a cold (Hornsund) Spitsbergen fjord in mid-summer 2002. *Polar Biology* 32, 549–559. doi: 10.1007/s00300-008-0549-2.

Shannon C, Weaver W (1949) *The Mathematical Theory of Communication.* Urbana, IL, University of Illinois Press, p. 117.

Shuman FR, Lorenzen CJ (1975) Quantitative degradation of chlorophyll by a marine herbivore. *Limnology and Oceanography* 20(4), 580–586.

Singh A, Krishnan KP (2019) The spatial distribution of phytoplankton pigments in the surface sediments of the Kongsfjorden and Krossfjorden ecosystem of Svalbard, Arctic. *Regional Studies in Marine Science* 31, 100815. https://doi.org/10.1016/j.rsma.2019.100815.

Stedmon CA, Markager S (2001) The optics of chromophoric dissolved organic matter (CDOM) in the Greenland Sea: an algorithm for differentiation between marine and terrestrially derived organic matter. *Limnology Oceanography* 46(8), 2087–2093.

Stedmon CA, Markager S, Kaas H (2000) Optical properties and signatures of Chromophoric Organic Dissolved Matter (CDOM) in Danish Coastal waters. *Estuarine, Coastal and Shelf Science* 51, 267–278.

Stramska M, Stramski D, Hapter R, Kaczmarek S, Stoń J (2003) Bio-optical relationships and ocean color algorithms for the north polar region of the Atlantic. *Journal of Geophysical Research* 108(C5), 3143. doi: 10.1029/2001JC001195.

Stramski D, Bricaud A, Morel A (2001) Modeling the inherent optical properties of the ocean based on the detailed composition of the planktonic community, *Applied Optics* 40(18), 2929–2945.

Svendsen H, Beszczynska-Møller A, Hagen JO, Lefauconnier B, Tverberg V, Gerland S, Ørbæk JB, Bischof K, Papucci C, Zajaczkowski M, Azzolini R, Bruland O, Wiencke C, Winther J-G, Dallmann W (2002) The physical environment of Kongsfjorden-Krossfjorden, an Arctic fjord system in Svalbard. *Polar Research* 21, 133–166.

Tripathy SC, Pavithran S, Sabu P, Naik RK, Noronha SB, Bhaskar PV, Kumar NA (2014) Is primary productivity in the Indian Ocean sector of Southern Ocean affected by pigment packaging effect? *Current Science* 107(6), 00113891.

Uitz J, Claustre H, Morel A, Hooker SB (2006) Vertical distribution of phytoplankton communities in open ocean: an assessment based on surface chlorophyll. *Journal of Geophysical Research* 111, C08005. doi: 10.1029/2005JC003207.

van de Poll WH, Kulk G, Rozema PD, Brussaard CPD, Visser JW, Buma AGJ (2018) Contrasting glacial meltwater effects on post-bloom phytoplankton on temporal and spatial scales in Kongsfjorden, Spitsbergen. *Elementa: Science of the Anthropocene* 6, 50. doi: 10.1525/elementa.307.

van de Poll WH, Maat D, Fischer P, Rozema P, Daly O, Koppelle S, et al. (2016) Atlantic advection driven changes in glacial meltwater: effects on phytoplankton chlorophyll-a and taxonomic composition in Kongsfjorden, Spitsbergen. *Frontiers Marine Science* 3, 200.

Vihtakari M, Welcker, J, Moe, B, Chastel O, Tartu S, Hop H, Bech C, Descamps S., Gabrielsen GW (2018) Blacklegged kittiwakes as messengers of Atlantification in the Arctic. *Scientific Reports* 8, 1178.

Walczowski W, Beszczynska-Möller A, Wieczorek P, Merchel M, Grynczel A (2017) Oceanographic observations in the Nordic sea and fram strait in 2016 under the IO PAN long-term monitoring program AREX. *Oceanologia* 59, 187–194.

Wang J, Cota GF, Ruble DA (2005) Absorption and backscattering in the Beaufort and Chukchi Seas. *Journal of Geophysical Research* 110, doi: 10.1029/2002JC001653.

Wang SQ, Ishizaka J, Yamaguchi H, Tripathy SC, Hayashi M, Xu YJ, Mino Y, Matsuno T, Watanabe Y, Yoo SJ (2014) Influence of the Changjiang River on the light absorption properties of phytoplankton from the East China Sea. *Biogeosciences* 11(7), 1759–73.

Welschmeyer NA (1994) Fluorometric analysis of chlorophyll a in the presence of chlorophyll b and pheopigments. *Limnology Oceanography* 39(8), 1985–1992.

Welschmeyer NA, Lorenzen CJ (1985) Chlorophyll budgets: Zooplankton grazing and phytoplankton growth in a temperate fjord and the Central Pacific Gyres. *Limnology Oceanography* 30(1), 1–21.

Woźniak B, Dera J, Ficek D, Majchrowski R, Kaczmarek S, Ostrowska M, Koblentz-Mishke OI (1999) Modelling the influence of acclimation on the absorption properties of marine phytoplankton. *Oceanologia* 41(2), 187–210.

14 Recent Advances in Seismo-Geophysical Studies for the Arctic Region under Climate Change Scenario

O. P. Mishra and Priya Singh
National Centre for Seismology

Neloy Khare
Ministry of Earth Sciences

CONTENTS

DOI: 10.1201/9781003265177-14

14.1 INTRODUCTION

It is a proven fact that the Arctic is delimited by the Arctic Circle (66°32'N) that approximates the southern boundary of the midnight sun (Figure 14.1a), contributing to changes in temperature, presence of mountain ranges, prevalence of large water bodies, complex geology and physiography, and different layers of permafrost that depict physical, geological, geographical and ecological characteristics of the Arctic region. The prodigiously large region of the Arctic extends to northern North America, northern Asia and northern Europe, accounting into eight countries with the existence of sea and ocean in between these continents.

The region is associated with considerable changeability in climate, meteorology and physical variability-related geography that control the terrestrial, freshwater and marine resources throughout the Arctic (Murray et al. 1998). The Arctic sea is covered with sea ice retreating at a rate which greatly exceeds the predictions of climate models. It is reported that sea-ice retreat has implications to a large-scale methane (CH_4) outbreak from the Arctic sea bed due to disintegration of unstable methane hydrates that remained sealed into the sedimentary by the permafrost cap, which in turn contributes towards economic benefit of sea-ice retreat due to easier Arctic navigation and oil exploration, but at the same time methane is a greenhouse gas of 23 times as powerful as CO_2 though shorter-lived in the atmosphere, which has the serious threat to the planet by impacting climate change scenario when the economic cost of climate change is greater with respect to per unit of additional warming (Wadham and Davis 2000; Wadham 2012).

Based on ecological parameters, the Arctic region is generally divided into the High Arctic and Low Arctic of which south of the Arctic is the subarctic, lying

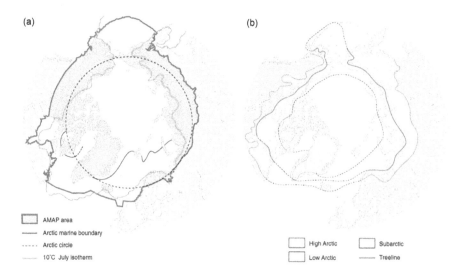

FIGURE 14.1 (a) The Arctic as defined by temperature (after Stonehouse 1989), and the Arctic marine boundary, also showing the boundary of the AMAP assessment area. (b) Arctic and subarctic floristic boundaries (after Bliss 1981; Linell and Tedrow 1981; Nordic Council of Ministers 1996).

between the closed canopy boreal forests to the south and the tree line to the north as shown in Figure 14.1b. The southern boundary of the subarctic generally corresponds with the southern limits of discontinuous and sporadic permafrost that remains present throughout most of the subarctic (Linell and Tedrow 1981). The marine boundary of the Arctic is characterized and is located along the convergence of cool, less saline, surface waters from the Arctic Ocean and warmer and saltier waters from ocean to the south. It is interesting to note that the eastern Canadian Arctic Archipelago is associated with marine boundary that exists nearly along a latitude of 630°N, which swings north Baffin Island and the west coast of Greenland whilst off the east coast of Greenland, the marine boundary lies at approximately 650°N. It is reported that warm Pacific water passing through the Berring strait meets the Arctic Ocean water at approximately 720°N from Wrangel Island in the west to Amundsen Gulf in the east (Stonehouse 1989). During the passage of warm Pacific water through the Bering Strait, the Pacific water started to undergo several modifications on the broad Chukchi shelf due to a series of ice-related processes consisted of freezing, melting, cooling and run-off, which further gets incorporated into the surface mixed layer or subducts and flows down the undersea canyons to contribute to the halocline (Murray et al. 1998). It is found to be a very difficult to define a distinct boundary between separating the Pacific water from Arctic water because of negligible hydro-geo-chemical contrast in two different types of water, which allowed to draw arbitrary boundary across the Bering Strait as the point where modification started to commence through ice-related processes. The Arctic marine area includes Arctic marine biology and sea-ice cover, the Arctic Monitoring and Assessment Programme (AMAP) marine area to the south of the region that extended to different zones, such as Davis Strait, the Labrador Sea and Hudson Bay, the Greenland, Iceland, Norwegian, Barents, Kara and White Sea, and the Bering Sea. As mentioned above, delineation of the Arctic is based on several factors, related to physical-geographical and politico-administrative setup of the regional extent that has been defined based on a compromise among various definitions, which has been used for AMAP assessment. The geographical coverage of the Arctic incorporates elements of the Arctic Circle: political boundaries, vegetation boundaries, permafrost limits, and major oceanographic features, and the Arctic region is essentially covered by the terrestrial and marine areas north of the Arctic Circle (66032′ N), and north of 620N in Asia and 600N in North America, and further modified to include the marine areas north of the Aleutian chain, Hudson Bay, and parts of the North Atlantic Ocean including Labrador Sea (Murray et al. 1998).

The Arctic region is characterized by low air temperature because of less solar radiation in comparison to that of other parts of the world. Regional and local climatic conditions in the Arctic are governed by the macro-scale topography of the Earth's surface with reference to the distribution of land, sea and mountain. Interestingly, the frequency and the preferred tracks of the persistent Pacific and Atlantic low-pressure systems and the position of the persistent high-pressure systems not only play an important role in the existence of the regional and local climates in the Arctic, but that also link the Arctic climatic system to the world climatic system. Climatic conditions in the Arctic consisted of maritime, which is evident for Iceland with average winter temperature between −20°C and 10°C, the Norwegian coast with average winter temperature of −20°C, and the adjoining parts of Russia with mean winter temperature of −110°C, whilst the mean summer temperature of 100°C has been recorded for the maritime region. Alaskan coast, a very narrow coast due to Alaskan mountains, and Coastal Ranges also fall under the maritime climatic condition whose winter is recorded to be moderate and summers are mild with average variability of temperature found to be from −50°C in winter to 100°C in summer. The continental subtypes, the other category of climatic condition in the Arctic, correspond to the interior of the Arctic from northern Scandinavia toward Siberia and from eastern Alaska toward the Canadian Arctic Archipelago with much lower precipitation and significant differences between summer and winter conditions with mean taken in July as 5–100°C, and mean for January from −200°C to −400°C, whilst the average temperature in the central Arctic varies between −30 and −350°C in winter and between 00°C and 20°C in summer (Prik 1959; Moore 1981). It is observed that variability in cold Arctic waters is small, but important in which the varying temperature condition in the western North Atlantic is opposite to that in the eastern North Atlantic. AMAP assessment report by Moore (1981) showed that the temperature state of the ocean is closely linked to atmospheric circulation having a positive feedback mechanism existing between the atmospheric and oceanic circulations. High atmospheric pressure is associated with low temperatures in the ocean, whilst low pressure is related to a warmer ocean. It is also demonstrated that changes in ocean climate influence transport mechanisms and ice cover, whilst in warm years, there is an increased transport of warm water masses to the Arctic, resulting in decreased ice cover. In cold years, transport of warm water to the Arctic is reduced and sea-ice coverage is greater (Ikeda 1990a, b, Aadlandsvik and Loeng 1991). It is also documented that the total precipitation in the Arctic is generally less than 500 mm with its range between 200 and 400 mm (Loshchicov 1965) with uneven distribution of the precipitation in such a way that the Arctic coast is associated with higher precipitation and the central Polar Basin and its vicinity have lower amount of precipitation. These observations indicate that meteorological and climatologically driven conditions have strong geophysical impacts and bearing on the Arctic region of the Earth.

The vast Arctic terrestrial landscape, covers an area of about $13.4 \times 10^6 \, km^2$ within the AMAP boundary with large tracks of land covered by glacial ice. These are the large masses of the ice that flow under their own weight and are formed where the mean winter snowfall exceeds the mean summer melting. Snow gets transformed into ice at suitable conditions of melting, refreezing and pressure as shown in Figure 14.2.

Greenland, often described as the largest island in the world, is actually comprised of numerous mountainous islands almost entirely covered with a permanent ice cap up to 3000 m thick (Stonehouse 1989; Bjerregaard 1995). It spans 24° of latitude (2670 km), is 1200 km across at its widest point and covers an area of some 2,186,000 km². The main tracts of ice-free land are in the southwest, the north (Peary Land) and the northeast (north of Scoresby Sund). Along the Greenland coast, outlet glaciers flow from the ice sheet to the sea. Glaciers terminating in the sea periodically break off, or calve, forming icebergs. Iceland is located south of the Arctic Circle (66°32′N). This mountainous and volcanically active island lies on the mid-Atlantic ridge. Its average height is approximately 500 m above sea level. One quarter of the country is less than 200 m above sea level. It has an area of 103,000 km²,

FIGURE 14.2   Topography and bathymetry of the Arctic (based on the ETOPO5 data set, NOAA 1988).

with 11% of its surface covered by glaciers and more than 50% of its land surface unvegetated (Stonehouse 1989). The Faeroe Islands, with a total area of 1399 km^2, are located 430 km southeast of Iceland. The terrain is mountainous with an average elevation of 300 m. Svalbard and Franz Josef Land are Arctic archipelagos of 63,000 and 10,000 km^2, respectively. These mountainous islands, and others lying to the north of Eurasia, are about 90% covered by ice. The Fenno Scandian Arctic area covers roughly 300,000 km^2, but most of this area is subarctic due to the warming influence of the Gulf Stream extension (Encyclopedia Britannica 1990). The Kola Peninsula (ca. 145,000 km^2) on the Russian mainland is also subarctic and contains many lakes. Permafrost is absent, except for sporadic occurrences at the tip of the peninsula, and the coasts are ice-free (Ives 1974; Luzin et al. 1994). The Russian

Arctic west of the Ural Mountains shows much variation in landscape, but large areas consist of flat, poorly drained lowlands with marshes and bogs. The Siberian coast is generally flat and includes the deltas of many large, north-flowing rivers. Ice-covered mountains are characteristic of the Russian peninsulas of Taimyr and Chukotka. In eastern Siberia, there are several mountain ranges (e.g., Verkhoyansk, Chersky and Momsky) with peaks reaching heights of over 2500 m. The entire area of Arctic Russia within the AMAP boundary is approximately $5.5 \times 106 \, km^2$. The numerous islands of the Russian Arctic cover an area of $135,500 \, km^2$. The largest of these is Novaya Zemlya, an archipelago with two main islands. The northern island is mountainous, with about half of the $48,000 \, km^2$ area covered by glaciers and a small ice cap. The southern island is smaller ($33,000 \, km^2$), largely ice-free and characterized by large coastal plains, especially in the southern parts. Alaska's Arctic, according to the AMAP definition, extends over an area of $1.4 \times 106 \, km^2$, and is dominated by rugged mountain ranges that stretch across the state in the south and north, reaching a maximum height of 6194 m at Mount McKinley. Extensive glaciers are found in the south central and south-eastern mountains. These ranges give way to foothills and low-lying coastal tundra plains in the southwest and along the northern coast of Alaska. Extending westward into the Pacific Ocean beyond the Alaska Peninsula are the volcanic Aleutian Islands (Stonehouse 1989). The Canadian Arctic landscape covers an area of approximately $4 \times 10^6 \, km^2$, comprised of the northern Canadian mainland in the south and the Arctic Archipelago to the north. At the most western boundary of the mainland is the Yukon Plateau, consisting of rolling uplands with valleys and isolated mountains. Southwest of this plateau are the Coast Mountains with extensive glaciers. To the northeast are the Mackenzie Mountains. These mountain ranges give way to the Interior Lowlands, comprised of plateaus ranging in height from 1200 m in the west to 150 m in the east. This region, which is transacted by the Mackenzie River, is characterized by extensive wetlands. The large Great Bear and Great Slave Lakes extend from the Interior Lowlands eastward into the Canadian Shield. The Shield extends to the east coast and contains numerous lakes and the vast expanse of Hudson Bay (National Wetlands Working Group 1988, Prowse 1990). The Canadian Arctic Archipelago extends far to the north of the mainland. Flat to rolling terrain is characteristic of the High Arctic islands in the western and central archipelago (e.g., Banks, Melville, Victoria, Bathurst and Prince of Wales Islands). The north-eastern islands (Baffin, Devon, Ellesmere and Axel Heiberg) contain rugged, ice-capped mountains up to 2000 m in height (Prowse 1990; Woo and Gregor 1992; Sly 1995).

In this chapter, an attempt has been made to review recent advancements made in seismo-geophysical research so far for the Arctic region to unravel the hidden mystery related to the seismotectonic and geophysical characteristics of the sub-surface layers vis-à-vis exploration of natural resources using geophysical tools for the region.

14.2 PHYSIOGRAPHICAL AND GEOLOGICAL SETUP OF THE ARCTIC REGION

The major physiographic regions (Figure 14.3a) and bedrock geology (Figure 14.3b) of the Arctic are adjudged as one of the complex geological set-ups with varying characteristics because of diverse dispositions of different types of rocks in the region. Greenland, and a vast region of the Canadian Arctic, from the Atlantic Ocean in the east to Great Bear Lake and Great Slave Lake in the west, is underlain by the Canadian Shield. This Precambrian, crystalline rock mass is exposed in some areas and covered by glacial deposits and thin soil in others. The Canadian Shield extends northward to include Baffin Island. The remaining islands in the Canadian Arctic Archipelago are primarily made up of Paleozoic and Mesozoic sedimentary rocks. Along the southwest coast of Hudson Bay are the Hudson Lowlands, comprised mainly of Lower Palaeozoic rock and covered by Quaternary sediments. To the west of the Shield are the Interior Plains made up of Devonian and Cretaceous sedimentary formations.

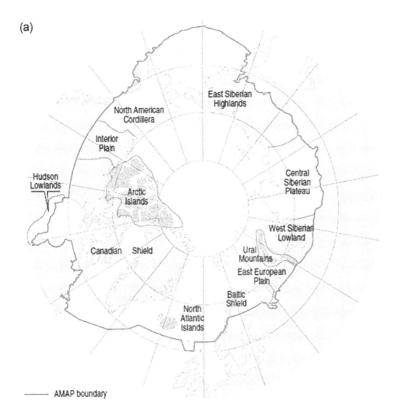

(a)

FIGURE 14.3 (a) Geologic and physiographic regions of the Arctic (after Linell and Tedrow 1981, by permission of Oxford University Press). (b) Bedrock geology of the Arctic (Geological Survey of Canada 1995).

(Continued)

(b)

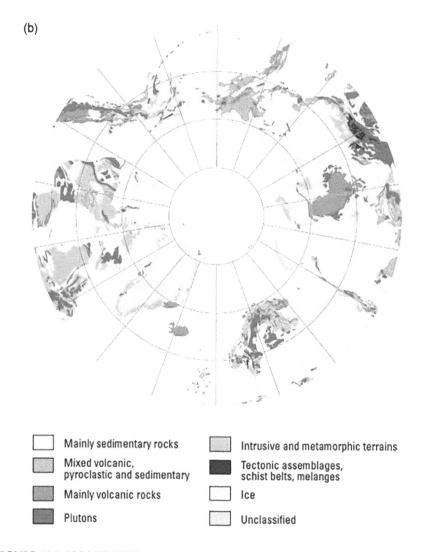

	Mainly sedimentary rocks		Intrusive and metamorphic terrains
	Mixed volcanic, pyroclastic and sedimentary		Tectonic assemblages, schist belts, melanges
	Mainly volcanic rocks		Ice
	Plutons		Unclassified

FIGURE 14.3 (CONTINUED)   (a) Geologic and physiographic regions of the Arctic (after Linell and Tedrow 1981, by permission of Oxford University Press). (b) Bedrock geology of the Arctic (Geological Survey of Canada 1995).

The North American Cordillera, a wide belt of high mountains and plateaus of Palaeozoic, sedimentary origin, extend along western Canada and Alaska. In Alaska, these mountains merge to the north with the Arctic Foothills, which descend still farther north into a coastal plain (Linell and Tedrow 1981; Prowse 1990; Natural Resources Canada 1994). Iceland has been created by volcanic activity along the mid-Atlantic ridge during the last 20 million years. New volcanic rock is constantly being added and about one-tenth of Iceland is covered by lava deposited since the last ice age (Einarsson 1980). The Aleutian Islands to the west of Alaska are also volcanic. Northern Fennoscandia and the Kola Peninsula are comprised of ancient, crystalline rocks forming the Baltic Shield. East of this region, from the White Sea to the Ural Mountains, lies the East European Plain, made up of sedimentary rock covered by a deep layer of glacial drift. The Ural Mountain complex, which includes Novaya Zemlya, is a region of folded Paleozoic bedrock, covered thinly with glacial deposits. The other Russian Arctic islands are also primarily formed by sedimentary formations of the Paleozoic Era. The West Siberian Lowland, comprised of till-covered sedimentary rocks, extends from the Urals to the Yenisey River. From here, eastward to the Lena River, is the Central Siberian Plateau. This till-covered region is underlain by the Anabar Shield and peripheral sedimentary rocks. To the north, the Taimyr Peninsula contains a folded mountain complex of sedimentary rocks, overlain by shallow soils. The region east of the Lena River is similarly comprised of folded sedimentary mountains, with some volcanic rocks. Glacial drift is discontinuous in this region (Linell and Tedrow 1981). Low plateaus and plains are characteristics of the region through which the Lena River passes of the northern margin of Siberia from the Taimyr Peninsula to the Kolyma River (Linell and Tedrow 1981). Role of seismo-geophysical tools has become paramount to explore sub-surface geology of the Arctic region with greater reliability to enrich our understanding about the seismo-geophysical potential and natural resources in the Arctic region.

14.2.1 Permafrost and Upper Soil Formation

Permafrost, or perennially frozen ground, is defined as a material that stays at or below 0°C for at least two consecutive summers (Woo and Gregor 1992). It may consist of soil, bedrock or organic matter. Spaces within the ground material may be filled with ice in the form of ice lenses, veins, layers and wedges. When very little or no ice is contained in the frozen substrate, this is referred to as dry permafrost (Figure 14.4a) (Linell and Tedrow 1981). Permafrost may reach depths of 600–1000 m in the coldest areas of the Arctic (Stonehouse 1989).

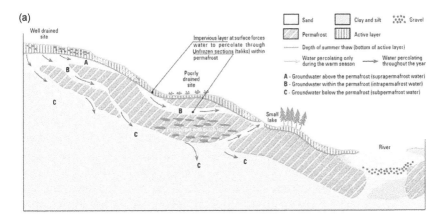

FIGURE 14.4   (a) Occurrence of groundwater in permafrost areas (after Mackay and Løken 1974; Linell and Tedrow 1981). (b) Circumpolar permafrost distribution (CAFF 1996).

(Continued)

(b)

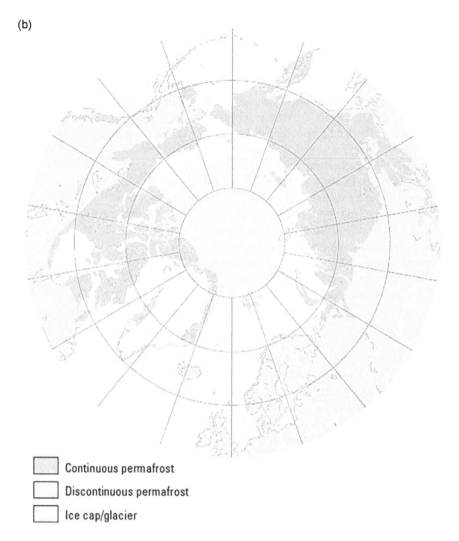

☐ Continuous permafrost

☐ Discontinuous permafrost

☐ Ice cap/glacier

FIGURE 14.4 (*CONTINUED*)   (a) Occurrence of groundwater in permafrost areas (after Mackay and Løken 1974; Linell and Tedrow 1981). (b) Circumpolar permafrost distribution (CAFF 1996).

   The distribution of permafrost is broadly determined by climate, particularly air temperature and the resultant energy balance between the air and the ground. Several secondary factors which affect permafrost occurrence include elevation, composition and colour of the ground surface, ground aspect, soil moisture, and extent and type of plant cover. Permafrost is generally not found under large bodies of water greater than three metres deep, the maximum depth to which winter ice develops. The insulating effects of glaciers and extensive snow cover also reduce permafrost development (Ives 1974). Figure 14.4b shows distribution of continuous and discontinuous permafrost north of 50°N. Perennially frozen ground occurs throughout the Arctic and extends into the forested regions to the south. Along the northern coastlines, frozen grounds meet the sea, with permafrost extending under some shelf seas. Permafrost influences soil development in the north. In general, Arctic soils are either poorly drained and underlain by solid, ice-rich permafrost, or well drained and situated over dry permafrost. Poorly drained soils are found in 85%–90% of the Low Arctic and in the few wet meadows of the High Arctic. Well-drained soils are common in the extensive, sparsely vegetated areas of the High Arctic, and are scattered throughout the Low Arctic in areas where water can escape, such as on steep slopes and beach ridges. Oxidation processes in these drier soils result in lower organic matter content compared to wetter soils (Rieger 1974; Everett et al. 1981).

During the summer, the upper layer of soil in the Arctic thaws and is termed the active layer. Its depth varies according to temperature, ground material, soil moisture content and plant cover, ranging from as little as several centimetres in far northern wet meadows to as deep as a few metres in warmer, more southern, dry areas with coarse-grained soils (Ives 1974) (Figure 14.4a). Soil-forming processes are largely restricted to the active layer, which is unstable due to frost action during repeated freezing and thawing. It is observed that frost action results in characteristic surface features, such as frost scars, stone circles, mud circles, solifluction lobes and stone stripes (Rieger 1974; Stonehouse 1989).

14.2.2 GROUNDWATER RESOURCES

Groundwater levels and distribution within the Arctic are greatly influenced by permafrost. Permafrost affects the amount of physical space in which groundwater can be held and the movement of water within drainage systems. There are three general types of groundwater: suprapermafrost, intrapermafrost and subpermafrost (Figure 14.4a). Suprapermafrost water lies above the relatively impermeable perma-frost table in the active layer during summer, and year-round under lakes and riv-ers that do not freeze. Intrapermafrost water resides in unfrozen sections within the permafrost, such as tunnels called 'taliks', located under alluvial flood plains and under drained or shallow lakes and swamps. Subpermafrost water is located beneath the permafrost table and its depth below the surface depends on the thickness of the permafrost. In this latter case, the permafrost acts as a relatively impermeable upper barrier. These three types of aquifers, which may be located in bedrock or in uncon-solidated deposits, may interconnect with each other or with surface water (Mackay and Løken 1974; van Everdingen 1990). Generally low levels of annual precipitation in the Arctic restrict the recharge of groundwater. In addition, infiltration of water to aquifers is restricted by permafrost year-round and by the frozen active layer for up to 10 months of the year. Frozen substrate does not entirely prevent water from seeping through to aquifers but slows the rate of infiltration by one or more orders of magnitude compared to unfrozen ground (van Everdingen 1990). The degree of groundwater recharge is influenced by the material comprising the substrate. Recharge is greatest in regions with coarse-grained, unconsolidated material and areas with exposed bedrock containing channels or fractures which allow passage of water. Infiltration, and therefore recharge, is least in areas covered by fine-grained deposits such as silt and clay (van Everdingen 1990). Groundwater is discharged via springs, base flow in streams and icings. These discharges can be fed by supra, intra or sub permafrost water. Perennial springs are generally fed by subpermafrost aquifers and less commonly by intrapermafrost water. Icings (also known as aufeis or naleds) are comprised of groundwater that freezes when it reaches the streambed during winter when base flow freezes. When fed by suprapermafrost water, icings generally stop growing before the end of winter. Icings formed by discharge from intra- or subpermafrost groundwater continue to build until temperatures climb above 0°C in spring. In cases of large icings, spring melt and run-off can result in significant redistribution of groundwater (van Everdingen 1990). Groundwater is quite extensive in the Arctic. For example, approximately two-thirds of the Yukon in the Canadian Arctic is underlain by aquifers (Hardisty et al. 1991). The largest groundwater aquifers in Iceland have been mapped in Elíasson (1994). These are generally found in highly permeable lava. Groundwater represents an important source of water in some Arctic countries.

14.3   SEISMO-GEOPHYSICAL STUDIES AND ITS IMPLICATIONS TO ENVIRONMENTAL VARIATION

The most exciting initiative for the recent polar studies was the International Polar Year (IPY) in 2007–2008 that witnessed a growing geo-scientific community who have made very impressive breakthrough in studying the structural dynamics of the polar region (Kanao et al. 2015). Significant advances in seismic and geophysical instrumentations, high-quality data acquisition, data processing and computation, advanced seismo-geophysical analytics, and integrated interpretational module led to achieve scientific objectives and targets in cryoseismology and cryogeophysics, as one of the important missions to explore the Arctic region for the benefit of science and society with reference to environmental variations in the region.

Kanao et al. (2015) defined the Arctic region where North Pole occupies the centre of the Arctic Ocean, which has been affected by environmental variation of the Earth in both short and long terms of time scale. The short-term environmental variation is associated with the undergoing global warming as the most significant phenomena to influence a rapid change in the cryospheric drivers, such as sea ice, ice sheet, ice shelves, ice caps and glaciers in the Arctic (IPCC 2007). In contrary, a long-term environmental variation during the Earth's history, affects the deformation of solid Earth overlain by the cryosphere. The different and diverse types of variations in the surface environments having spatial and temporal changes can be estimated aptly through comprehensive investigations using seismo-geophysical and geological methods. Recent studies on seismo-geophysical and geo-engineering in the Arctic region aimed to address several unaddressed issues related to polar seismicity and its relationship with structure and dynamics of the crust and overlying ice sheets, sea ice, and other cryospheric evolution, seismic wave propagation to crustal and sub-crustal deep layers and tectonics, crustal and lithospheric thickness mapping and modelling with reference to plate reconstructions, resource mapping (water/methane/sea bed minerals), seismic attenuation structure and exploration of Arctic mid-ocean ridge system, which are described successively in following sections.

14.3.1 SEISMICITY AND CRYOSEISMIC EVENTS (ICE QUAKES) IN THE ARCTIC REGION

Unlike the Antarctic, the Arctic region is devoid of permanent seismic stations beyond 660°N, but several temporary seismographic stations of Eastern Asia including in Far East of Russia helped us to analyse nature and seismicity pattern of the Arctic region. Following the 1995 North Sakhalin earthquake (Mw 7.6), a big research project for studying stagnant slab and its relationship with mantle dynamics using broadband seismic station and global positioning system (GPS) was taken up by the Russian Academy of Science in collaboration with international scientists of different countries (Kanao 2018a). Recording of seismicity of the Arctic region (Figure 14.5a and b) and its detailed analyses are one of the most challenging issues to understand the seismic potential and nature of seismogenic strengths of various sub-surface layers beneath the Arctic region.

It is well understood that seismological processes and related phenomena are attributed by seismicity that can be translated into structure of the Earth and its dynamics in the Arctic, which has not yet been adequately studied because of the lack of adequate seismographic stations there in and around the North Pole of the Arctic region. A well-defined distribution map of the permanent GSN stations in the Arctic region has been demarcated by red colour. Data from IRIS/DMS and PASSCAL showed recording of teleseismic events (Mw ≥ 5.5) during 1990–2004 for different epicentre distance ranges in the Arctic (Figure 14.5b). It provides capability to study deeper structure of the earth beneath the Antarctic region.

Recent advances in seismo-geophysical research unravelled the fact that conduction of continuous long-term monitoring of seismic and the dynamic interaction between the cryosphere and geosphere in surface layers of the Earth will provide comprehensive information about the unresolved issues of tectonics and resource potential of the Arctic region with special emphasis on Eurasian continent and surrounding oceans to establish its relationship with regional evolution during the Earth's history in terms of amalgamation and disposal of super-continents (Kanao et al. 2015). A serious effort has been made by Storchak et al. (2015) to compile earthquake data from the International Seismological Centre (ISC) to assess the long-term accumulation and improvements in seismic event data for the polar region as shown in Figure 14.5c that sheds light on the fact that a majority of earthquakes in the Arctic region are concentrated conspicuously along the plate boundaries from the Northern Atlantic Ridges, the Barents Sea, the Gakkel Ridge within the Arctic Ocean, the Laptev Sea in the northern part of Siberia, to the Far East region of Russia. It is debated that seismicity induced by the subducting plates and deglaciation processes are associated with glacial isostatic adjustment (GIA), which are very much evident around the Alaska and the Baltic Sea, Lopland, northern Europe, principally related to GIA (Kozlovskaya 2013). GIA-based seismic events are related to the transform faults at the ocean bottom and coastal margins of the continent whose hypocentres are confined to the boundary between the continent and the ocean, a coastal line at the margin of the continental ice sheets, ice shelf, and glaciers, which generates tectonic events (inner crustal events) rather than ice quakes or cryoseismic events.

Based on the data from Global Seismographic Network (Butler and Anderson 2008), it has been reported that more than 200 'glacial earthquakes' occurred along the continental margins of which about 95% found occurred at Greenland and remaining events were located in Alaska and Antarctica, which has strong seasonality and increasing frequency in the 21st century (Dahl-Jensen et al. 2010) as shown in Figure 14.6.

More interestingly, local earthquakes reported for Greenland, Baffin Island and Eastern Canada demonstrate close correspondence with crustal and sub-crustal seismic strengths of the regions under study. Study on seismicity and its characterization provides deep insights into tectonic provinces (Figure 14.7a–d), and their history of

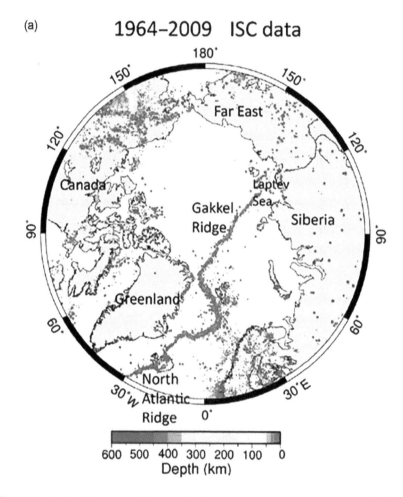

FIGURE 14.5 (a) Seismicity of the Arctic Ocean and surrounding regions in 1964–2009. Hypocentres are determined by a new ISC location algorithm (after Bondár and Storchak 2011; adapted from Kanao 2018a).

(Continued)

(b)

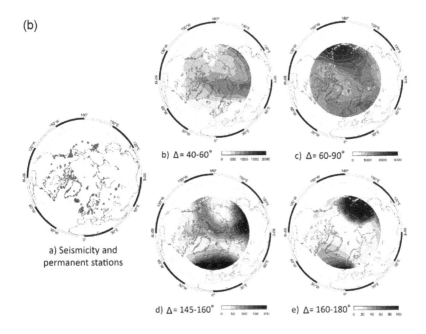

(c)

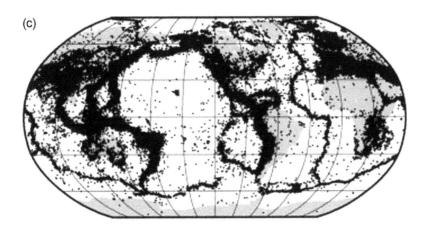

FIGURE 14.5 (*CONTINUED*) (b) A distribution map of the permanent GSN stations in the Arctic region (red colour; data compiled from IRIS/DMS and PASSCAL). Distribution of teleseismic event numbers at each location in the Arctic counted from a list including earthquakes with magnitude greater than or equal to 5.5 in the period of 1990–2004, for the different hypocentre distance ranges. Grey contour scales indicate the accumulated earthquake numbers for the period of 1990–2004 (adapted from Kanao et al. 2015). (c) Hypocentres, magnitudes and arrival times from ~5.1 million seismic events during 1960–2013 are included in the ISC Bulletin (after Storchak et al. 2015).

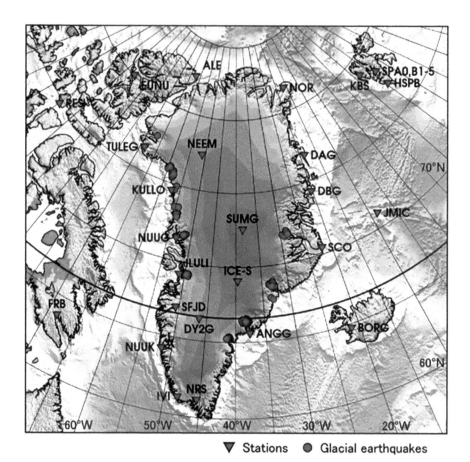

FIGURE 14.6 Distribution of glacial earthquakes around Greenland and vicinity in 1993–2008 (Blue circles). Seismic stations including both permanent and temporal are shown by red triangles. Hypocentres are taken from Nettles and Ekström (2010). (Adapted from Kanao et al. 2012.)

evolution, recent dynamism, heterogeneous structures of the crust and sub-crust in the Arctic region that comprises Eurasian Arctic, the wide area of Siberia, Baikal Rift Zone (BRZ), Far East Russia, Arctic Ocean along with Northern Canada and Greenland.

It is intriguing to note that the role of ice melting on genesis and suppression of earthquakes is well documented by many researchers in their studies carried out for the Polar Regions, including Greenland, Canadian region and the Himalayan glacier region, referred to as the third pole of the world (Wood 2000; Wu and Johnston 2000; Mishra 2014; Olivieri and Spada 2015; Mishra et al. 2021; Khare et al. 2018). They argued that the spatio-temporal distribution of earthquakes is principally associated with the absence of the thick layer of ice due to significant melting of ice sheets under climate change scenario, which has been referred to as glacial melt mass change induced earthquakes by Mishra et al. (2021) for the glaciated areas of the Himalayan

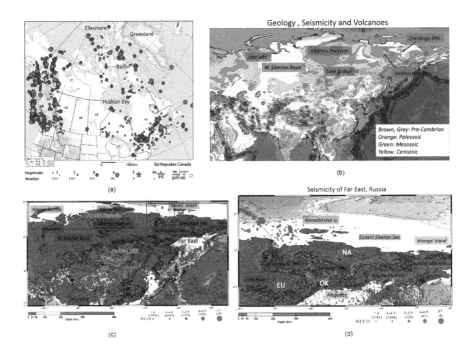

FIGURE 14.7 (a) Seismicity in Canada and surrounding areas in the Arctic in February 2014, based on the compiled data at Natural Resources Canada (after Kanao 2018a). (b) Distribution of tectonic provinces, seismicity and volcanoes in Eurasian continent and surrounding regions (after database of Cornell University; the world geology map, seismicity is after the ISC, 2011). Tectonic provinces are classified as follows: Pre-Cambrian (brown; Archaean, grey; Proterozoic), Paleozoic (orange), Mesozoic (green), and Cenozoic (yellow) (adapted from Kanao et al. 2015). (c) Regions of the Arctic Eurasia, Baikal Rift, Far East with Seismicity of different magnitudes and depths from ISC catalogue (1964–2002) (Adapted from Kanao et al. 2015). (d) Seismicity of Far East, Russia. Seismicity information is taken by ISC (1964–2002). The red lines indicate the plate boundaries between North American Plate (NA), Okhotsk micro-Plate (OK) and Eurasian Plate (EU) (adapted from Kanao et al. 2015).

region. Nonetheless, Olivieri and Spada (2015) did not find a distinct correlation between seismic activity and ice melting in Greenland. Moreover, vertical distribution of earthquakes suggests two pertinent mechanisms involved in genesis of earthquakes in which post-glacial rebound promotes moderate-size crustal earthquakes at a regional scale whilst the continuous ice melting promotes local, shallow, low-magnitude seismicity.

Seismicity of Arctic mid-ocean ridge of deep focus earthquakes (Figure 14.8) helped to determine seismic structure beneath the Arctic Ocean. The plausible reasons of variability in seismic activities along the ridge and in the European Arctic are put forth due to ultraslow spreading processes, characterized by generation of moderate to large earthquakes in magma-rich tectonic belts distributed along the rift system. The pronounced seismic activities are found along the Gakkel Ridge to the northern part of the Svalbard Islands and also to the Franz Josef Land archipelago. Recent studies highlighted that global seismic networks are not the complete solution

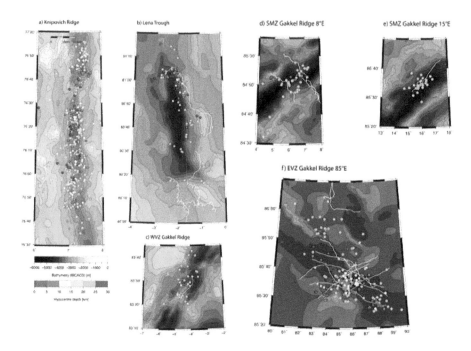

FIGURE 14.8 Local seismicity maps with the bathymetry, showing epicentre locations of earthquakes (white circles) of local earthquakes that have a location error of less than 5 km (length of the semi-major axis of the 95% error ellipse). Grey circles denote earthquakes without hypocentre depth determination. Colours indicate hypocentre depth range; yellow triangles denote OBS locations; yellow lines in all other figures show the drift paths of the seismic arrays on ice floes. (Adapted from Schindwein et al. 2015.)

for detecting microseismic activities as because of several constraints of poor locations of seismic events at the remote plate boundaries, transform faults, different types of transportable seismographs and hydrophones on sea ices, icebergs, and in ocean bottoms. It has, therefore, become possible to detect and monitor seismicity by making use of arrival times and waveforms of the tertiary (T) phase that travels through the ocean, which would help us to estimate fine crustal structure at the sea bottom frontiers in bipolar regions, such as the area beneath the Arctic and the Southern Oceans.

14.3.2 CRUSTAL AND LITHOSPHERIC STRUCTURE OF THE ARCTIC REGION

During and after the IPY, several efforts have been made to study crustal structures of the different tectonic provinces of the Arctic Ocean using seismological and geophysical methods. The recording of microseismicity and data from the ocean bottom seismograph recorders, including Gakkel Ridge–Amerasia Basin (Dziak et al. 2009; Schlindwein et al. 2015; Antonovskaya et al. 2015; Grad et al. 2015). A series of seismological research have been carried out for studying the seismogenic crustal and sub-crustal sources beneath the Arctic region. Extensive geophysical studies carried out to estimate crustal and sub-crustal structures for the southern Svalbard, between the Mid-Atlantic Ridge and the Barren Sea using wide-angle reflection seismic method with additional deployment of Ocean Bottom Seismometer (OBS) to measure seismic energy of strong water multiples as an important ingredient for estimating shallow part of the crust along with continent-ocean transition (COT) and the continental structures down to the Lithosphere-asthenosphere system (LAB) with support of gravity and density modelling (Grad et al. 2015). They found that about 4–9 km thick oceanic crust formed at the Knipovich Ridge as a distinct and narrow COT and 30–35 km thick continental crust beneath the Barents Sea has also been reported. Interestingly, change in seismic velocity with depth suggests lherzo-lite composition of the uppermost oceanic mantle, and dunite composition beneath the continental crust.

14.3.2.1 Crustal Structure of Siberia

It is worth to mention that active source experiments have also been carried out by setting up of the long-distance seismic profiles by the peaceful Nuclear Explosions (PNEs) in a wide area of Siberia to investigate its lithospheric thickness of the Siberia (Melnik et al. 2015). Studies by Melnik et al. (2015) highlighted the fact that it is very difficult to ascertain the variability of the asthenospheric thickness underneath the lithosphere by PNE surveys. The velocity variation in the upper mantle is found associated with a vertical layered structure or otherwise horizontal heterogeneity. The principal constraint of investigating deep structures beneath Siberia using PNE survey is the scanty distribution of PNE short intervals of about 1000 km. However, high-resolution PNE data enabled us to assimilate the upper mantle models using seismic velocity at the Moho discontinuity, which in turn allowed us to analyse hori-zontal heterogeneity of the three-layered density structure with varying velocities from 8.0 to 8.7 km/s. The layer with very high velocity suggests the presence of a high-density eclogite composition and the bottom of the middle layer corresponded to the base of the lithosphere followed by the presence of the third layer of low-seismic velocity of 8.5 km/s suggestive of asthenosphere.

The continental flood basalt beneath the Western Siberia Basin is interpreted to be associated with the P-T boundary, and Yakutian kimberlite region is regarded as the main tectonic terrains of the Western Siberia Basin, which has been clearly demarcated into the thick sediments and the thin crust that are signifying isostasy together with the presence of the 'eclogite layer' within the upper mantle associated with high value of seismic attenuation due to severe magmatic activity during which the physical/chemical condition of the mantle might have been deformed after 250 Ma. The crustal structure of the Siberia based on Lg and Sn phases of seismic wave propagation determined by Chiu and Snyder (2015) revealed that most parts of the Amerasia Basin have a crustal thickness intermediate between the typical of thin continental and oceanic crust as shown in Figure 14.9.

14.3.2.2 Crustal Thickness Mapping and Plate Reconstruction for the High Arctic

Integrated geophysical studies enabled to unravel the hidden mystery about the plate tectonic history in detail along with the distribution of oceanic and continental lith-osphere within the Arctic region extending to the Amresia Basin under the High

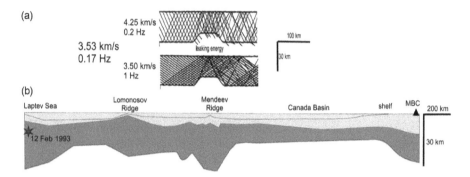

FIGURE 14.9   (a) Theoretical ray propagation models (modified from Kennett 1986) showing 'leakage' of Sn wave energy and disruption/attenuation of typical Lg wave energy where crustal thicknesses decrease over short lateral distances. (b) Crustal thickness model along a ray path between the Laptev Sea and seismic station MBC based on the probable path of earthquake waves on 12 February 1993. The crustal model is based on the recent ARCTIC-2000 (Lebedeva-Ivanova et al. 2006) and older Prince Patrick Island (Harrison and Brent 2005) seismic refraction survey results. The crust exhibits strong lateral variations and heterogeneity along this path, and thus, higher frequency waves would be expected to leak at the characteristic frequency of 0.17 Hz. (Adapted from Chiu and Snyder 2015.)

Arctic using data generated from the Arctic Gravity Project (ArcGP) (Alvey et al. 2008), which helped in assessing the crustal thickness and continental lithosphere thinning factors with conspicuous distribution of bathymetry, free-air gravity data, magnetic anomalies and sediment thickness for the region as shown in Figure 14.10. A very large negative residual thermal gravity anomaly of about −380 mGal and its subsequent correction for both oceanic and continental lithosphere resulted in realistic estimate of Moho depth. The gravity anomaly inversion was used to estimates crustal thickness at different ages with a reference crustal thickness of 35 km (Figure 14.11). Such approach is adjudged as the most apt method for discriminating various plate tectonic regimes either remotely located or not adequately investigated.

The integrated study resulted in three plate reconstruction models (two end member models and one hybrid model) that furnish information on the age and location of oceanic lithosphere for opening of the Amerasia Basin and the style of rifting with reference to different ridges (Figure 14.12).

Alvey et al. (2008) showed that the first end member model, the Mendeleev Ridge, is rifted from the Canadian Margin, whilst the second one end member model, the Mendeleev Ridge, is rifted from the Lomonosov Ridge of Eurasian Basin. Most importantly, the hybrid model is a composite model that contains elements of both end member models, which is very much consistent with the location of the ocean-continent transition from continental lithosphere thinning factors derived from gravity inversion.

The estimated value of 20 km crustal thickness for Late Cretaceous Makarov Basins from gravity inversion was found very much corroborative to that obtained from seismic refraction (Lebedeva-Ivanova et al. 2006). On the other hand, Alvey et al. (2008) combines plate reconstruction modelling to predict oceanic lithospheric age, and gravity inversion. It was done by incorporating lithosphere thermal gravity

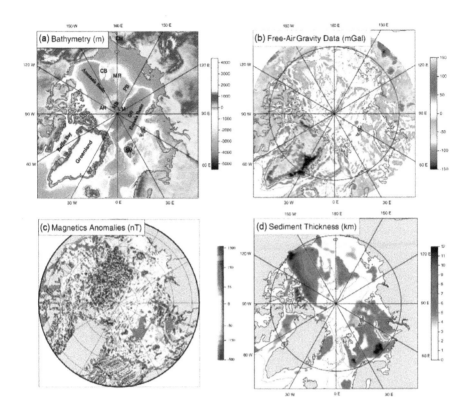

FIGURE 14.10 Maps based on integrated geophysical results: (a) Bathymetry-topography (m) from IOC, IHO and BODC (2003); (b) free-air gravity anomalies (mGal) from ArcGP Project (free-air data for regions below sea level and Bouguer including ice thickness correction for regions above sea level) (Forsberg and Kenyon 2004); (c) magnetic anomaly data (nT) (Verhoef et al. 1996; Glebovsky et al. 1998); and (d) sediment thickness data (km) (Laske et al. 1997). AR = Alpha Ridge, CB = Chukchi Borderlands, CH = Chukotka, GR = Gakkel Ridge, LM = Lomonosov Ridge, MR = Mendeleev Ridge, PB = Podvodnikov Basin, SV = Svalbard. (Adapted from Alvey et al. 2008.)

anomaly correction to predict crustal thickness. It also provides a strategy for testing plate reconstruction models against seismic estimates of crustal thickness. It is highlighted that prediction of accurate crustal thicknesses and ocean basin formation ages requires sufficient and prior knowledge of sediment thickness from seismic reflection data.

14.3.2.3 Crustal Structure of the Baikal Rift

Crustal thicknesses of the BRZ and other parts of the Arctic region have been estimated using integrated methodologies of geosciences, comprising seismic, electromagnetic and gravity tools of geophysics for getting better insight into the crustal and mantle dynamics. The location of BRZ is very intriguing in sense that it is located in the central part of the large Eurasia Continent and is distantly located from the subduction zone of the Western Pacific to the east and the India-Himalaya collision zone to the south.

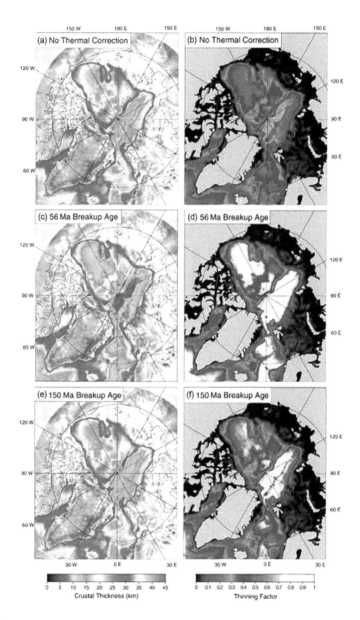

FIGURE 14.11 Maps showing the estimates of crustal thickness at different ages with a reference crustal thickness of 35 km based on (a) predicted crustal thickness obtained from gravity inversion without using a lithosphere thermal gravity anomaly correction (TGC) with the corresponding continental lithosphere thinning factor shown in (b). Predicted crustal thickness obtained from gravity inversion with a TGC treating the entire region as continental with a uniform rift age of 56 Ma and volcanic addition prediction corresponding to a volcanic rifted margin (c) with the corresponding lithosphere thinning factor (d). Predicted crustal thickness obtained from gravity inversion with a TGC treating the entire region as continental and volcanic addition that corresponds to a volcanic rifted margin of age of 150 Ma (e) with its corresponding lithosphere thinning factor (f). (Adapted from Alvey et al. 2008.)

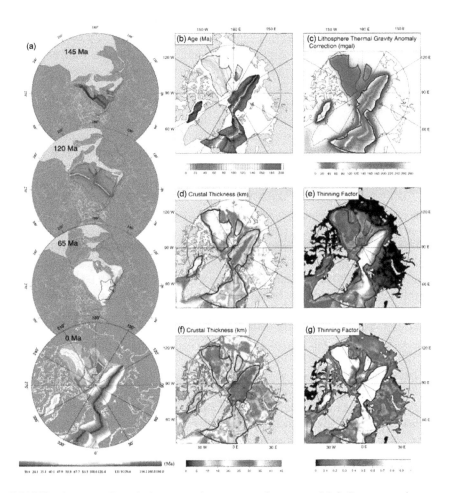

FIGURE 14.12 (a) Set of plate tectonic reconstructions – model 3. Reconstructions are shown in a Eurasian plate fixed reference frame. At around 145 Ma the initiation of the break-up of Mendeleev Ridge from the Canadian continental margin took place. The older ocean basin observed is the South Anyui ocean basin, which will be subducted during the opening of the Amerasia Basin. 120 Ma – the Amerasia Basin has opened fully, halted by the grounding of the Chukchi Borderlands against the Russian continental margin at this time. 65 Ma – after a period of compression between the North American Block and the Eurasian Block, a period of extension occurs resulting in the rifting of Mendeleev Ridge from the Lomonosov Ridge/ Barents Sea Margin. This results in the subduction of part of the older Amerasia Basin. The opening of the Labrador Sea is initiated at this time (Chalmers and Laursen 1995). 0 Ma – the current situation with predicted ocean age. (b) Predicted age (Ma) for the oceanic lithosphere within the Amerasia Basin with isochrons for the North Atlantic and Eurasia Basin as defined by Gaina et al. (2005, 2002). (c) Lithosphere thermal gravity anomaly correction (mGal) at surface. (d) Predicted crustal thickness (km) obtained from the gravity inversion without a sediment correction using the predicted age grid shown in (b). (e) Continental lithosphere thinning factors corresponding to (d). (f) Predicted crustal thickness (km) obtained from the gravity inversion using a sediment thickness correction and the predicted age grid shown in (b). (g) Continental lithosphere thinning factors corresponding to (f). OCTs used within the plate reconstruction model are indicated by a black line. (Adapted from Alvey et al. 2008.)

The exact geotectonic presence of the BRZ is found to be in between the northern Siberian craton and the southern Paleozoic-Mesozoic mobile (Kanao 2018b, c). It is also pertinent to mention that crustal structure and seismicity at BRZ have been studied in detail by different researchers using different tools of seismology (Zhao et al. 2006; Ananyin et al. 2009; Kanao et al. 2015).

One of the best classical studies by Zhao et al. (2006) presented detailed seismic deep structure of BRZ in Siberia of the Arctic region using 3D P-wave teleseismic tomography assimilated by inverting arrival times of P-wave of the teleseismic earthquakes wave for a given dense distribution of seismic rays in lateral and vertical directions (Figures 14.13 and 14.14a–c).

Zhao et al. (2006) showed that upper mantle under the BRZ found appreciably associated with low-velocity anomalies, whilst the lithosphere shows high velocity under the Siberia (Figure 14.15). The low velocity has been interpreted by the authors that it corresponds to plume due to mantle upwelling, which plays a significant role in the initiation and evolution of BRZ. They described that the lithosphere of the stable Siberian craton is represented by a high-velocity anomaly having a thickness of 150–180 km as shown in Figure 14.15. The lithosphere of the complex Mongolian fold belt is imaged by strong lateral heterogeneities in which the Baikal mantle plume played an important role in the rift formation having control of other factors such as older (pre-rift), orienting lithosphere linearly with respect to the plume, showing favourable orientation of the far-field forces exerted by the India-Asia collision.

It has been reported that low-velocity anomalies correspond to the upper mantle with its variability to the extent of up to 2% beneath the BRZ.

As mentioned above, the low-velocity feature extends down to the surface up to the mantle transition zone, which explains the evolution of the BRZ, whilst the lithosphere of the complex Mongolian fold belt is found associated with strong heterogeneity (Zhao et al. 2006). Thus, the distribution of structural heterogeneities beneath the Baikal rift comprises a branched chain of Late Cenozoic half-grabens extending over a distance of about 1500 km in Siberia (Molnar and Tapponier 1975; Logachev and Florensov 1978; Zonenshain and Savostin 1981; Zhao et al. 2006).

14.3.3 Seismic Explosion Experiment and Numerical Modelling

There is a record advancement in controlled source experiments on ice caps in polar region to collect high-quality seismic data, which have been used in numerical modelling to estimate thickness of ice sheets and the seismic structure of the crust and upper mantle in the adjacent area of the Arctic region (e.g., central eastern Greenland) along with unravelling several hidden factors that influenced the signal-to-noise ratio of the recorded seismic data as clearly evident from the recorded seismograms. The seismic structure of Greenland is known through active seismic experiments in the coastal regions with airgun sources on experimental basis. The structure of the interior Greenland as determined with limited receiver function measurements of the Moho depth (Voss and Jocket 2007), has provided the estimate of structure with great degree of uncertainty (Kumar et al. 2007). The presence of a high-velocity lowermost layer created confusion on whether the layer corresponds to the part of the

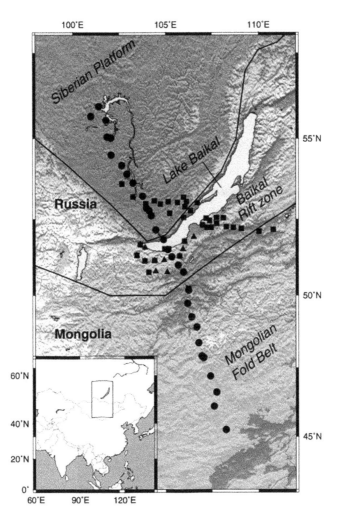

FIGURE 14.13 Map of the study area. Solid squares show the locations of 36 portable seis-mic stations deployed by University of California (UC) and University of Wisconsin (UW) during July to October 1991. Solid circles and triangles denote 28 portable stations deployed by UC and 12 portable stations installed by UW, respectively, from July to September 1992. The two curved lines show the approximate boundaries between the Siberian craton, the BRZ and the Mongolian fold belt (Logachev et al. 1983). Insert shows the location of the present study area. (Adapted from Zhao et al. 2006.)

mantle or to the lowermost crust (Artemieva and Thybo 2013). Nonetheless, active controlled source seismic explosion experiment has serious challenges because of the presence of a thick 2–3.5 km ice sheet overlying the basement rock. The latest experi-ment for acquiring the refraction seismic profile on the top of ice cap was designed by installing a total of 10 broadband seismometers for a period of 3 years on the ice caps and 13 seismometers on bed rock outside the ice cap for a period of 2 years (Shulgin and Thybo 2015).

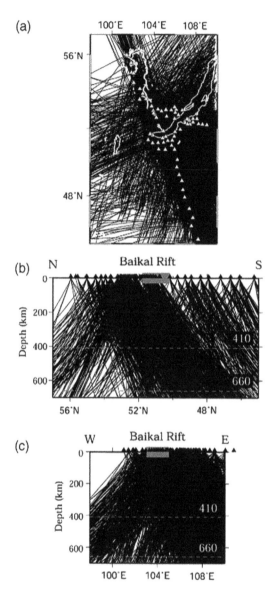

FIGURE 14.14 Distribution of the 1782 teleseismic rays used in this study in plain view (a) and in north–south (b) and east–west (c) vertical cross-sections. Triangles denote the seismic stations used. The bold bar in panels (b) and (c) denotes the location of the BRZ. (Adapted from Zhao et al. 2006.).

The above-mentioned, labour-intensive and challenging seismic explosion experiment has been conducted along the profile of 320 km length on the icy terrain to acquire data by drilling of 50 boreholes up to a depth of 80 m with aid of 1 tonne of water per hour under high pressure. Each borehole was loaded with about 100 kg explosive of Trinitrotoluene (TNT) and sealed with water in plastic bags.

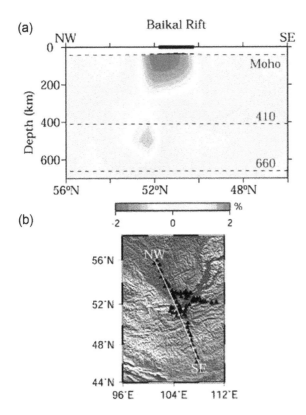

FIGURE 14.15   (a) Vertical cross-section of P-wave velocity image along a profile as shown in panel (b). The model CRUST2.0 was used for the crustal correction. Red and blue colours denote slow and fast velocities, respectively. The velocity perturbation scale is shown below the cross-section. The three dashed lines denote the Moho, 410 and 660 km discontinuities. The bold bar at the top of the cross-section denotes the location of the BRZ. Solid triangles in panel (b) show the locations of seismic stations. (Adapted from Zhao et al. 2006.)

The explosive charge sizes used 1 ton at the ends. On an average 500 kg explosive was used along the profile and about 100 kg at 35–85 m depth in individual boreholes to acquire high-quality of seismic data through different shooting strategy for hours on an ice cap. This experiment was found much more superior than that by Jacob et al. (1994) in sense of high signal-to-noise ratio, frequency characteristics, and characteristics of the controlled source data acquired under arduous field conditions.

Recorded seismic data showed that amplitudes of different seismic phases have clear variability along the profile associated with the largest shots at both ends of the profile. The amplitude undulations are found largest in the direction where the ice thickness decreases. Direct comparison of the amplitudes is problematic for the ice wave from all sources as shown in Figure 14.16.

Analysis of the data with respect to charge and detonation depth indicated a linear relation between the peak amplitudes and charge load for the measured amplitudes of the ice and Pg phases at 20 and 60 km distances. Shulgin and Thybo (2015) compared the data recorded on the Greenland ice cap with the shot records from a

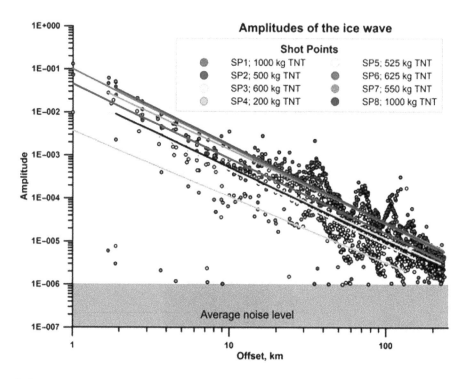

FIGURE 14.16 Plot of the amplitude of the ice wave as a function of offset from the shot point in log–log space. The colours correspond to the individual shot points marked in the legend. Lines are the best-fit for each shot point (black line is for shot point 1). Grey zone shows the average noise level. (Adapted from Shulgin and Thybo 2015.)

conventional onshore experiment that explains different phases of seismic wave as shown in Figure 14.17.

On the basis of the numerical modelling of a realistic ice sheet model, it was inferred that the near-surface seismic source produced a very characteristic wave train with a group velocity smaller than the S-wave speed in the ice treated as an ice sheet guided. S-wave developed by the superposition of post-critical reflections between the free surface and the ice bed, is referred to as Le, analogous to Lg wave, a crustally guided wave (Toyokuni et al. 2015). The analyses showed that the crustal Sg-coda wave has variability from 3.1 to 2.6 km/s and that corresponds to the characteristics waveform observed from the Greenland ice sheet.

14.3.4 STUDIES USING ELECTROMAGNETICS AND BIOGEOCHEMISTRY

As mentioned above, the Polar Regions have several unresolved geo-scientific issues that find ample scope of applications of geophysical methods. Electromagnetic, especially magnetotelluric (MT), studies are carried out in the region of high-altitude glaciated terrain because of its several advantages despite several sources of errors in the measurements. High impedance contrast allowed to overcome very high contact resistance between electrodes and the snow and ice cover at the surface; due to vertical propagation, horizontal polarized plane wave in the proximity to the geomagnetic poles.

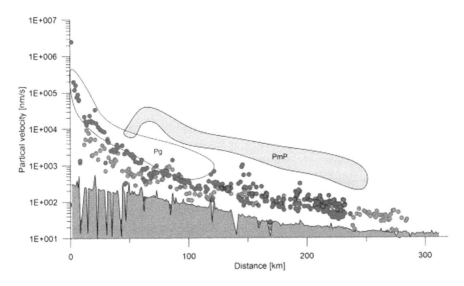

FIGURE 14.17 Comparison of the shot efficiency for ice and conventional environments (onshore shooting in the Kenya Rift). The amplitude data are taken for shot point 1: blue – ice wave; green – Pg; red – PmP; and grey shading – mean noise level. The shaded areas are results for the KRISP90 experiment in Kenya, modified after Jacob et al. (1994): light blue – Pg amplitudes; and pink – PmP amplitudes. (Adapted from Shulgin and Thybo 2015.)

It is due to generation of blizstatic, a localized random electric fields caused by the spin drift of moving charged snow and ice particles that produce significant noise in the electric fields during strong winds (Hill 2020). Electromagnetic studies are mainly carried out to study near-surface hydrologic systems, and glacial and ice sheet dynamics, whilst MT helps deciphering deeper electrical structure, and volcanic and tectonic problems at larger scale beneath the polar region.

MT has better control to provide more information about the nature of ice sheets and glacier dynamics because of large-scale variations in the bulk resistivity of glaciers and ice sheets due to variable ice density, temperature, impurity characteristics and the presence of organic and inorganic materials. The beauty of applying electromagnetic method in the polar region is based on the fact that heat flow from below generally facilitates basal ice sheet melting. If heat flow becomes high enough to raise temperature to the melting point for the given pressure (depth) conditions, it allows to discriminate ice sheets with active basal melting from the ice sheet without basal melting (Jordon et al. 2018a, b). This in turn changes in resistivity at the base of the ice column due to phase changes from solid ice to liquid water. The extensive zone of low resistivity has been interpreted as porous geologic materials containing hyper-saline fluids that occur through the processes of freeze concentration termed as cryoconcentration. MT and Transient Electromagnetic (TEM) have been used in the Svalbard Archipelago, Norway to explore potential zones for geothermal resources, crustal structure, and potential for CO_2 sequestration in the region (Beka et al. 2017a, b).

The Arctic is of great interest with respect to groundwater discharge because climate change data predict that the Arctic will warm faster than any other environs. The impacts of climate change on the physical and biogeochemical aspects of ground water discharge in the Arctic have become very significant. It is documented that the Arctic region is most neglected with reference to studies made on groundwater discharge and its impacts on aquatic ecosystems (Lecher 2017). Carbon and nitrogen are found to be most common solutes with respect to groundwater discharge that gets transported and brings a significant change with the onset of climate change. A recent study by Lecher (2017) suggested a conceptual model of groundwater discharge in the Arctic as shown in Figure 14.18.

It is observed that groundwater discharge could transport methane from the saturated zone of active layer to the lake from where it can easily escape to the atmosphere (Lecher et al. 2017). It is further supported by isotopes in methane (CH_4) that suggests large contributions of groundwater to the methane inventory of near-shore area where the lake to air methane flux is highest (Figure 14.19). It explains significant source of methane to the atmosphere in the Arctic.

Under climate change scenario, methane transport is expected to increase in the atmosphere because the Arctic is moving towards a more groundwater-dominated system due to increase in permafrost thawing with warmer temperature by deepening the active layer and contribute water to it (Bense et al. 2009; Lecher et al. 2017).

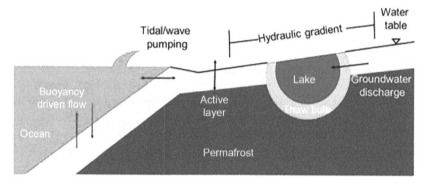

FIGURE 14.18 A conceptual model of groundwater discharge in the Arctic, which can be driven by hydraulic gradients, tidal/wave pumping and buoyancy differences (Lecher 2017).

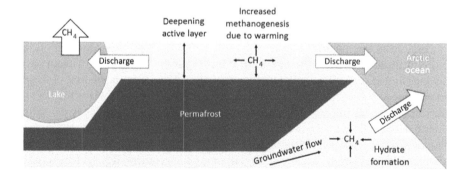

FIGURE 14.19 A conceptual model of methane–groundwater interactions. The diagram is shown as a cross section (Lecher 2017).

14.4 BIOPHYSICAL REMOTE SENSING OF THE ARCTIC REGION

The Arctic region is considered erroneously as a barren and unproductive land with minimal vegetation cover (Laidler and Treitz 2003). Satellite remote sensing offers potential for extrapolating or scaling up biophysical measures derived from local sites, to landscape and even regional scales. The Arctic region has been investigated through various remote sensing studies that involve vegetation mapping or assessing tundra characteristics, and analysis of the Arctic environments using medium resolution satellite data. Spatial and temporal distribution of vegetation helps to understand the nature and extent of climate change. Landsat data based Thematic Mapping (TM) is being used for broad vegetation mapping and analysis. High spatial resolution remote sensing provides necessary sampling scale to link field data to remotely sensed reflectance data; consequently, these data help improve the representation of biophysical variables over sparsely vegetated regions of the Arctic (Laidler and Treitz 2003).

14.5 EPILOGUE

Different sets of studies using diverse tools of seismo-geophysical domain have ushered a new and ample avenue for young researchers to take up several unaddressed issues related to the ice dynamics, seismogenesis, crustal, and sub-crustal imaging, and natural resource estimates for the Arctic region as their research career. Integrated seismo-geophysical studies would certainly allow us for conducting detailed mapping of natural resources, thickening of ice sheets and discharge of groundwater resources by deploying high-resolution sensors. Detailed processing of diverse data with the aid of powerful analytics may generate a novel idea for adopting sustainable strategy to deal with impacts of climate change in the Arctic region.

ACKNOWLEDGEMENTS

The authors are grateful to the Secretary, Ministry of Earth Sciences, India, for consistent motivation to work on integrated science with multiple disciplines to understand the role of different geo-scientific tools to assess the impacts of the Climate Change. Discussion with colleagues of National Centre for Seismology, New Delhi, is gratefully acknowledged.

REFERENCES

Aadlandsvik, B., Loeng, H., 1991. A study of the climate system of the Barents Sea, *Polar Research* 10(1), 45–49.

Alvey A., Gaina, C., Kusznir, N.J., Torsvik, T.H., 2008. Integrated crustal thickness mapping and plate reconstructions for the high Arctic, *Earth and Planetary Science Letters* 274, 310–321.

Ananyin, L.V., Mordvinova, V.V., Gots, M.F., Kanao, M., Suvorov, V.D., Tatkov, G.I., Thubanov, T.A., 2009. Velocity structure of the crust and upper mantle in the Baikal Rift Zone on the long-term observations of broad-band seismic stations, *Doklady Earth Science* 428, 1067–1071.

Antonovskaya, G., Konechnaya, Y., Kremenetskaya, E.O., Asming, V., Kværna, T., Schweitzer, J., Ringdal F., 2015. Enhanced earthquake monitoring in the European Arctic. *Polar Science* 9, 158–167. doi: 10.1016/j.polar.2014.08.003.

Artemieva, I.M., Thybo, H., 2013. EUNAseis: a seismic model for Moho and crustal structure in Europe, Greenland, and the North Atlantic region, *Tectonophysics* doi: 10.1016/j.tecto.2013.1008.1004.

Beka, T.I., Bergh, S.G., Smirnov, M., Birkelund, Y. 2017a. Magnetotelluric signatures of the complex tertiary fold–thrust belt and extensional fault architecture beneath Brøggerhalvøya, Svalbard, *Polar Research* 36(1):1409586. doi: 10.1080/17518369.2017.1409586.

Beka, T.I., Senger, K., Autio, U.A., Smirnov, M., Birkelund, Y., 2017b Integrated electromagnetic data investigation of a Mesozoic CO_2 storage target reservoir-cap-rock succession, Svalbard, *Journal of Applied Geophysics* 136, 417–430.

Bense, V.F., Ferguson, G., Kooi, H., 2009. Evolution of shallow groundwater flow systems in areas of degrading permafrost, *Journal of Geophysical Research* 2009, 36.

Bjerregaard, P., 1995. Health and environment in Greenland and other circumpolar areas, *Science of the Total Environment* 160/161, 521–527.

Bliss, L.C., 1981. North American and Scandinavian tundras and polar deserts. In: L.C. Bliss, O.W. Heal and J.J. Moore (eds.), *Tundra Ecosystems: A Comparitive Analysis*. pp. 8–24. Cambridge University Press, Cambridge.

Bondár, I., Storchak, D., 2011. Improved location procedure at the International Seismological Centre, *Geophysical Journal International* 186, 1220–1244.

Butler, R., Anderson, K., 2008. Global Seismographic Network (GSN), IRIS Annual Report, pp. 6–7.

CAFF, 1996. Proposed protected areas in the circumpolar Arctic (1996). Conservation of Arctis Flora and Fauna, Directorate for Nature Management, Trondheim, Habitat Conservation Report No. 2.

Chalmers, J.A., Laursen, K.H., 1995. Labrador Sea — the extent of continental and oceanic-crust and the timing of the onset of sea-floor spreading, *Marine and Petroleum Geology* 12, 205–217.

Chiu, K., Snyder, D.B., 2015. Regional seismic wave propagation (Lg & Sn phases) in the Amerasia Basin and High Arctic, *Polar Science* 130–145.

Dahl-Jensen, T., Larsen, T.B., Voss, P.H., the GLISN Group, 2010. Greenland Ice Sheet Monitoring Network (GLISN): A Seismological Approach. GEUS Report of Survey Activities (ROSA) for 2009, pp. 55–58.

Dziak, R.P., Park, M., Lee, W.S., Matsumoto, H., Bohnenstiehl, D.R., Haxel, J.H., 2009. Tectono-magmatic activity and ice dynamics in the Bransfield Strait back-arc basin, Antarctica, *Proceedings of the International Symposium on Polar Sciences* 16, 59–68.

Einarsson, T., 1980. *Geology (in Icelandic)*. Mal og Menning, Reykjavik. p. 240.

Elíasson, J., 1994. Northern hydrology in Iceland. In: T.D. Prowse, C.S.L. Ommaney and L.E. Watson (eds.). *Northern Hydrology: International Perspectives*. pp. 41–80. Environment Canada, Saskatoon, Saskatchewan, NHRI Science Report No. 3.

Encyclopedia Britannica, 1990. *The New Encyclopedia Britannica*. Chicago. Vol. 14, pp. 288–292.

Everett, K.R., Vassiljevskaya, V.D., Brown, J., Walker, B.D., 1981. Tundra and analogous soils. In: L.C. Bliss, O.W. Heal and J.J. Moore (eds.). *Tundra Ecosystems: A Comparative Analysis*. pp. 139–179. Cambridge University Press, Cambridge.

Forsberg, R., Kenyon, S., 2004. Gravity and geoid in the Arctic region—the northern polar gap now filled, in Proceedings of the GOCE Workshop, p. 6, ESA-ESRIN Frascati, Italy, ESA Publications Division, Noordwijk, http://earth.esa.int/workshops/goce04/goce_proceedings/57_forsberg.pdf.

Gaina, C., Roest, W.R., Muller, R.D., 2002. Late Cretaceous–Cenozoic deformation of northeast Asia, *Earth and Planetary Science Letters* 197, 273–286.

Geological Survey of Canada, 1995. Generalized Geological Map of the World, and Linked Databases.

Glebovsky, V.Y., Kovacs, L.C., Maschenkov, S.P., Brozena, J.M., (1998). Joint compilation of Russian and US Navy aeromagnetic data in the central Arctic seas. In: N. Roland and F. Tessesnsons, (eds.). *ICAM III; Third International Conference on Arctic Margins.* Polarforshung, pp. 35–40.

Grad M., Mjelde, R., Krysinski, L., Czuba, W., Libak, A., Guterch, A., IPY Project Group, 2015. Geophysical investigations of the area between the Mid-Atlantic Ridge and the Barents Sea: from water to the lithosphere-asthenosphere system, *Polar Science* 9(1), 168–183.

Hardisty, P., Schilder, V., Dabrowski, T., Wells, J., 1991. Yukon and Northwest Territories groundwater database. In: T.D. Prowse, C.S.L. Ommaney and L.E. Watson (eds.). *Northern Hydrology: International Perspectives.* pp. 465–482. Environment Canada, Saskatoon, Saskatchewan, NHRI Science Report No. 3.

Harrison, J.C., Brent, T.A., 2005. Basins and fold belts of Prince Patrick Island and adjacent area, Canadian Arctic Islands, *Geological Survey of Canada, Bulletin* 560, 208. doi: 10.4095/220345. CDROM.

Graham, H.J., 2020. On the use of electromagnetics for earth imaging of the polar regions, *Surveys in Geophysics* 5–45, doi: 10.1007/s10712-019-09570-8.

Ikeda, M., 1990a. Decadal oscillation of the air-sea-ocean system in the northern hemisphere, *Atmosphere-Ocean* 28, 106–139.

Ikeda, M., 1990b. Feedback mechanism among decadal oscillations in the northern hemisphere atmospheric circulation, sea ice and ocean circulation, *Annals Glaciology* 14, 120–123.

International Seismological Centre, 2011. On-line Bulletin, http://www.isc.ac.uk. (International Seismological Centre, Thatcham, United Kingdom).

IOC, IHO, BODC, 2003. Centenary Edition of GEBCO Digital Atlas, published on CD-ROM on behalf of the Intergovernmental Oceanographic Commission and the International Hydrographic Organization as part of the General Bathymetric Cart of the Oceans British Oceanographic Data Centre, Liverpool, UK.

IPCC, 2007. Intergovernmental Panel on Climate Change (IPCC), 2007. Climate Change 2007: The Physical Science Basis. Contributions of Working Group1 to the Fourth Assessment Report of the Intergovernmental Panel on Climate Change. Cambridge University Press, p. 996.

Ives, J.D., 1974. Permafrost. In: J.D. Ives and R.G. Barry (eds.). *Arctic and Alpine Environments.* pp. 159–194. Methuen and Co. Ltd., London.

Jacob, A.W.B., Vees, R., Braile, L.W., Criley, E., 1994. Optimization of wide-angle seismic signal-to-noise ratios and P-wave transmission in Kenya, *Tectonophysics* 236, 61–79.

Jordan, T.A., Martin, C., Ferraccioli, F., Matsuoka, K., Corr, H., Forsberg, R., Olesen, A., Siegert, M., 2018a. Anomalously high geothermal fux near the South Pole, *Scientific Reports* 8, 16785. doi: 10..1038/s41598-018-35182-0.

Jordan, T.M., Williams, C.N., Schroeder, D.M., Martos, Y.M., Cooper, M.A., Siegert, M.J., Paden, J.D., Huybrechts, P., Bamber, J.L., 2018b. A constraint upon the basal water distribution and thermal state of the Greenland Ice Sheet from radar bed echoes, *Cryosphere* 12, 2831–2854 https://doi.org/10.5194/tc-12-2831-2018.

Kanao, M., 2018a. *Studies on Seismicity in the Polar Region in Book 'Polar Seismology - Advances and Impact',* doi: 10.5772/intechopen.78554.

Kanao, M., 2018b. *Introduction: Progress of Seismology in Polar Region in Book 'Polar Seismology - Advances and Impact',* doi: 10.5772/intechopen.78550.

Kanao, M., 2018c. *Seismological Studies on the Deep Interiors of the Earth Viewed from the Polar Region in Book 'Polar Seismology - Advances and Impact',* doi: 10.5772/ intechopen.78552.

Kanao, M., Tsuboi, S., Butler, R., Anderson, K., Dahl-Jensen, T., Larsen, T., Nettles, M., Voss, P., Childs, D., Clinton, J., Stutzmann, E., Himeno, T., Toyokuni, G., Tanaka, S., Tono, Y., 2012. Greenland ice sheet dynamics and glacial earthquake activities. In: J. Müller and L. Koch (eds.). *Ice Sheets: Dynamics, Formation and Environmental Concerns*. pp. 93–120. Nova Science Publishers, Inc., Hauppauge, NY. Chap. 4.

Kanao, M., Zhao, D., Wiens, D.A., Stutzmann, É., 2015. Recent advance in polar seismology: Global impact of the International Polar Year, *Polar Science* 9(1), 1–4.

Kanao, M., Suvorov, V.D., Toda, S., Tsuboi, S., 2015. Seismicity, structure and tectonics in the Arctic region, *Geoscience Frontiers* 665–677.

Kennett, B.L.N., 1986. Lg waves and structural boundaries, *Bulletin of the Seismological Society of America* 76, 1133–1141.

Khare, N., Mishra, O.P., Taguchi, S., 2018. Recent advances in climate science of polar region, *Polar Science* 18, 1–4.

Kozlovskaya, E., 2013. Monitoring of slow seismic events from Arctics using the data of the POLENET/LAPNET broadband temporary array. In: IAHS-IASPO-IASPEI.IUGG Joint Assembly, S105S2.03, Gothenburg, Sweden.

Sebastian, S.A., Pomeroy, J.W., Marsh, P., 2017. Diagnosis of the hydrology of a small Arctic basin at the tundra-taiga transition using a physically based hydrological model, *Journal of Hydrology* 550, 685–703.

Kumar, P., Kind, R., Priestley, K., Dahl-Jensen, T., 2007. Crustal structure of Iceland and Greenland from receiver function studies, *Journal of Geophysical Research: Solid Earth* 112.

Kyle, P.R., Moore, J.A., Thirlwall, M.F., 1992. Petrologic evolution of anorthoclase phonolite lavas at Mount Erebus, Ross Island, Antarctica, *Journal of Petrology* 3, 849–875.

Laidler, G.J., Treitz, P., 2003. Biophysical remote sensing of arctic environments, *Progress in Physical Geography* 27(1), 44–68.

Laske, G., Masters, T.G., Anonymous, 1997. A global digital map of sediment thickness. Eos, Transactions. American Geophysical Union, p. 483.

Lebedeva-Ivanova, N.N., Zamansky, Y.Y., Langinen, A.E., Sorokin, M.Y., 2006. Seismic profiling across the Mendeleev Ridge at 82 N: evidence of continental crust. *Geophysical Journal International* 165, 527–544.

Lecher, A.L., 2017. Groundwater discharge in the arctic: a review of studies and implications for biogeochemistry. *Hydrology* 4, 41; doi: 10.3390/hydrology403004.

Lecher, A.L., Chuang, P.-C., Singleton, M., Paytan, A., 2017. Sources of methane to an Arctic lake in Alaska: an isotopic investigation, *Journal of Geophysical Research: Biogeosciences* 122, 753–766.

Linell, K.A., Tedrow, J.C.F., 1981. *Soil and Permafrost Surveys in the Arctic*. Clarendon Press, Oxford, p. 279.

Logatchev, N.A., Zorin, Y.A., Rogozhina, V.A., 1983. Baikal rift: active or passive? E comparison of the Baikal and Kenya rift zones. *Tectonophysics* 94(1–4), 223–240.

Loshchicov, V.S, 1965. Snow cover on the ice of the central Arctic. *Probl. Arktiki* 17: 36–45. (In Russian).

Luzin, G.P., Pretes, M., Vasiliev, V.V., 1994. The Kola Peninsula: geography, history and resources. *Arctic* 47: 1–15.

Mackay, D.K., Løken, O.H., 1974. Arctic hydrology. In: J.D. Ives and R.G. Barry (eds.). *Arctic and Alpine Environments*. pp. 111–132. Methuen and Co. Ltd., London.

Melnik E.A., Suvorov, V.D., Pavlov, E.V., Mishenkina, Z.R., 2015. Seismic and density heterogeneities of lithosphere beneath Siberia: evidence from the Craton long-range seismic profile. *Polar Science* 119–129.

Mishra, O.P., 2014. Intricacies of the Himalayan seismotectonics and seismogenesis: need for integrated research. *Current Science* 106(2).

Mishra, O.P., Khare, N., Das, S.B., Kumar, V., Singh, J., Singh, P., Ghatak, M., Shekhar, S., Tiwari, A., Gera, S.K., Mahto, R., Gusain, P., 2021. Glacial mass change induced earthquakes in the Himalayan region of South Asia: one of the proxies of climate change? In: N. Khare (eds.). *Understanding Present and Past Arctic Environments.* Elsevier.

Molnar, P., Tapponnier, P., 1975. Cenozoic tectonics of Asia: effects of a continental collision: features of recent continental tectonics in Asia can be interpreted as results of the India-Eurasia collision. *Science* 189(4201), 419–26.

Moore, J.J., 1981. Mires. In: L.C. Bliss, O.W. Heal and J.J. Moore (eds.). *Tundra Ecosystems: A Comparative Analysis.* pp. 35–37. Cambridge University Press, Cambridge.

Murray, J.L., Hacquebord, L., Gregor, D.J., Loeng, H., 1998. Physical/geographical characteristics of the arctic. In: J.L. Murray (ed.). *AMAP Assessment Report: Arctic Pollution.* pp. 9–24. Oslo, AMAP.

Natural Resources Canada, 1994. *Geology and Canada.* Department of Energy, Mines and Resources, Ottawa, p. 32.

Nettles, M., Ekström, G., 2010. Glacial earthquakes in Greenland and Antarctica. *Annual Review of Earth and Planetary Sciences* 38, 467–491.

NOAA, 1988. *Data Announcement 88-MGG-02, Digital relief of the surface of the Earth.* NOAA, National Geophysical Data Center, Boulder, Colorado.

Nordic Council of Ministers, 1996. The Nordic Arctic Environment – Unspoilt, Exploited, Polluted? Nordic Council of Ministers, Copenhagen. *Nord* 26, 240.

Olivieri, M., Spada, G., 2015. Ice melting and earthquake suppression in Greenland. *Polar Science* 9, 94–106.

Prik, Z.M., 1959. Mean position of surface pressure and temperature distribution in the Arctic. *Tr. Arkticheskogo Nauch.-Issled, Inst.* 217, 5–34.

Prowse, T.D., 1990. Northern hydrology: an overview. In: T.D. Prowse and C.S.L. Ommanney (eds.). *Northern Hydrology: Canadian Perspectives.* pp. 1–36. Environment Canada, Saskatoon, Saskatchewan, NHRI Science Report No. 1.

Rieger, S., 1974. Arctic soils. In: J.D. Ives and R.G. Barry (eds.). *Arctic and Alpine Environments.* pp. 749–769. Methuen and Co. Ltd., London.

Schlindwein, V., Andrea, D., Edith, K., Christine, L., Florian, S., 2015. Seismicity of the Arctic mid-ocean Ridge system, *Polar Science* 146–157.

Shulgin, A., Hans, T., 2015. Seismic explosion sources on an ice cap – Technical considerations. *Polar Science* 107–118.

Sly, P.G., 1995. Human impacts on the Hudson Bay region: present and future environmental concerns. In: M. Munawar and M. Luotola (eds.). *The Contaminants in the Nordic Ecosystem: Dynamics, Processes and Fate,* pp. 171–263. SPB Academic Publishing, Amsterdam.

Stonehouse, B., 1989. *Polar Ecology.* Blackie, London, p. 222.

Storchak, D.A., Masaki, K., Emily, D., James, H., 2015. Long-term accumulation and improvements in seismic event data for the polar regions by the International Seismological Centre. *Polar Science* 5–16.

Toyokuni, G., Takenaka, H., Kanao, M., Tsuboi, S., Tono, Y., 2015. Numerical modeling of seismic waves for estimating influence of the Greenland ice sheet on observed seismograms. *Polar Science* 9(1), pp 80–93.

van Everdingen, R.O., 1990. Groundwater hydrology. In: T.D. Prowse and C.S.L. Ommanney (eds.). *Northern Hydrology. Canadian Perspectives,* pp. 77–101. National Hydrology Research Institute, Environment Canada, Saskatoon, NHRI Science Report No. 1.

Verhoef, J., Roest, W.R., Macnab, R., Arkani-Hamed, J., 1996. Magnetic anomalies of the Arctic and North Atlantic oceans and adjacent land areas.

Voss, M., Jokat, W., 2007. Continent-ocean transition and volumi-nous magmatic underplating derived from P-wave velocity modelling of the East Greenland continental margin. *Geophysical Journal International* 170, 580e604.

Voss, M., Jokat, W., 2007. Continent-ocean transition and volumi-nous magmatic underplating derived from P-wave velocity modelling of the East Greenland continental margin. *Geophysical Journal International* 170, 580e604.

Voss, M., Jokat, W., 2007. Continent-ocean transition and volumi-nous magmatic underplating derived from P-wave velocity modelling of the East Greenland continental margin. *Geophysical Journal International* 170, 580e604.

Voss, M., Jokat, W., 2007. Continent-ocean transition and volumi-nous magmatic underplating derived from P-wave velocity modelling of the East Greenland continental margin. *Geophysical Journal International* 170, 580e604.

Voss, M., Jokat, W., 2007. Continent-ocean transition and volumi-nous magmatic underplating derived from P-wave velocity modelling of the East Greenland continental margin. *Geophysical Journal International* 170, 580e604.

Voss, M., Jokat, W., 2007. Continent-ocean transition and volumi-nous magmatic underplating derived from P-wave velocity modelling of the East Greenland continental margin. *Geophysical Journal International* 170, 580e604.

Voss, M., Jokat, W., 2007. Continent-ocean transition and volumi-nous magmatic underplating derived from P-wave velocity modelling of the East Greenland continental margin. *Geophysical Journal International* 170, 580e604.

Voss, M., Jokat, W., 2007. Continent-ocean transition and volumi-nous magmatic underplating derived from P-wave velocity modelling of the East Greenland continental margin. *Geophysical Journal International* 170, 580e604.

Voss, M., Jokat, W., 2007. Continent-ocean transition and volumi-nous magmatic underplating derived from P-wave velocity modelling of the East Greenland continental margin. *Geophysical Journal International* 170, 580e604.

Voss, M., Jokat, W., 2007. Continent-ocean transition and voluminous magmatic underplating derived from P-wave velocity modelling of the East Greenland continental margin. *Geophysical Research Letters* 170, 580–604.

Wadhams, P. 2012. Arctic ice cover, ice thickness and tipping points. *AMBIO* 41, 23–33.

Wadhams, P., Davis, N.R., 2000. Further evidence of ice thinning in the Arctic Ocean. *Geophysical Research Letters* 27(24): 3973–3975, doi: 10.1029/2000GL011802.

Woo, M.K., Gregor, D.J., 1992. *Arctic Environment: Past, Present and Future.* McMaster University, Department of Geography, Hamilton, Ont, p. 164.

Wood, W. 2000. Attitude change: persuasion and social influence. *Annual Review of Psychology* 51: 539–570.

Wu, P., Paul, J., 2000. Can deglaciation trigger earthquakes in N. America? *Geophysical Research Letters* 27, 1323–1326.

Zhao, D., Lei, J., Inoue, T., Yamada, A., Gao, S.S., 2006. Deep structure and origin of the Baikal Rift Zone. *Earth and Planetary Science Letters* 243, 681–691.

Zonenshain, L.P., Savostin, L.L., 1981. Geodynamics of the Baikal rift zone and plate tectonics of Asia. *Tectonophysics* 76, 1–45.

Zonenshain, L.P., Savostin, L.L., 1981. Geodynamics of the Baikal rift zone and plate tectonics of Asia. *Tectonophysics* 76, 1–45.

Zonenshain, L.P., Savostin, L.L., 1981. Geodynamics of the Baikal rift zone and plate tectonics of Asia. *Tectonophysics* 76, 1–45.

15 Decadal Arctic Sea Ice Variability and Its Implication in Climate Change

Alvarinho Luis
National Centre for Polar and Ocean Research

Neelakshi Roy
Kumaun University

CONTENTS

15.1 INTRODUCTION

Seawater freezes to form a thin veneer of sea ice at high latitudes in winter. The sea ice inhibits air-sea momentum and heat transfer, and its highly reflective surface gives rise to positive ice-albedo feedback. It is one of the climate indicators because it is sensitive to warming and plays an essential role in amplifying climate change. Sea ice has received considerable attention in the recent past, largely because the Arctic sea ice is in an exponential decline (Stroeve et al. 2012), compared to that in the early 1980s. There have been 10 lowest minimum Arctic sea ice covers after 2007.

DOI: 10.1201/9781003265177-15

Moreover, the September sea ice extent (SIE) decreased by 45% from 1979 to 2016, and if current trends continue, some Arctic shelf seas are forecasted to be ice-free during summer in the 2020s (Onarheim et al. 2018). The sea ice decline, which was found to be faster than that forecasted by model simulations (Stroeve et al. 2007), could be explained by internal variability wherein the loss is related to the increase in the concentration of the greenhouse gases over the past few decades. On the other hand, for the Antarctic, it showed a statistically significant increasing trend of ~1% over the past 40 years, since the passive microwave measurements of sea ice began in 1978 (Parkinson 2019).

The sea ice decrease has potential a broad range of climate consequences. With the sea ice cover retreating the year after year, more shortwave radiation penetrates the low-albedo water and warms the ocean, leading to further ice reductions as a result of the ice-albedo feedback (Screen and Simmonds 2010). The ice-albedo feedback accounts for nearly half of the surface temperature increase associated with greenhouse warming in the Arctic (Hall 2004; Graversen et al. 2014). The ice continuously impacts its surroundings by restraining heat and momentum exchanges between the atmosphere and ocean. As the ice forms/melts, the salinity of the underlying ocean increases/decreases, impacting the overturning and ocean circulation. Furthermore, as the sea ice drifts, it releases cold, low-salinity water, affecting the stratification (Aagaard and Carmack 1989).

The decreasing Northern hemisphere (NH) sea ice cover is consistent with Arctic warming (Walsh 2013; Hartmann et al 2013) induced by increases in the atmospheric greenhouse gases due to anthropogenic activities. Several studies carried out using multichannel satellite passive-microwave data for the past three decades have revealed a precipitous declining sea ice trend, since satellite observations began in October 1978 (Parkinson and Cavalieri 1989; Cavalieri et al. 1999). The loss of the perennial sea ice cover by about 50% of its extent (Comiso 2006; Stroeve et al. 2012) produces feedbacks that tend to enhance the warming through Arctic amplification (AA) (Screen and Simmonds 2010).

Potential drivers responsible for dwindling sea ice in the NH are either ocean (Zhang et al. 1999) and atmosphere (Rigor et al. 2002) or both. The dominant ones are albedo (Randall et al. 1998), increases in water vapour from ice-free ocean leading to more clouds, surface winds and poleward atmospheric and oceanic energy transport. The summer sea ice melt exposes areas of the dark ice-free ocean that absorbs 93% of the solar radiation reaching the surface, whereas white ice reflects 50%–80% to space (Piestone et al. 2014). Because the Arctic planetary albedo has decreased between 1979 and 2011 from 0.52 to 0.48, an amount of $6.4 \pm 0.9 \, \text{W/m}^2$ of solar energy heats the Arctic Ocean (Pistone et al. 2014). About 75% of the air temperature fluctuations on sea ice occur due to variations in moist-static energy transport or local ocean heat release to the atmosphere (Olonscheck et al. 2019). The increase in cyclone frequency, mostly to the north of Siberia, coincides with the large decrease in Arctic ice cover for 1989–1990 (Maslanik et al 1996). The cyclones modulate sea ice in two ways: (i) by controlling the exchange of fluxes of sensible and latent heat between the surface and atmosphere (Murray and Simmonds 1995), and (ii) the enhanced surface wind speed intensifies the surface stresses on to the sea ice and shifts the sea ice distribution.

The see-saw pattern of alternating atmospheric pressure between polar and mid-latitudes (centred on ~45°N) represented by Arctic Oscillation (AO) varies from positive to negative phase. In the positive (negative) phase it produces a strong (weak) polar vortex, with the mid-latitude jet stream shifted northward (southward). From the 1950s to the 1980s, the AO flipped between positive and negative phases, but it entered a strong positive pattern between 1989 and 1995. So the precipitous sea ice decline since the mid-1990s may have been partly triggered by the strongly positive AO mode during the preceding years (Rigor et al. 2002; Rigor and Wallace 2004).

With the availability of four decades of SIE from passive microwave sensors, the goals of this chapter are to place before the reader its variability for different decades: 1979–88, 1989–98, 1999–2008, and 2009–2018 and to explore the role of the forcing parameters (local and remote) to account for the SIE variability. Finally, we link our findings with the literature through a stimulating discussion.

15.2 MATERIAL AND METHODS

The SIE data used in this chapter were obtained from surface brightness temperature (Tb) recorded by the multichannel passive microwave satellites since 1978. The continuous and sometimes overlapping observations by the passive sensors (Table 15.1) whose swath facilitates a complete coverage of the Earth surface in 1–2 days provide measurements in all-weather and day-and-night conditions. The passive microwave sensor measures Tb emitted energy by a surface, which varies with frequency and polarization for a given surface. Compared to sea ice, the seawater has its Tb higher in the vertical channel than that in the horizontal, so the boundary between the two can be delineated. The sea ice data have been rigorously inter-calibrated between the SSM/I and SSM/IS sensors (Cavalieri et al. 1999, 2012), to produce a homogeneous dataset for trend studies. Most algorithms use some form of a polarization difference or polarization ratio and a linear mixing formula with Tb to estimate the concentration of sea ice, which is expressed as fractional coverage of ice from 0% to 100%, within the field of view of the sensor.

TABLE 15.1

A Series of Passive Microwave Satellite Sensors Launched over the Different Years for Recording Sea Ice Concentration in the Polar Regions

Platform and Instrument	Period
Nimbus-7 Scanning Multichannel Microwave Radiometer (SMMR)	26 October 1978–20 August 1987
The Defense Meteorological Satellite Program (DMSP) F8 Special Sensor Microwave Imager (SSM/I)	21 August 1987–18 December 1991
DMSP-F11 SSM/I	19 December 1991–29 September 1995
DMSP-F13 SSM/I	30 September 1995–31 December 2007
DMSP-F17 Special Sensor Microwave Imager and Sounder (SSMIS)	1 January 2008–31 March 2016
DMSP-F18 SSMIS	1 April 2016–December 2019

   The NASA team SSMI sea ice algorithm (Cavalieri et al. 1984) and the Bootstrap algorithms (Comiso 1995) were developed to derive sea ice concentration from Tb using empirically derived algorithms. The algorithms use different combinations of frequency and polarization channels but generally share the common approach of determining the coefficients for pure surface types (so-called 'tie points') (100% ice, 100% water) and interpolating between these points to find the ice concentration that corresponds to a set of Tb values. The NASA team uses polarization ratios at 19 GHz and gradient ratios of vertically polarized channels at 19 and 37 GHz to obtain the tie points for sea ice and open water. The advantage of the NASA team algorithm is that it takes into account changes in surface temperatures through the use of ratios of differences of Tb. The polarization ratios are, however, affected by the physical characteristics of some types of sea ice cover (Mätzler et al. 1984).

   The Bootstrap algorithm uses dual polarization at 37 GHz and vertical polarization at 19 GHz to resolve surface areas where there is layering in the snow and ice. The disadvantage is that the algorithm requires prior knowledge of the emissivities (or Tb) of 100% ice and that of 100% water daily. In the post-processing stage, spurious sea ice retrievals due to weather and coastal effects (mixed land/ocean grid cells) are discarded and the missing data are filled in with spatial and/or temporal interpolation. Generally, both the above algorithms provide similar results in the Arctic during winter, when the emissivity is relatively stable, and surfaces are dry.

   Different corrections were applied before generating the final product. While the nominal grid cell size of the gridded products is 25 km×25 km, the −3dB footprint of the 19.35 GHz SSM/I and Special Sensor Microwave Imager and Sounder (SSMIS) passive microwave channel is 72 km×44 km, which causes land-to-ocean spillover. To remove spurious coastal ice, land spill overcorrection was applied (Cavalieri et al. 1997). The rationale behind this land spillover approach is that ice will have retreated from most coasts in late summer, so that coastal ice recorded by the passive microwave instruments is probably a false detection. To rule out the possibility of removing ice where it does exist, the algorithm searches for the presence of open water in the vicinity of the grid cell to be corrected. The method uses monthly data from 1992 as a basis for correcting SSM/I data and data from 1984 for correcting the Scanning Multichannel Microwave Radiometer (SMMR) data. Daily sea ice concentration from low-resolution passive microwave data is not highly accurate because the ice edge, whether it is a diffuse marginal or compact ice zone, is not represented well. So the process involves the use of a 4 km multi-sensor analysed sea ice extent (MASIE) product from the same day to resolve the ice edge with greater precision and accuracy. Many errors due to missing data and transient weather effects were removed after averaging daily data over a month.

   Spurious ice from residual weather effects was eliminated by applying Polar Stereographic Valid Ice Masks Derived from National Ice Center (NIC) Monthly Sea Ice Climatologies. There are 12 NIC valid ice masks, one for each month. Before January 2015, the processing used the passive microwave-based maximum ice-extent climatology for masks. Another reason for the change to the new NIC valid ice masks is that the old maximum extent masks would sometimes remove or add ice in months when ice is changing rapidly such as in May and June. The masks were scrutinized to make manual adjustments to them so that ice extent was not erroneously added or removed. The algorithm parameters called tie points sometimes need adjustments to have a consistent time series, since sensors and spacecraft orbit are slightly different. The tie points were derived from regressions of Tb during satellite overlap periods.

The accuracy of Arctic SIC at a grid cell in the source data is usually cited as within ±5% of the actual sea ice concentration in winter, and ±15% during the summer when melt ponds are present on the sea ice, but some comparisons with operational charts report much larger differences (Agnew and Howell 2002; Partington et al 2003). Accuracy tends to be best within the consolidated ice pack where the sea ice is relatively thick (greater than 20 cm) and ice concentration is high. Accuracy decreases as the proportion of thin ice increases (Cavalieri 1995). Melt ponds in particular can lead to SIC underestimation by as much as 40% (Rösel et al. 2012). Generally, the SIC estimates are reliable for values that are at 15% or higher, with sensor footprints up to ~45 km × ~70 km.

We used the SIE product G02135 distributed online (Fetterer et al. 2017) which is processed through the NASA team algorithm. Produced by the National Snow and Ice Data Center (NSIDC) using the NASA team algorithm, the data are distributed in near real time. The product has been extensively evaluated in earlier studies (e.g. Agnew and Howell 2003; Steffen and Schweiger 1991; Emery et al. 1994). We used SIE which is the total area under ice cover within a pixel with ice concentration exceeding 15%. Because microwave emissivity is sensitive to melt, most studies focus on SIE rather than a sea ice area to examine temporal variability. We computed monthly anomalies relative to 1981–2010 climatology following Zwally et al. (2002). We also provide SIE trend expressed in percentage which was computed relative to 1981–2010 mean SIE. To account for sea ice volume (SIV) changes, we also used the Pan-Arctic Ice Ocean Modelling and Assimilation System (PIOMAS V.2.1) dataset (Schweiger et al. 2011). PIOMAS uses a computational model of sea ice and ocean elements to assimilate empirical data such as sea ice concentration and sea surface temperature (SST). The output from the parallel ocean and ice model incorporates major features such as concentration, extent, thickness and sea ice motion in the polar oceans. The results agree with buoy observations of ice motion, satellite observations of ice extent and ice thickness derived from submarine observations. The model biases are within 8% in Arctic ice motion, within 9% in Arctic ice thickness (Zhang and Rothrock 2003). In this study, monthly SIV data were obtained for each month spanning 1979–2018. SIV anomalies were calculated by deducting the monthly mean 1981–2010 climatology.

The reanalysis data consisting of SST (°C), air temperature (Ta, °C), mean sea level pressure (MSLP, mb), wind speed (m/s), and a sum of latent and sensible heat flux (W/m²) were obtained from the fifth-generation ECMWF atmospheric reanalysis of the global climate (ERA5, Hersbach and Dee 2016). When the ocean gains heat, the sign of the heat flux is positive. Reanalyses model output is available since 1979 at a horizontal resolution of 31 km and 137 vertical levels from the surface to 0.01 hPa which uses a fixed prediction model and data assimilation system to generate global atmospheric variables. It is noted that non-physical trends and variability may be present in the records due to changes in the observing system and that the climatology of some variables (surface energy fluxes) is not well represented. The anomalies were calculated relative to the benchmark period of 1981–2010.

The associations were highlighted in the literature between sea ice cover and low-frequency global climate oscillations such as the Pacific Decadal Oscillation (PDO) (Lindsay and Zhang 2005), Atlantic Meridional Oscillation (AMO) (Park and Latif 2008), Southern Oscillation (SOI) (Gloerson 1995) and AO (Rigor et al. 2002). PDO is akin to a long-lived El Niño-like pattern of Pacific climate variability (Zhang et al. 1997). The standardized values for the PDO index were derived as the leading principal component of monthly SST anomalies in the North Pacific Ocean, poleward of 20°N (Mantua et al. 1997). The AMO represents changes in SST of the North Atlantic Ocean. The AMO has affected air temperatures, rainfall and frequency of major hurricanes over much of the NH. It was computed using de-trended Kaplan SST, with an area-weighted average over the North Atlantic (0°N–70°N). The SOI represents large-scale fluctuations in air pressure occurring between the western and eastern tropical Pacific during El Niño (warm SST along the tropical Pacific) and La Niña (cold SST along the tropical Pacific) episodes. The AO represents the atmospheric circulation over the NH high latitude which is portrayed by a westerly circumpolar vortex that extends from the surface to the stratosphere (Thompson and Wallace 2000). A positive phase of AO is marked by high pressure over the polar region with the mid-latitude jet stream blowing strongly and consistently from west to east, confining the cold Arctic air to the polar region. When the AO index is negative, low pressure exists over the polar region, weaker zonal winds and greater movement of frigid polar air into middle latitudes. It is noted that the AO shows a close relationship with the North Atlantic Oscillation (NAO). Pearson correlation coefficient (r_{xy}) was used to assess the degree of association between time series/indices (x) and SIE (y), whose significance was judged at 95% and 99% significance level using a two-tail Students' test with the null hypothesis of a zero trend, using:

$$t = \frac{r_{xy}\sqrt{n-2}}{\sqrt{1-r_{xy}^2}}; n : \text{degrees of freedom}$$

We used trend analysis on the time series. The trend represents the least square linear fit ($y = mx + c$; where y is the dependent variable (sea ice, for example) and x is the independent variable (for example, time), c is the intercept, and m is the slope) to the SIE. The trend is presented in km²/year and %/decade which is a slope divided by the average of SIE over the reference period (1981–2010) (Tables 15.2–15.4). The different seasons referred to in this chapter are winter (January–March), spring (April–June), summer (July–September), autumn (October–December), and winter (July–September), analogous to other studies (cf. Cavalieri and Parkinson 2012). We begin by highlighting the variabilities and trends in the SIE spanning four decades spanning the 1979–2018 period and then turn our attention to the drivers of these SIE changes and implications of continued sea ice loss.

15.3   GENERAL FEATURES OF THE ARCTIC

The Arctic is a semi-enclosed ocean, almost surrounded by land (Figure 15.1a). Though the sea ice is mobile around the Arctic basin, it tends to confine to the cold

TABLE 15.2

Yearly and Seasonal Arctic Sea Ice Extent Trends with Standard Deviations for 1979–2019. NI Means No Ice; The Trends Were Found to Be Not Significant; Units: km²/yr for Trend, and %/decade

Period Sectors	Units	Yearly (Jan–Dec)	Winter (Jan–Mar)	Spring (Apr–Jun)	Summer (Jul–Sep)	Autumn (Oct–Dec)
Northern Hemisphere	km²/yr	−55,632±10023	−45,000±11,131	−41,000±10,160	−76,000±19,144	−64,000±15,071
	%/decade	−4.77±0.001	−2.99±0.56	−3.09±0.58	−9.88±1.9	−6.02±1.10
Okhotsk and Japan Seas	km²/yr	−3288±1421	−7733±3599	−4428±2017	NI	−1209±821
	%/decade	−8.79±3.13	−7.84±3.17	−12.06±4.49	NI	−8.43±5.82
Bering Sea	km²/yr	−1638±1332	−2042±2790	−2349±2326	−57±21	−2104±1093
	%/decade	−5.48±2.85	−2.98±2.90	−6.57±4.78	−27.04±8.33	−13.56±6.79
Hudson Bay	km²/yr	−3040±1113	−89±136	−2200±927	−4341±1623	−5531±2298
	%/decade	−3.91±1.52	−0.07±0.01	−1.90±0.85	−21.90±8.59	−10.60±4.73
Baffin Bay and Labrador Sea	km²/yr	−6376±1434	−6621±2340	−5416±1576	−4340±982	−5549±1363
	%/decade	−11.64±2.62	−5.18±1.83	−5.25±1.53	−23.46±5.31	−10.42±2.56
Gulf of Saint Lawrence	km²/yr	−623±344	−2088±1093	−416±312	NI	5.79±58
	%/decade	−10.10±5.36	−11.54±5.69	−8.36±6.49	NI	0.36±4.12
Greenland Sea	km²/yr	−6168±1928	−9265±2947	−5535±1936	−4517±1995	−5352±1919
	%/decade	−9.62±2.76	−11.24±3.18	−7.25±2.34	−11.64±5.20	−9.11±3.06
Kara and Barents Seas	km²/yr	−15,428±3026	−13,703±2843	−15,836±3386	−14,710±3035	−17,464±3614
	%/decade	−13.23±2.59	−8.26±1.71	−10.43±2.23	−32.49±6.70	−16.92±3.50
Canadian Archipelago	km²/yr	−916±480	0.07±0.67	−195±159	−2411±1401	−1060±511
	%/decade	−1.32±0.61	9.01±0.001	−0.26±0.22	−4.45±2.24	−1.47±0.62
Central Arctic	km²/yr	−1770±682	−781±329	−415±226	−3793±1562	−2092±872
	%/decade	−0.55±0.12	−0.24±0.04	−0.13±0.04	−1.19±0.31	−0.65±0.18

TABLE 15.3
Sea Ice Extent Trend for Four Decades Computed using Yearly Data; Trend Values Are Shown in Bold and Italics Are Significant at 95%, while Those in Bold Ones Are Significant at 99% Confidence

Sub-Region	Units	1979–88	1989–98	1999–08	2009–18
Seas of Japan	km²/yr	*−17,217 ± 936*	1353 ± 598	**−16,666 ± 855**	−442 ± 452
and Okhotsk	%/decade	*−46.02 ± 2.50*	**3.61 ± 1.59**	**−44.55 ± 2.29**	−1.18 ± 1.21
Bering Sea	km²/yr	3837 ± 431	18 ± 569	−2313 ± 655	**−26,370 ± 1272**
	%/decade	12.83 ± 1.44	0.06 ± 1.90	−7.73 ± 2.19	**−88.19 ± 4.25**
Hudson Bay	km²/yr	1212 ± 380	**−12,314 ± 661**	147 ± 454	6371 ± 488
	%/decade	1.56 ± 0.49	**−15.82 ± 0.85**	0.19 ± 0.58	8.18 ± 0.63
Baffin Bay and	km²/yr	−4797 ± 1400	1785 ± 685	*−14,057 ± 786*	−254 ± 847
Labrador Sea	%/decade	−8.60 ± 2.51	3.20 ± 1.23	*−25.20 ± 1.41*	−0.45 ± 1.52
Saint	km²/yr	**2765 ± 146**	−3140 ± 164	138 ± 123	539 ± 212
Lawrence	%/decade	**44.84 ± 2.37**	**−50.91 ± 2.66**	2.24 ± 2.00	8.73 ± 3.44
Greenland	km²/yr	−287 ± 885	64 ± 764	223 ± 465	**−13,678 ± 660**
Sea	%/decade	−0.45 ± 1.38	0.09 ± 1.19	0.35 ± 0.72	**−21.34 ± 1.03**
Kara and	km²/yr	−15,608 ± 1616	7066 ± 1445	*−39,447 ± 2211*	−15,281 ± 1533
Barents Seas	%/decade	−12.69 ± 1.31	5.74 ± 1.17	*−32.06 ± 1.79*	−12.42 ± 1.24
Arctic Ocean	km²/yr	315 ± 211	−389 ± 101	−2905 ± 290	−5419 ± 495
	%/decade	0.09 ± 0.06	−0.12 ± 0.03	−0.91 ± 0.09	−1.69 ± 0.15
Canadian	km²/yr	−1713 ± 249	−2903 ± 254	−3241 ± 292	2896 ± 315
Archipelago	%/decade	−2.47 ± 0.34	−4.18 ± 0.37	−4.67 ± 0.42	4.18 ± 0.45
Northern	km²/yr	−20,624 ± 2207	−21,853 ± 2625	**−1,17,092 ± 5206**	*−52 655 ± 3397*
Hemisphere	%/decade	−1.77 ± 0.19	−1.87 ± 0.22	**−10.04 ± 0.45**	*−4.52 ± 0.29*

TABLE 15.4
Yearly and Seasonal Arctic Sea Ice Volume Trends (km³/year) with Standard Deviations for the NH; Trend Values Are Significant at 99% Using a Two-Tailed Student t Test

Decades	Yearly (Jan–Dec)	Winter (Jan–Mar)	Spring (Apr–Jun)	Summer (Jul–Sep)	Autumn (Oct–Dec)
1979–1988	−54 ± 122	−63 ± 133	30 ± 128	−104 ± 141	−81 ± 126
1989–1998	−242 ± 83	−238 ± 107	−303 ± 120	−244 ± 92	−185 ± 83
1999–2008	−506 ± 179	−551 ± 200	−530 ± 190	−452 ± 159	−492 ± 173
2009–2018	−156 ± 75	−79 ± 84	−76 ± 79	−251 ± 93	−218 ± 86
1979–2018	−306 ± 51	−336 ± 57	−307 ± 52	−279 ± 47	−301 ± 51

Arctic waters. Sea ice converges and piles up into thick ridges whose longer life cycle leads to ice remaining frozen longer during the summer melt. Some Arctic sea ice remains through the summer and continues to grow the following autumn. Of the total sea ice of $\sim15\times10^6\,km^2$ formed in winter (January–March), about $7\times10^6\,km^2$ survive at the end of the summer season (July–September) which depends on the geography of the basin. The Central Arctic, Baffin Bay, Greenland Sea and the Barents Sea which are not bound by land on all sides retain some ice year-round, so a fully sinusoidal seasonal cycle exists. On the other hand, the Seas of Okhotsk and Japan and the Gulf of St. Lawrence host sea ice only during July–September (summer), while Bering remains ice-free in August. Sectors such as Hudson Bay, Laptev Sea, East Siberian, Chukchi Sea, Beaufort Sea, Canadian Archipelago, Kara and the Barents Sea and Central Arctic show the rapid late fall and early winter freeze-up for the whole region with early summer melting. It is pointed out that the maximum sea ice that occurs in February ($\sim34,000\,km^2$, on average) corresponds to 3% of the total surface area of the Japan Sea (Nihashi et al. 2017).

Generally, the Arctic sea ice is 2–3 m thick, which in some regions thickens to 4–5 m due to the ridging effect. The Arctic region to the north of the Atlantic Ocean remains ice-free since it opens to the warmer water from the south. The waters along the eastern shelves of Canada and Russia remain frozen due to cold air moving off the land from the west. Moreover, the southward-flowing cold-water currents facilitate sea ice to grow along the eastern Canadian coast. The snowfall tends to be low, except near the ice edge. However, the Atlantic sector of the Arctic, between Greenland and Scandinavia, receives precipitation, which is high in winter, facilitated by storms forming in the Atlantic Ocean. Water from several rivers in Russia and Canada and Pacific Ocean inflow through Bearing Strait provides fresher, less dense water to the Arctic Ocean which allows more ice formation.

15.4 RESULTS

Linear trends are presented for yearly and seasonal SIE for 1979–1988, 1989–98, 1999–2008 and 2009–2018. Following the literature (Cavalieri and Parkinson 2012), the NH is divided into nine sectors: Seas of Okhotsk and Japan, Hudson Bay, Baffin Bay and the Labrador Sea, the Gulf of Saint Lawrence, the Greenland Sea, the Kara and Barents Seas, the Arctic Ocean, and the Canadian Archipelago (Figure 15.1a). The Okhotsk and Japan Seas are connected by the Tatar Strait region, located in the northeastern Japan Sea. The sea ice cover in the Japan Sea peaks in February ($\sim34,000\,km^2$ on average), which disappears by March. We note that there are four more seas (Beaufort, Chukchi, East Siberian and Laptev Seas) which contribute only $\sim1\%$ to the total NH sea ice variability (Stroeve and Notz 2018); these are excluded from the study.

Figure 15.1b and c shows changes in the maximum (March) and minimum (September) SIE, since the advent of satellite-based measurements in 1979. These were processed online at NSIDC (https://nsidc.org/data/seaice_index/). From an extent of $18.2\times10^6\,km^2$ in March 1979, it dropped by 11.6% to $14.52\times10^6\,km^2$ in March 2019 which is the peak winter. Likewise, from September 1979, the SIE dropped by 42.6% from 7.05×10^6 to $4.32\times10^6\,km^2$ for peak summer. In the following sections, we focus on regional SIE changes.

15.4.1 NORTHERN HEMISPHERE TREND

A typical annual cycle in the SIE for NH is portrayed in Figure 15.2a, where the SIE trend for March (peak boreal winter) was found to be $-41,742 \pm 8108 \, km^2/year$, while that for September (peak boreal summer) was estimated as $-82,490 \pm 16,712 \, km^2/year$. The yearly SIE showed a trend of $-55,632 \pm 10,023 \, km^2/year$ ($-4.77\% \pm 0.95\%/decade$) (Figure 15.2a) over the 41 years. The trend computed for each month (averaged over the period of 1979–2018) showed a high standard deviation from July to October (insert in Figure 15.2a). The largest negative (positive) excursion from the trend line occurred in September 2012 (September 1980), which had the smallest

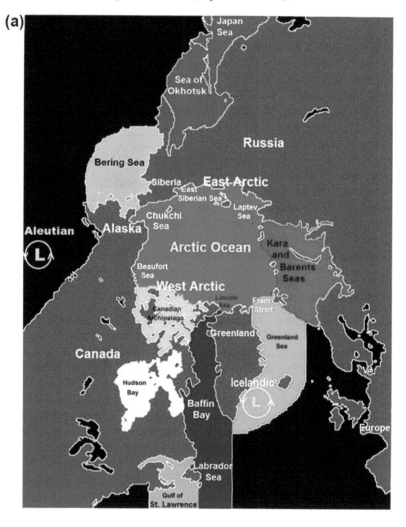

FIGURE 15.1 (a) The study area indicating the different sectors in colour. The Aleutian Low and Icelandic lows are two main low-pressure systems in the high-latitude NH whose approximate semi-permanent locations are shown. (b) Changes in the maximum SIE for March spanning 1979–2019, and (c) changes in the minimum SIE for September over the 1979–2019 period.

(Continued)

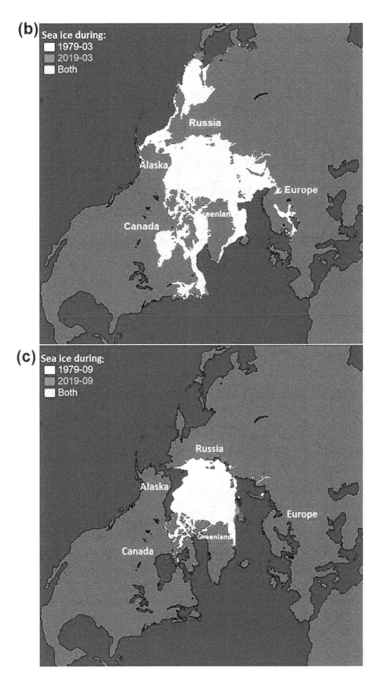

FIGURE 15.1 (*CONTINUED*) (a) The study area indicating the different sectors in colour. The Aleutian Low and Icelandic lows are two main low-pressure systems in the high-latitude NH whose approximate semi-permanent locations are shown. (b) Changes in the maximum SIE for March spanning 1979–2019, and (c) changes in the minimum SIE for September over the 1979–2019 period.

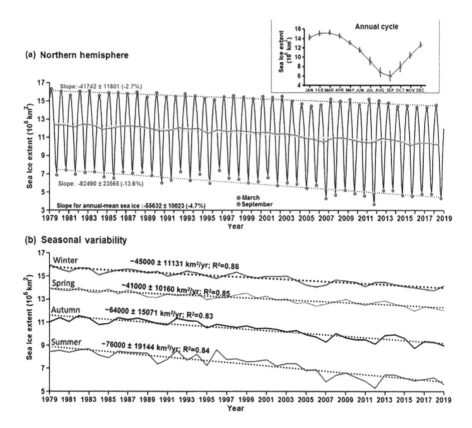

FIGURE 15.2 (a) Annual cycle of the NH SIE. The insert shows the monthly cycle with standard deviations bars. (b) Seasonal variation of the SIE for the NH.

(largest) SIE of $3.57 \times 10^6 km^2$ ($7.67 \times 10^6 km^2$) for the 41 years (Figure 15.2a). The overall trend was 14% and 7.4% more negative for the 41 years compared to the corresponding trend for 28 years (1979–2006; Parkinson and Cavalieri 2008) and 32-year period (1979–2010; Cavalieri and Parkinson 2012). Though all the trends were negative, they exhibited an increasingly negative tendency from winter ($-2.99\% \pm 0.61\%$/decade), spring ($-3.09\% \pm 0.63\%$/decade), autumn ($-6.02\% \pm 1.21\%$/decade) and summer ($-9.88\% \pm 2.09\%$/decade). A comparison of seasonal trends from the 32-year analysis revealed that for 1979–2019, the magnitude of the negative trend was higher for winter by 12.5%, for spring by 7.6%, for summer by 8.4% and for autumn by 11.7%.

15.4.2 REGIONAL TREND

Table 15.2 shows the yearly and seasonal trends for the nine sectors over the period of 1979–2018. The three regions that contribute 50% to the overall negative SIE trend of the NH are the Seas of Kara and Barents ($-15,428 \pm 3026 km^2$/

year), the Baffin Bay/Labrador Sea (-6376 ± 1434 km²/year) and the Greenland Sea (-6168 ± 1928 km²/year). The yearly trends for the Seas of Kara and Barents, Baffin Bay, Greenland Sea, Gulf of St Lawrence and Canadian Archipelago indicated that these regions lost more sea ice by 14%, 16%, 5%, 24% and 17%, respectively, when compared with the corresponding yearly SIE trends for 1979–2010 (Cavalieri and Parkinson 2012). The SIE trend reversed from 300 to -1638 km²/year (at 95% confidence level) for the Bering Sea. On the other hand, some regions had less negative yearly SIE trends (significant at a 99% confidence level) for the 41-year period compared to the 32-year period. These included the Seas of Okhotsk and Japan, Hudson Bay, Baffin Bay, Greenland Sea and the Arctic Ocean.

In winter, highly negative SIE trends were found for Hudson Bay, Gulf of Saint Lawrence, Kara and Barents Seas by 89%, 10%, 2% and 12%, respectively, compared to the 1979–2010 period. Likewise, the Central Arctic Ocean experienced the highest loss to an extent of 290% (from -200 ± 100 to -781 ± 329 km²/year), while the Canadian Archipelago exhibited increases in the SIE by 7% (significant at 99% confidence). On the other hand, other regions where sea ice loss was lower when compared to the 1979–2010 period were Okhotsk and Japan Seas by 16%, the Bering Sea by 18% and the Baffin Bay/Labrador Sea by 30%. In spring, less negative SIE trends were observed for the Arctic Ocean, Canadian Archipelago, Greenland Sea, the Baffin Bay/Labrador Sea, Hudson Bay, and Seas of Okhotsk and Japan by 80%, 51%, 8%, 32%, 29% and 9.6%, respectively. In contrast, the Bering Sea, Gulf of Saint Lawrence and Kara and Barents Seas exhibited a more negative SIE trend (significant at 99% confidence level) to an extent of 49%, 38% and 13%, respectively, compared to the 1979–2010 period. While the Kara and Barents Seas and Canadian Archipelago experienced a sea ice loss of 6.5% and 9.5% relative to the 1979–2010 period, all other regions showed less negative SIE trends in summer, with a 90% increase in SIE for the Arctic Ocean. In autumn, a drop in the SIE was found for the Bering Sea by 40%, the Kara and Barents Seas by 26% and the Gulf of Saint Lawrence by 97%, compared to the 1979–2010 period.

In brief, the Gulf of Lawrence and the Greenland Sea showed the highest sea ice decline in winter, the Seas of Okhotsk and Japan in spring, the Kara and Barents Seas in summer and autumn. The Canadian Archipelago and Gulf of Saint Lawrence experienced positive trends in winter and autumn, respectively. A highly negative trend was observed for the Hudson Bay in winter and the Central Arctic in winter, spring and autumn.

The linear trends for each of the decades are shown in Table 15.3. For 1979–88, the Seas of Japan and Okhotsk exhibited a highly negative trend ($-17,217 \pm 936$ km²/year). Interestingly in the following decade, this region showed a positive SIE trend (1353 ± 598 km²/year), which switched to yet another highly negative trend ($-16,666 \pm 855$ km²/year). For 2009–18, the loss in SIE was lowest at -442 ± 452 km²/year. For the Bering Sea, the SIE trend dropped from -3837 ± 431 km²/year for 1979–88 to $-26,370 \pm 1272$ km²/decade for 2009–18. Across the decades, a negative trend ($-12,314 \pm 661$ km²/year) for Hudson Bay for 1989–98 was found. The Baffin Bay exhibited negative trends for 1999–08 (-4797 ± 1400 km²/year) and 1999–08

(−14,057 ± 786 km²/year), while the SIE for Saint Lawrence showed a highly negative trend (−3140 ± 164 km²/year) for 1989–98. The Greenland Sea also showed a highly negative trend for 2009–18, while the Kara and Barents Seas lost sea ice across all the decades, except for 1989–98. The Arctic Ocean exhibited the lowest trends across the decades, while the Canadian Archipelago showed negative trends in the range of −2%/decade for 1979–88 to −4%/decade for 1989–98 and 1999–08, and a positive trend (4%/decade) for 2009–18. Overall, the NH showed a negative trend across the decades, with the highest loss (−10%) for 1999–08.

In brief, negative trends were found for the Seas of Japan and Okhotsk and Kara and Barents for 1979–88 and 1999–08. The Hudson Bay, Saint Lawrence and Canadian Archipelago lost sea ice, while the ice cover in the Seas of Japan and Okhotsk, Bering Sea, Baffin Bay and Greenland showed increases for 1989–98. The SIE depleted from all seas, except Hudson Bay, Saint Lawrence, Greenland Sea, and Kara and Barents Seas for 1999–08. Overall, the SIE showed negative trends for NH across the decades with the highest retreat (−10%/decade) for 1999–2008.

15.4.3 REGIONAL SEA ICE EXTENT BY MONTH

The SIE trends for each of the Arctic sectors and the NH for the four decades are depicted in Figure 15.3. The Seas of Okhotsk and Japan showed a negative trend all through the year for 1979–88 and 1999–2008, with a high wintertime sea ice loss of −30,884 ± 2287 km²/year and −34,210 ± 2135 km²/year, respectively. For 2009–18, the sector lost ice from March to June, with a high loss (−14,395 km²/year) in May. Weak positive trends were observed from March to May for 1989–98 and from January to February for 2009–18. Overall, the SIE exhibited a drop in sea ice (−9%/decade), with the highest loss in April (−9217 km²/year) for 1979–2018. The Bering Sea lost ice all through the year, with the highest drop in April (−70,877 km²/year) for 2009–18. It is interesting to note that the SIE increased in April for 1979–88 by 22,677 km²/year. On the other hand, the Seas of Okhotsk and Japan lost sea ice by −5% for 1979–2018.

The sea ice loss/gain appeared post-May for Hudson Bay. A bimodal pattern was observed in the SIE trends in July and November for 1989–98, while a similar pattern marked by increases in sea ice was found for 1979–88 and 2008–18. Overall, the Hudson Bay lost sea ice by −4%/decade for 1979–2018. The SIE extent for the Baffin Bay and the Labrador Sea revealed that from a positive trend in January to June for 1979–88, the sector lost more sea ice from January to July, with the highest drop in winter for 1989–98 (−32,311 ± 2091 km²/year). For 1999–2008, the sector showed an overall negative SIE trend, which switched to positive for 2009–18, with the highest trend recorded in February. This sector lost −11%/decade for 1979–2018. For the Gulf of Saint Lawrence, the increases in sea ice that occurred from January to May for 1979–88 (45%/decade) reversed for 1989–98 (−51%/decade); the trend switched to slightly positive for 2009–18 (8%/decade). Overall, this sector indicated a drop in SIE (−10%/decade) for 1979–2018. The Greenland Sea showed a highly negative trend in SIE for 2009–18, which resulted in an overall loss of −10%/decade for 1979–2018.

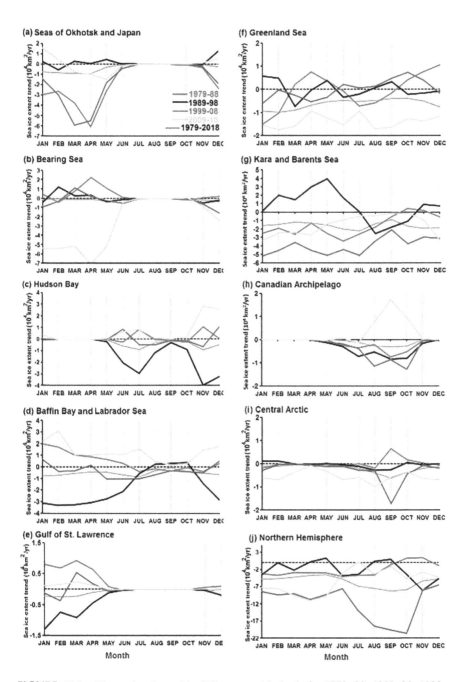

FIGURE 15.3 The regional trend in SIE on monthly basis for 1979–88, 1989–98, 1999–2008 and 2009–18.

Negative SIE trends for the Kara and Barents Seas were observed during most of the year across the decades, except for 1999–2008, where a highly positive trend was encountered during January–July. The overall SIE trend revealed a sea ice loss of −13% for 1979–2018. The Canadian Archipelago exhibited negative SIE trends for all the decades, except for 2009–18, which showed increases in SIE (7257 km^2/year). The overall SIE decline was found to be −1% for 1979–2018. For the Central Arctic, weak SIE trends were observed across the decades from January to August, except for 2009–18 which exhibited variable trends across the months. A highly negative (positive) trend was observed for 1999–2008 (1979–88) in September. Negative trends over the year were observed for 2009–18. Overall, the SIE depleted by −0.55%/decade, with an appreciable ice loss in summer (−1%/decade).

In brief, the Seas of Okhotsk and Japan, Bearing Sea and the Gulf of Saint Lawrence showed higher trends during December to May, while the Hudson Bay, Canadian Archipelago and Central Arctic showed higher trends during June to November. For 2009–18, the seas of Bearing, Greenland, Kara and Barents Seas, and Central Arctic Ocean exhibited sea ice loss, while the Hudson Bay, Canadian Archipelago, Baffin Bay and the Labrador Sea, and the Gulf of St. Lawrence showed a positive tendency in SIE.

Though our calculations showed that the SIE changes are related to seasons, for completeness it is necessary to provide insights on how the sea ice thickness varied with seasons using the SIV dataset over the last four decades (Table 15.4). For summer, the highest loss of SIV occurred for 1999–2008 (−551 ± 200 km^3/year), followed by −238 ± 107 km^3/year for 1989–98. Likewise, for autumn, we found a similar tendency for SIV, with the highest depletion for 1999–2008 (−530 ± 190 km^3/year) and 1989–98 (−303 ± 120 km^3/year). We found that the SIV showed an increase for 1979–99 for autumn. For spring, the SIV declined to third lowest for 1999–2008 (−492 ± 173 km^3/year), followed by −218 ± 86 km^3/year for 2009–18 and −185 ± 83 km^3/year for 1989–98. The winter period showed the fourth-lowest SIV for 1999–2008 (−452 ± 159 km^3/year). We note that the highest loss of SIV shifted from summer and autumn for 1999–2008 to winter and spring for 2009–2018.

15.4.4 REGIONAL DRIVERS FOR SEA ICE CHANGE

We examined the role of local forcings such as wind, MSLP, ERA5-based SST and ERA5-based turbulent heat flux on SIE. The rationale for choosing these parameters was as follows. Due to ocean circulation (thermohaline and wind-driven) and solar radiation, the SST changes spatially relative to the atmospheric layer just above sea/ice surface; this facilitates latent and sensible heat transfer between ocean/sea ice and atmosphere and cools/warms the surface atmospheric boundary layer leading to changes in atmospheric pressure which influences the regional wind speed. The histogram for correlation for the boreal summer, autumn, winter and spring between the parameter and SIE at zero-lag correlation are shown in Figure 15.4. The correlation coefficients exceeding 0.64 and 0.76 were found to be significant at 95% and 99% confidence level.

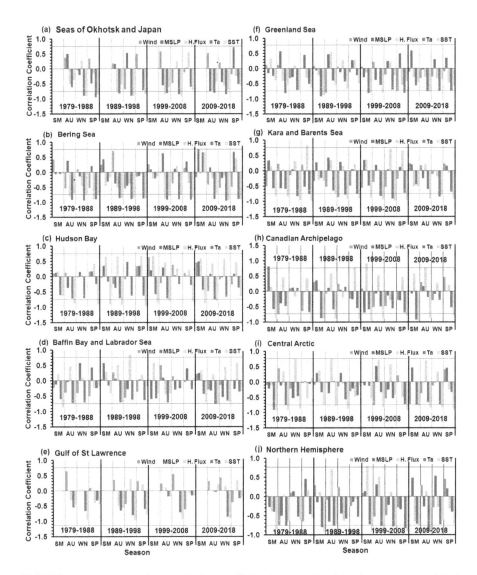

FIGURE 15.4 The zero-lag correlation coefficient between local forcing parameters (wind speed, mean sea level pressure (MSLP), heat flux, air temperature (Ta) and sea surface temperature (SST)) and SIE for different Arctic sectors demarcated in Figure 15.1a. The x-axis represents month from seasons for four decades.

For the Seas of Okhotsk and Japan, increases in Ta and SST caused decreases in the SIE, as revealed by a significantly negative correlation between them for winter and spring for 1979–88. For 1989–98, the influence of wind speed, heat flux and Ta reduced the SIE for winter, while the increases in Ta and SST facilitated a decline in the SIE for spring. For 1999–2008, increases in the heat flux and Ta depleted the SIE for autumn, while Ta and SST favour reduction in the SIE for winter. The warmer SST caused decreases in the SIE for spring. For 2009–18, warmer air due to high heat flux and low MSLP favoured depletion of SIE for autumn, while warmer SST and Ta caused decreases in the SIE for winter. It is noted that there was no sea ice for summer. For the Bering Sea, heat flux, a warmer atmosphere and higher SST exhibited a negative role on the SIE for autumn and winter, while Ta and SST opposed increases in the SIE for winter for 1979–88. For 1989–98, the increases in Ta and SST were effective in reducing the SIE for autumn, winter and spring. Increases in Ta, SST and heat flux for autumn suggested lower SIE, while Ta and SST were effective in reducing the SIE for winter and spring. For 2009–18, positive correlations between SIE and MSLP, Ta and SST were found for summer. However, for autumn, winter and spring the increases in Ta and SST dictated the SIE decline.

For the Hudson Bay, the SST influenced the SIE during summer, autumn and spring, while Ta played an important role in depleting the SIE for autumn and winter for 1979–88. For 1989–98, increases in the SIE had positive feedback on the heat flux for summer, autumn and spring. However, increases in SST and Ta were counteractive to SIE for the above reasons. A similar scenario was observed for 1999–2008, except that the positive feedback on heat flux was seen only for summer. The SST was effective in reducing the SIE for summer, autumn and spring for 2009–18, while Ta overwhelmed the decreases in SIE for autumn. For the Baffin Bay, the SST favoured the SIE decline for summer and autumn; likewise, a warmer atmosphere depleted the SIE in autumn and winter for 1979–88. For 1989–98, it was found that Ta and SST negatively influenced the SIE for autumn, while for 1999–2008, Ta showed a significantly negative correlation with the SIE for autumn. For 2009–18, we found that SST exhibited a negative relationship with the SIE for summer, autumn and winter, while higher Ta reduced the SIE for autumn and winter.

For the Gulf of Saint Lawrence, winds in autumn enhanced the SIE possibly through the ridging effect; however, Ta and SST worked in tandem to reduce the SIE for 1979–88. For 1989–98, increases in Ta in autumn depleted the SIE, while wintertime SST and Ta reduced the SIE. The SIE also showed a negative influence on SST for spring. Likewise, for 1999–2008 and 2009–18, warmer atmosphere and increases in SST opposed sea ice growth for winter. For Kara and the Barents Sea, the SST significantly influenced the depletion of the SIE for 1979–88. Warmer atmosphere reduced the SIE for winter and spring. We also found that the heat flux was also an important contributor to the SIE decline in spring. Higher SST and increases in Ta contributed to reducing the SIE for autumn, winter and spring for 1989–98 and 1999–2008. For winter and spring, the SIE induced positive feedback which enhanced the heat flux to the atmosphere for 1999–2008. The warmer SST reduced the SIE for autumn, winter and spring for 2009–18. Warmer atmosphere induced negative anomalies in the SIE for winter and spring.

For the Greenland Sea, increases in Ta induced negative anomalies in the SIE for autumn and winter, while increases in SST depleted the SIE for winter and spring for 1979–88. For summer, Ta and SST played a negative role in reducing the SIE for 1989–98. Likewise, we found that the wind speed in autumn and SST supported the SIE decline for winter. For 1999–2008, increases in the summertime Ta and SST reduced the SIE. Similarly, warmer SST for autumn, wintertime warm Ta and enhanced wind speed for spring induced negative anomalies in the SIE. For 2009–18, we found higher Ta reduced the SIE for autumn, winter and spring, while the increases in the SST for autumn rendered a lower SIE.

For Central Arctic, we found that Ta and SST for summer and autumn played a role in the SIE decline for 1979–88. For winter, the heat flux and SST were drivers for lowering the SIE. For 1989–98, increases in SST induced negative sea ice anomalies for summer, autumn and winter. We note that positive feedback of SIE on the heat flux was found for autumn post-1979–88. For 1999–2008, the higher wind speed was responsible for the SIE decline for autumn, while increases in the SST registered a drop in SIE. Similarly, higher Ta and SST for spring lowered the SIE for spring. For 2009–18, it was revealed that higher wind speed, warmer Ta and SST led to decreases in the SIE. For autumn and winter, Ta and SST were responsible for a drop in the SIE, while SST was found to be the lone driver for negative anomalies in sea ice for spring.

For the Canadian Archipelago, SST played a significant role in inducing lower SIE for summer, autumn and spring for 1979–88. Increases in Ta were also found to deplete the SIE for spring. Summertime SIE was reduced due to increases in Ta and SST for 1989–98. For autumn and spring, the SST negatively influenced SIE. We found a positive correlation between the heat flux and SIE for spring. For 1999–2008, the warmer SST facilitated heat transfer to the overlying atmosphere, resulting in a lower MSLP during summer. The SST in autumn and spring harmed the SIE, and springtime Ta also facilitated a drop in the SIE. For 2008–18, we encountered negative correlations between SST/Ta and SIE for summer. Warmer SST promoted heat flux transfer to the atmosphere for autumn, while increases in SST led to the depletion in the SIE for spring.

In summary, a significant influence of SST on the SIE was revealed for autumn, winter and spring. Atmospheric warming also contributed to the SIE changes across the decades. However, the SIE in the Greenland Sea and the Gulf of Saint Lawrence experienced a lower impact of SST/Ta. Overall, the SIE for the NH was significantly influenced by SST and Ta for summer and autumn for 1999–2008 and 2009–2018. Changes in the wind speed and heat flux played a secondary role in the SIE modulation.

15.4.5 Remotely Linked Processes for Sea Ice Change

The zero-lag correlations between indices of PDO/AMO/SOI/AO and the SIE on a decadal basis are shown in Figure 15.5. For the Seas of Okhotsk and Japan, we found a negative impact of SOI on autumn SIE, while the PDO was significantly facilitating a drop in the wintertime SIE for 1979–88. For the Bering Sea, PDO facilitated increases in the wintertime SIE for 1979–88. The negative association of AMO with the SIE ($r = -0.82$) was highly significant (99%) for summer, while AO induced increases in the SIE in autumn for 1989–98. For 1999–2008, the PDO favoured depletion of the SIE for autumn and winter, while the SOI induced increases in the SIE for winter and spring. For 2009–18, the AO facilitated increases in the SIE for summer.

For Hudson Bay, we found a positive association of the PDO with SIE for summer ($r = 0.75$) for 1979–88. The AMO was negatively correlated with the SIE for summer and spring for 1989–98. For 2009–18, the PDO favoured increases in the SIE, while SOI reduced the SIE for summer. The AMO strongly favoured decreases in the SIE

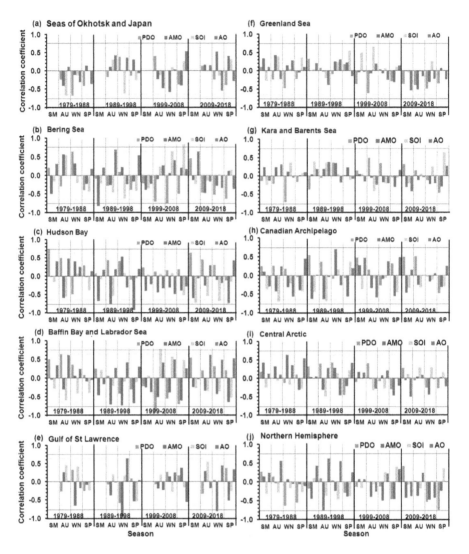

FIGURE 15.5   The same as in Figure 15.4, but for remote forcing parameters (PDO, AMO, SOI and AO).

for spring ($r = -0.74$). For the Baffin Bay, the PDO supported increases in the SIE for autumn for 1979–88, while the AMO caused a decline in the SIE for autumn, winter and spring for 1989–98. For 1999–2008, PDO caused a reduction in the SIE ($r = -0.87$), while the SOI favoured increases in the SIE for autumn. The PDO was negatively associated with the SIE ($r = -0.7$) for spring.

For the Gulf of Saint Lawrence, the AMO opposed increases in the SIE for winter across the decades, except for 1999–2008. The wintertime AO favoured the increases in the SIE for 1989–98. We found that the SOI favoured increases in the SIE for spring for Kara and Barents Seas for 2009–18. For the Greenland Sea, the SOI showed a positive association with the SIE for autumn for 1999–2008. We found that none of these indices significantly influenced the SIE for Central Arctic. As for the Canadian Archipelago, we found only the SOI opposed the increases in the SIE for autumn for 1979–88 and 1989–98.

15.5   DISCUSSION

Various studies have emphasized the possible causes for the Arctic sea ice loss and the role of several physical processes is delineated (Serreze and Barry 2011). Among them is a robust phenomenon known as AA – the double pace of Arctic surface warming compared to the lower latitudes (Screen and Simmonds 2010). The NH air temperature has risen by about $0.08 \pm 0.02°C$/year (~3.1°C for 1979–2018) (Kumar et al. 2020). They also showed that the SST over the NH has exhibited a rapid warming trend of $0.089 \pm 0.01°C$/year (~3.5°C for 1979–2018) compared to the NH land surface temperature ($0.072 \pm 0.01°C$/year (~2.8°C for 1979–2018). The increase in atmospheric warming has been attributed to the rising concentration of greenhouses, particularly that of CO_2 that accounts for 77% of man-made greenhouse gas emissions. CO_2 has been increasing at the rate of 21.5 ppm/year (Figure 15.6) due to the large-scale combustion of fossil fuels and deforestation. The rapid and accelerated loss in the NH SIE can be largely attributed to the warming atmosphere and rising SST, as seen from the zero-lag correlation (Figure 15.4). Notz and Stroeve (2016) established the relationship between the September SIE and cumulative atmospheric CO_2 emissions, and highlighted the importance of restricting future CO_2 emissions to limit global warming to below 2°C and retain the Arctic summer ice cover.

The timing of the strongest SIE trends varies from region to region on a seasonal basis (Figure 15.3). Our results showed that for the Seas of Okhotsk and Japan, higher sea ice loss for April was associated with the positive trends in Ta (0.25°C/year) and SST (0.07°C/year), both of which showed a significantly negative association with SIE for 1979–88. Likewise, for 1999–2008, a positive Ta trend (0.5°C/year) and SST (0.07°C/year) was responsible for enhanced sea ice depletion for March. It has been reported that the interannual variability of SIE is highly correlated with a surface temperature in autumn and winter (Ohshima et al. 2006), and Ta has risen progressively

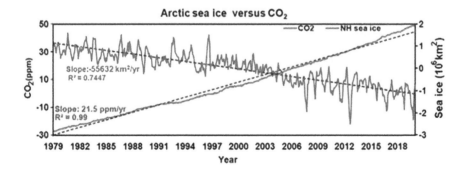

FIGURE 15.6   Relationship between CO_2 concentration and NH SIE.

over the past 50 years (Kim 2012). In these surface-warming conditions, it is inferred that the Seas of Okhotsk and Japan lost sea ice annually by $-8.79\% \pm 3.13\%$ over the last 40 year period (Table 15.2).

What mechanism(s) is associated with the SST warming in the Okhotsk Sea? The overall thermal regime in the northern Sea of Okhotsk is determined by the effect of the impact of the atmospheric warming or cooling on surface waters, formation or melting of sea ice, convection in the water column and the effect of currents, particularly the advection of warm water from the Pacific Ocean. The north thermal regime depends to a large extent on the inflow of the Western Kamchatka Current whose advection of warm ocean waters has shown a significant variation in recent decades (Figurkin et al. 2008). It was characterized by enhanced flow from the 1980s to the 1990s, followed by a decreased advection during the late 1980s, early 1990s and 2000s. In the northwest, the Amur River discharge is another factor influencing the SST regime in the basin. In late autumn, river discharge leads to lower salinities and strong stratification which encourages the formation of sea ice (Ohshima et al. 2010). In spring, the more river discharge and its relatively warm waters extend far south, so the SIE was lowest (Table 15.2). From 1983 to 2005, the river flow was low, after which the discharge increased, corresponding to a warm period. In the south, the summer temperature regime is defined by the intensity of the warm Soya Current, whose intensity is greatest in years when winter ice cover is larger. Warming of the northern Okhotsk Sea generally corresponds to cooling in the south.

The Bering Sea showed a large drop in the SIE in spring (-88%) for 2009–18 (Figure 15.3b) which was marked by positive trends in SST ($0.3°C$/year) and Ta ($0.42°C$/year), which is confirmed by a strongly negative correlation between the local drivers and the SIE (Figure 15.4b). Conversely, increase in the SIE for 1979–88 was facilitated by the cooling trend in Ta ($-0.17°C$/year) and SST ($-0.06°C$/year). Why the sea ice is short-lived in the Bearing Sea? For about half of the year, the Bering Sea is free of ice, but begins to form in November or the sea ice is advected into the Bering Sea through Bering Strait mainly by anomalies in easterly winds associated with the Aleutian Low, which was particularly strong during the 1980s (Francis and Hunter 2007). Although the ice is generally limited to the shelves, its seasonal extent advances. Along the leeward side of the coasts and islands freezing takes place in polynyas, and then the ice is advected southward (or south-westward) where it is eventually melted by the warmer water. During freezing, the seawater ejects brine (high-salinity (>34 psu) water) at the polynyas and the ice freshens the water (~30 psu) at the melting edge. The amount of production and advection of ice is determined by which storm track dominates in a given winter; the greatest ice production occurs in the years when the Aleutian Low is intensified and storms migrate along the principal storm track. The ice begins to retreat by late March or April. The Aleutian Low is one of the two main low-pressure systems (the other being the Icelandic Low) in the high-latitude NH. During summer, the Aleutian Low is typically weak due to long periods of daylight and high solar insolation. A distinct change occurs in atmospheric pressure fields during winter wherein high MSLP (Siberian High) dominates Asia, while the Aleutian Low strengthens and dictates the weather over the North Pacific and the Bering Sea.

In the Hudson Bay, high sea ice losses were found for November and July for 1989–98. The reason was that SST/Ta showed a warming trend 0.18/0.11°C/year and 0.07/0.48°C/year for July and November, respectively. On the other hand, the heat flux transfer to the atmosphere showed a negative trend (−0.96/−0.11 W/m²/year) for July/November due to more heat absorption by the ocean devoid of sea ice. The positive trend in the SIE for 2009–18 (Figure 15.3c) was due to the cooling in SST/Ta (−0.07/−0.47°C/year) for July/November.

In the Baffin Bay and the Labrador Sea, high wintertime sea ice depletion for 1989–98 was attributed to high heat flux transfer to the atmosphere (−0.3 W/m²/year) which increased Ta (0.3°C/year), in conditions when the AMO acted favourably. For spring, the warmer atmosphere played a major role in decreasing the SIE, while for autumn Ta and SST showed warming trends. With Ta and SST showing a cooling trend for winter, spring and autumn facilitated a positive SIE trend for 1979–88 and 2009–18. For the Gulf of Saint Lawrence, the sea ice loss was attributed to the positive trend in TA/SST of 0.22/0.08°C/year (not significant at 95%) and the transfer of atmospheric heat to the ocean in winter for 1989–98. On the other hand, for the winter of 1979–88, a decreasing trend in Ta/SST (not significant at 95%) facilitated increases in SIE (Figure 15.4e).

In the Baffin Bay, Ta has increased by 2–3°C/decade since the late 1990s (Peterson and Pettipas 2013), resulting in earlier melt onset and delayed ice freeze (Stroeve et al. 2014). Accordingly, a rapid decline of SIE across all seasons has been reported in the Bay (Stroeve and Notz 2018). The interannual variation of the SIE was found to be correlated with winter Ta (Tang et al. 2004) with a periodicity of 1.7, 5 and 10 years. On the other hand, oceanic processes related to the West Greenland Current determine the decadal variation of sea ice for the Labrador Sea (Clem and Renwick 2015).

In the Greenland Sea, the seasonal cycles of sea ice loss/gain have been influenced by significant changes in Ta/SST and heat flux. The high year-round sea ice loss for 2009–18 was due to higher heat flux in the atmosphere that enhanced Ta and SST. The basin is influenced by warm Atlantic water supplied by the Norwegian Atlantic Current which flows poleward (Chatterjee et al. 2018). We also found large changes in the sea ice trend across the seasons over the four decades were observed in the Kara and Barents Seas. Shapiro et al. (2003) pointed out that the winter SIE has declined since 1850, and the retreat observed during the recent decades (Figure 15.3g) has been the largest decrease in the Arctic in conformity Parkinson and Cavalieri (2008). During winter and spring for 1989–98, the wind speed, heat flux, Ta and SST showed negative trends of −0.040 m/s/year, −0.03 W/m²/year, −0.27°C/year and −0.03°C/year which facilitated increases in the SIE. On the other hand, a maximum sea ice loss across the seasons for 1999–2008 was driven by warmer SST/Ta which showed a significant correlation (>95% confidence) with the SIE (Figure 15.4g).

The Barents SIE variations are linked to several processes, which includes large-scale atmospheric circulation anomalies (Maslanik et al. 2007; Deser and Teng 2008; Zhang et al. 2008), cyclone activity (Simmonds and Keay 2009; Sorteberg and Kvingedal 2006), local winds and ice advection from the Arctic Ocean (Hilmer et al. 1998; Kwok 2009), and locally generated oceanic heat anomalies (Schlichtholz 2011) or those advected into the Barents Sea (Vinje 2001; Francis and Hunter 2007). The Atlantic water which enters between Norway and Bear Island in the Barents

Sea's main oceanic heat source, which during its passage through the Barents Sea, loses most of its heat to the atmosphere (Häkkinen and Cavalieri 1989; Årthun and Schrum 2010).

The Canadian Archipelago showed increases in the SIE for August–October for 2009–18 (2×10^4 km²/year) which was associated with cooling Ta/SST tendency (−0.12/0.04°C/year), and the positive trend in the wind speed (0.06 m/s/year) which possibly contributes to the ridging effect on the sea ice. We also observed that the increase in the sea ice cover increased the heat flux (0.2 W/m²/year) through positive feedback. In contrast, from May to October the sea ice depleted across the other decades due to atmospheric and ocean warming which facilitated of transfer of heat to the ocean. For the Central Arctic, we observed increases in the summertime SIE for 1979–88 which was driven by SST/Ta cooling (−0.01/−0.04°C/year), with a negative trend in the heat flux (−0.24 W/m²/year). The summertime sea ice loss for 1999–2008 was influenced by the positive trend in Ta/SST (0.05/0.05°C/year). Likewise, a strongly Ta warming in summer and winter and supported by warmer SST facilitated the sea ice loss across all seasons for 2009–18.

From the analysis of Arctic surface air temperature data for the period of 1979–97, Rigor et al. (2000) reported that the seasonal and spatial trends are generally consistent with the SIE trends. For December–February, they found a warming trend over the Eurasian landmass and the Laptev and East Siberian Seas to the north of Russia, whereas the eastern Siberia and north of the Canadian Archipelago experienced cooling trends. During September–November, Alaska and the Beaufort Sea showed cooling trends. For March–May, the warming trends were detected over the entire Arctic region. During June–August, however, there are no significant trends in the temperature data (Rigor et al. 2000), which is interesting because from June to September are the months for which we find the strongest negative trends for the ice cover as a whole.

The role of climate indices is much more complex to decipher since their influence changes seasonally and regionally. To understand how the indices changed over the decades, we present in Figure 15.7, the time series of PDO, AMO, SOI and AO. The PDO was highly positive for all seasons for 1979–88, which weakened by 50% for 1989–98. It switched to negative, with lowest values for autumn and summer for 1999–2008. Since 2014 winter, it is prominently positive in the entire record. Although the PDO is not considered as a strong driver of winter sea ice variability, the analysis suggests that the PDO phase modulates the atmospheric response to sea ice variability (Screen and Francis 2016). The PDO phase influences the wintertime atmospheric circulation, mostly over the Pacific Ocean, with its influence extending into the Arctic (Screen and Francis 2016). Anomalous southerly winds occur during negative PDO over the central North Pacific facilitated by a weakened Aleutian Low, which adverts air warmed by wintertime sea ice loss in the Sea of Okhotsk and the Bering Sea into the Central Arctic. Likewise, during the negative PDO phase, anomalous westerly and southerly flow south and east of Greenland adverts air into the Central Arctic that has been warmed by sea ice loss in the Labrador Sea, Baffin Bay and the Greenland Sea.

The AMO showed a consistent increase from −0.13 for 1979–88 to 0.16 for 2009–18, with high values occurring in summer since 1999; it entered into the positive

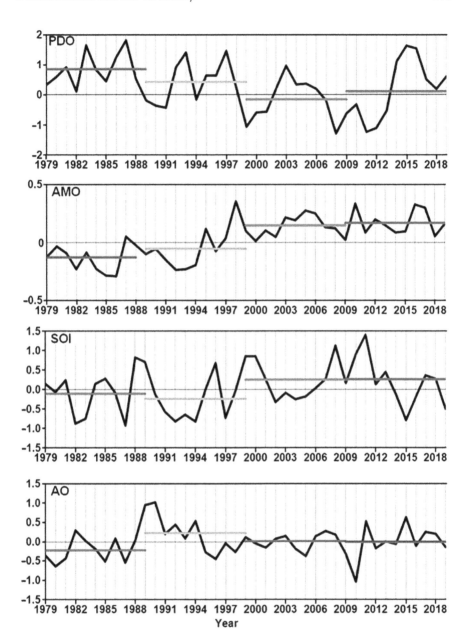

FIGURE 15.7 Time series of PDO, Atlantic Multidecadal Oscillation (AMO), Southern Oscillation Index (SOI) and AO over 1979–2019. Coloured horizontal lines indicate averages of the indices over each of the decades: 1979–88, 1989–98, 1999–2008 and 2009–18.

phase in 1997 winter. The shift in AMO was accompanied by anomalous oceanic heat transport toward the Arctic, whose role dominates over the sea ice multidecadal variability in the Atlantic sector of the Arctic (Årthun et al. 2012; Miles et al. 2014; Onarheim et al. 2014; Zhang 2015). The phase of AMO can alter the impact of sea ice loss on the horizontal wave propagation in winter, wherein the Arctic sea ice retreat is linked to a trough-ridge-trough response over Pacific-North America only during a negative AMO phase. Osborne et al. (2017) argued that this is a result of enhanced meridional temperature gradient promoted by the sea ice loss, just south of the climatological maximum, in the mid-latitudes of the central North Pacific. This causes a southward shift in the storm tracking the North Pacific, which reinforces the Aleutian Low allowing the circulation anomalies to proliferate into North America. While the climate response to sea ice decline is sensitive to AMO-linked SST anomalies in the North Pacific, there is little sensitivity to larger-magnitude SST anomalies in the North Atlantic. It is pointed out that the AMO controls the frequency of atmospheric blocking highs over the Euro-Atlantic sector by changing the baroclinicity and transient eddy activity (Häkkinen et al. 2011; Peings and Magnusdottir 2014). The reduced (enhanced) blocking highs over the Euro-Atlantic sector can further weaken (enhance) the vertical wave propagation, leading to a strengthened (weakened) stratospheric polar vortex (Li et al. 2018; Nishii et al. 2011).

The SOI was in the negative phase (on average) for 1979–88 and 1989–98, which switched to positive after 1998 for most of the period, except for 2015 (Figure 15.7). From negative in 1979–88 to positive phase in 1989–98, the AO has weakened in the last two decades, but the Arctic sea ice continued to decline. The positive (negative) AO phase produces a strong polar vortex, with the mid-latitude jet stream shifted northward (southward). According to Liu et al. (2004), positive phases of the AO and SOI/El Niño-Southern Oscillation (ENSO) produce congruent (opposite) sea ice changes in the western (eastern) Arctic. However, the magnitude of the AO-induced sea ice changes and that of SOI/ ENSO is much smaller than the regional ice trends. During El Niño/La Niña events, the alterations in the regional Ferrel cell lead to anomalous poleward/equatorward mean meridional heat flux into the sea ice areas in the northeast Pacific or northwest America sector/northeast America or northwest Atlantic sector, which increases/decreases air temperature and restricts/promotes sea ice growth in the Chukchi and southern Beaufort Seas/the Northwest Passage. Their analysis also indicated that more (less) ice in the western (eastern) Arctic is a result of a combined effect of the anomalous mean surface heat flux and ice advection during the positive AO phases.

From the 1950s to the 1980s, the AO flipped between positive and negative phases, but it entered a strong positive pattern between 1989 and 1995. The acceleration in the sea ice retreat since the mid-1990s may have been partly triggered by the strongly positive phase of AO, apart from other factors, during the preceding years (Rigor et al. 2002; Rigor and Wallace 2004) that flushed older, thicker ice out of the Arctic Ocean via Fram Strait, leading to thinning of the winter ice pack which in turn preconditioned the summer ice pack for faster melt. The winter AO index is marked by cyclonic MSLP anomalies over the high latitudes, especially over the Norwegian and Greenland Seas. The negative anomalies imply a strong cyclonic circulation resulting in the flow into and out of the Fram Strait and southward along the east coast of Greenland. The large-scale variability in the extratropical surface temperature is closely related to changes in the position of storm tracks and jet stream (Francis and Vavrus 2012), which is largely regulated by the leading atmospheric modes of AO.

15.6 IMPACTS OF PLUMMETING ARCTIC SEA ICE

The Arctic sea ice cover is not only a receptive pointer of climate change, but it also has strong feedback effects on the other components of the climate system. Associated with ongoing Arctic warming (Martin et al. 1997; Screen and Simmonds 2010), if a 50% drop in the SIE in summer and an even higher decline in SIV continues, then the sea ice cover is likely to drop as well. But if the sea ice changes are related more closely to oscillatory behaviours in the climate system, then it is likely that there will also be fluctuations between the epoch of sea ice decrease and that of sea ice increase. And with the precipitous decline in the Arctic sea ice cover, the system is losing the ability to control and stabilize the earth's climate system. Turning the Arctic from white to blue ocean is expected to change the region's albedo, and the fact that the ocean surface reflects roughly 10% of solar radiation, the future Arctic will experience high albedo-driven warming. Warm Ocean will induce low-/high-pressure systems only to accelerate winds. Cloud formation and drizzle in the warm air layer increases down-welling longwave radiation and initiates melt onset. As solar heating increases with earlier melt onset dates in summer, positive ocean heat-content anomalies increase in magnitude (Timmermans 2015). In autumn, larger storms supported by heat from the open ocean would cause wave-induced mixing of the Arctic Ocean, which would release subsurface heat absorbed during the summer, making it difficult for ice to form in the autumn. Advection of warm air temperature anomalies generated over open Arctic waters in summer is increasing melt over the Greenland ice sheet to release a large volume of freshwater, affecting the stratification in the Arctic Ocean.

The retreat of summer sea ice during the last decade from the shelves allowed the water to warm up. The warmer water extends to the seabed and thaws the offshore permafrost that exists since the last Ice Age. Underneath it is a thick sediment layer holding large amounts of solid methane hydrates. With the release of the overlying pressure, the hydrates transform into gaseous methane, which bubbles up through the water column and eventually gets into the atmosphere. This release is pushing up the global methane levels after being stagnant through the early years of this century. Methane traps heat 23 times more efficiently per molecule than CO_2.

The acceleration of thinning and retreat of Arctic sea ice is also related to 'strong positive feedback mechanisms associated with ice-albedo and open water formation efficiency' (Lindsay and Zhang 2005; Holland et al. 2006). These positive feedback mechanisms leave the ice pack exposed to forcing from other processes: natural and anthropogenic. For example, higher downward longwave radiation associated with a warmer atmosphere, and higher water vapour and cloudiness over the Arctic Ocean (Francis and Hunter 2006) have become a key factor driving low summer SIEs since the mid- to late 1990s. Evidence suggests that the atmospheric circulation has continued to change the wintertime sea ice distribution since the mid-1990s (Maslanik et al. 2007; Francis and Hunter 2007). It is noted that the effects of sea ice decrease may be sensitive to the study region. Sea ice loss in the Sea of Okhotsk, Hudson Bay and the Labrador Sea, which fall in the vicinity of the main storm tracks, seems to have a larger role to influence the mid-latitude storm track (Kidson et al. 2011; Screen 2013).

Arctic sea ice plays a vital role in the global ecosystems (Post et al. 2013) and its vanishing will not only contribute to rising sea level but render many species homeless. It is also well-argued that vanishing sea ice will have a major impact on mid-latitude weather and climate (Vihma 2014). Ursus maritimus (Polar bear), for example, are utterly reliant on Arctic sea ice to live as they use the ice for foraging. As the ice disappears, they lose the ability to sustain themselves. Molnár et al. (2020) found that most polar bear populations are at risk of disappearing by 2100 due to the plummeting SIE. An increasing number of polar bears are forced by the longer summer periods to rest along Arctic coastlines, where their powerful sense of smell attracts them to human waste, stocked food and animal carcasses, thereby bringing them into greater conflict with humans. The reliance on Inuit and coastal Cree on sea ice for travelling and hunting is reflected in their detailed knowledge of its processes, characteristics and annual cycles (McDonald et al. 1997). The sea ice determines the ecology of the ice biota and influences the pelagic systems under the ice and at ice edges (Legendre et al. 1992a).

15.7 CONCLUSIONS

We set out to analyse decadal variability using satellite-based monthly anomalies of SIE which were computed relative to 1981–2010. We found a decreasing coverage of Arctic sea ice in line with Arctic warming (Walsh 2013; Hartmann et al. 2013) via ice-albedo feedbacks as highlighted elsewhere (Screen and Simmonds 2010). The Seas of Okhotsk and Japan, Bearing Sea and the Gulf of Saint Lawrence showed higher trends during December to May, while the Hudson Bay, Canadian Archipelago and Central Arctic showed higher trends during June to November. For 2009–18, the seas of Bearing, Greenland, Kara and Barents Seas and Central Arctic Ocean exhibited sea ice loss, while the Hudson Bay, Canadian Archipelago, Baffin Bay and the Labrador Sea, and the Gulf of St. Lawrence showed increases in the SIE.

A significant influence of SST on the SIE was revealed for autumn, winter and spring. Atmospheric warming also contributed to the SIE changes across the decades. However, the SIE in the Greenland Sea and the Gulf of Saint Lawrence experienced a lower impact of SST/Ta. Overall, the SIE for the NH was significantly influenced by SST and Ta for summer and autumn for 1999–2008 and 2009–2018. Changes in the wind speed and heat flux played a secondary role in the SIE modulation. The physical mechanism underlying the Arctic sea ice decline includes thermodynamic processes involving alterations in air temperature, radiative and turbulent energy fluxes, and ocean heat storage, and dynamical processes such as modulations in winds and ocean currents.

The Bering Sea is free of ice for half of the year, where it begins to form in November or the sea ice is advected into the Bering Sea through Bering Strait mainly by anomalies in easterly winds associated with the Aleutian Low, which was particularly strong during the 1980s. Anomalous southerly winds occur during negative PDO over the central North Pacific facilitated by a weakened Aleutian Low, which adverts air warmed by wintertime sea ice loss in the Sea of Okhotsk and the Bering Sea into the Central Arctic. During the negative PDO phase, anomalous westerly and southerly flow south and east of Greenland advects air into the Central Arctic that has been warmed by sea ice loss in the Labrador Sea, Baffin Bay and the Greenland Sea. The analysis suggests that the life cycle of Arctic sea ice in most sectors is reduced by a host of factors, some of which are regional while others are dictated by large-scale atmospheric circulation induced by the phases of AO, PDO and AMO.

As a way forward, it is recommended that monitoring of sea ice should continue, as we have noted its precipitous decline across the seasons due to several factors outlined in this work. While most of the changes are communicated to the sea ice through the atmospheric changes, the ocean component is also essential in explaining the seasonality of the trend patterns. The record of 40 years of satellite data used here is relatively short, and the ability of global coupled climate models to realistically represent the complex polar climate system is in doubt. Considerable efforts are required to reconcile the observations and modelling efforts to evolve a better scenario of variations in the high latitudes for prediction and early warning.

ACKNOWLEDGEMENTS

We thank NCPOR Director, Dr M. Ravichandran, for motivation and providing the research facilities. Sea ice extent data were downloaded from the NSIDC site: https://nsidc.org/data/g02135. PDO was downloaded from http://research.jisao.washington.edu/pdo/PDO.latest.txt, AMO from https://www.esrl.noaa.gov/, SOI from http://www.cpc.ncep.noaa.gov/ and AO from https://legacy.bas.ac.uk/met/gjma/sam.html. The PIOMAS data were downloaded from http://psc.apl.uw.edu/research/projects/arctic-sea-ice-volume-anomaly data/). The CO_2 was obtained from https://www.esrl.noaa.gov/gmd/ccgg/trends/. We thank the guest editor, Dr. Neloy Khare, for waiting patiently for the ms, the editing of which was delayed due to the lockdown imposed by COVID-19.

REFERENCES

Årthun M, Eldevik T, Smedsrud LH, Skagseth Ø, Ingvaldsen RB, 2012. Quantifying the influence of Atlantic heat on Barents Sea ice variability and retreat. *J. Clim.* 25(13): 4736–4743. doi: 10.1175/JCLI-D-11-00466.1.

Årthun M, Schrum C, 2010. Ocean surface heat flux variability in the Barents Sea. *J. Mar. Syst.* 83: 88–98.

Aagaard K, Carmack EC, 1989. The role of sea ice and other fresh water in the Arctic circulation. *J. Geophys. Res.* 94. doi: 10.1029/JC094iC10p14485.

Agnew TA, Howell S, 2002. Comparison of digitized Canadian ice charts and passive microwave sea-ice concentrations. Geoscience and Remote Sensing Symposium, 2002. IGARSS '02. *2002 IEEE International.* pp. 231–233. doi: 10.1109/IGARSS.2002.1024996.

Cavalieri DJ, Parkinson CL, DiGirolamo N, Ivanoff A, 2012 Intersensor calibration between F13 SSMI and F17 SSMIS for global sea ice data records. *IEEE Geosci. Remote Sens. Lett.* 9: 233–236. doi: 10.1109/LGRS.2011.2166754.

Cavalieri DJ, Parkinson CL 2012 Arctic sea ice variability and trends, 1979–2010. *Cryosphere* 6: 881–889.

Cavalieri DJ, ParkinsonCL, GloersenP, Comiso JC, 1999. Zwally HJ1999Deriving long-term time series of sea ice cover from satellite passive-microwave multisensor data sets. *J. Geophys. Res.* 104(C7): 15803–15814. doi: 10.1029/1999JC900081.

Cavalieri DJ, Parkinson CL, Gloersen P, Zwally HJ, 1997. *Arctic and Antarctic Sea Ice Concentrations from Multichannel Passive-Microwave Satellite Data Sets: October 1978-September 1995- User's Guide. NASA TM 104647.* Goddard Space Flight Center, Greenbelt, MD, p. 17.

Cavalieri DJ, St. Germain KM, Swift CT, 1995. Reduction of weather effects in the calculation of sea ice concentration with the DMSP SSM/I. *J. Glaciol.* 41: 455–464.

Cavalieri DJ, Gloersen P, Campbell WJ, 1984. Determination of sea ice parameters with the Nimbus 7 SMMR. *J. Geophys. Res.* 89: 5355–5369.

Chatterjee S, Raj RP, Bertino L, Skagseth Ø, Ravichandran M, Johannessen OM, 2018. Role of Greenland Sea gyre circulation on Atlantic Water temperature variability in the Fram Strait. *Geophys. Res. Lett.* 45: 8399–8406. doi: 10.1029/2018GL079174.

Clem KR, Renwick JA, 2015. Austral spring Southern Hemisphere circulation and temperature changes and links to the SPCZ. *J. Clim.* 28(18): 7371–7384. doi: 10.1175/JCLI-D-15-0125.1.

Comiso JC, 2006. Abrupt decline in the Arctic winter sea ice cover. *Geophys. Res. Lett.* 33: L18504. doi: 10.1029/2006GL027341.

Comiso JC, 1995. SSM/I sea ice concentrations using the bootstrap algorithm. *NASA Ref. Pub.* 1380, 49.

Deser C, Teng H, 2008. Evolution of arctic sea ice concentration trends and the role of atmospheric circulation forcing, 1979–2007. *Geophys. Res. Lett.* 35: L02504, doi: 10.1029/2007GL032023.

Emery WJ, Fowler C, Maslanik J, 1994. Arctic sea ice concentrations from special sensor microwave imager and advanced very high-resolution radiometer satellite data. *J. Geophys. Res.* 99: 18329–18342. doi: 10.1029/94JC01413.

Fetterer F, Knowles K, Meier WN, Savoie M, Windnagel AK, 2017. Updated daily. Sea Ice Index, Version 3. Boulder, Colorado USA. NSIDC: National Snow and Ice Data Center. https://doi.org/10.7265/N5K072F8. Accessed on 20 February 2020.

Figurkin AL, Zhigalov IA, 2008. Vanin NS2008 Oceanographic conditions in the Sea of Okhotsk in the early 2000s. *Izvestiya TINRO* 152, 240–252 (Russian).

Francis J, Vavrus S, 2012. Evidence linking Arctic amplification to extreme weather in mid-latitudes. *Geophys. Res. Lett.* 39: L06801. doi: 10.1029/2012GL051000.

Francis JA, Hunter E 2007 Drivers of declining sea ice in the Arctic winter: a tale of two seas. *Geophys. Res. Lett.* 34: L17503. doi: 10.1029/2007GL030995.

Gloersen P, 1995. Modulations of hemispheric sea ice cover by ENSO events. *Nature* 373: 503–506.

Graversen RG, Langen PG, Mauritsen T, 2014. Arctic amplification in CCSM4: Contributions from the lapse rate and surface albedo feedbacks. *J. Clim.* 27: 4433–4450 doi: 10.1175/JCLI-D-13-00551.1.

Häkkinen S, Rhines PB, Worthen DL, 2011. Atmospheric blocking and Atlantic multidecadal ocean variability. *Science* 334(6056): 655–659. doi: 10.1126/science.1205683.

Häkkinen S, Cavalieri DJ 1989 A study of oceanic surface heat fluxes in the Greenland, Norwegian, and Barents Sea. *J. Geophys. Res.* 94(C5): 6145–6157.

Hall A, 2004. Role of surface albedo feedback in climate. *J. Clim.* 17: 1550–1568 doi: 10.1175/1520-0442(2004)017<1550:TROSAF>2.0.CO;2.

Hartmann DL, et al. 2013. Observations: atmosphere and surface. In *Climate Change 2013: The Physical Science Basis*, T. F. Stocker et al., Eds., Cambridge University Press, pp. 159–254.

Hersbach H, Dee D 2016, ERA5 reanalysis is in production, *ECMWF Newsletter* 147: 7. https://www.ecmwf.int/en/newsletter/147/news/era5-reanalysis-production (last access: 14 May 2020).

Hilmer M, Harder M, Lemke P, 1998. Sea ice transport: a highly variable link between the Arctic and North Atlantic. *Geophys. Res. Lett.* 25(17): 3359–3362.

Holland MM, Bitz CM, Tremblay B, 2006. Future abrupt reductions in the summer Arctic sea ice. *Geophys. Res. Lett.* 33: L23503. doi: 10.1029/2006GL028024.

Kidson J, Taschetto AS, Thompson DWJ, England MH, 2011. The influence of Southern Hemisphere sea-ice extent on the latitude of the mid-latitude jet stream. *Geophys. Res. Lett.* 38: L15804. doi: 10.1029/2011GL048056.

Kim ST, 2012. A review of the Sea of Okhotsk ecosystem response to the climate with special emphasis on fish populations. *ICES J. Mar. Sci.* 69(7): 1123–1133. doi: 10.1093/icesjms/fss107.

Kumar A, Yadav J, Mohan R, 2020. Global warming leading to alarming recession of the Arctic sea-ice cover: insights from remote sensing observations and model reanalysis. *Heliyon* 6(7): e04355. doi: 10.1016/j.heliyon.2020.e04355.

Kwok R, 2009. Outflow of Arctic Ocean sea ice into the Greenland and Barents Seas: 1979–2007. *J. Clim.* 22: 2438–2457.

Li F, Orsolini YJ, Wang HJ, Gao YQ, He SP, 2018. Modulation of the Aleutian–Icelandic low seesaw and its surface impacts by the Atlantic Multidecadal Oscillation. *Adv. Atmos. Sci.* 35(1): 95–105. doi: 10.1007/s00376-017-7028-z.

Liu J, Curry JA, Hu Y, 2004. Recent Arctic Sea Ice Variability: Connections to the Arctic Oscillation and the ENSO. *Geophys. Res. Lett.* 31: L09211. doi: 10.1029/2004GL019858.

Lindsay RW, Zhang J 2005. The thinning of Arctic sea ice, 1988–2003: have we passed a tipping point? *J. Clim.* 18(22): 4879–4894. doi: 10.1175/JCLI3587.1.

Martin S, Munoz E, Drucker R, 1997. Recent observations of a spring-summer surface warming over the Arctic Ocean. *Geophys. Res. Lett.* 24(10): 1259–1262.

Maslanik J, Drobot S, Fowler C, Emery W, Barry R, 2007. On the Arctic climate paradox and the continuing role of atmospheric circulation in affecting sea ice conditions. *Geophys. Res. Lett.* 34: L03711. doi: 10.1029/2006GL028269.

Mantua NJ, Hare SR, Zhang Y, Wallace JM, Francis RC, 1997. A Pacific interdecadal climate oscillation with impacts on salmon production. *Bull. Am. Meteor. Soc.* 78: 1069–1079. doi: 10.1175/1520-0477(1997)078<1069:APICOW>2.0.CO;2.

Mätzler C, Ramseier RO, Svendsen E 1984 Polarization effects in sea-ice signatures. *IEEE J. Oceanic Eng.* 9: 333–338.

Mcdonald M, Arragutainaq L, Novalinga Z, eds. 1997. *Voices from the Bay: Traditional Ecological Knowledge of Inuit and Cree in the Hudson Bay Bioregion.* Ottawa: Canadian Arctic Resources Committee and Sanikiluaq, NWT.

Legendre L, Ackley SF, Dieckmann GS, GuUiksen B, Horner SR, Hoshiai T, Melnikov I, Reeburgh WS, Spindler M, Sullivan CW, 1992. Ecology of sea ice biota 2: global significance. *Pol. Biol.* 12: 429–444.

Miles MW, Divine DV, Furevik T, Jansen E, Moros M, Ogilvie AEJ, 2014. A signal of persistent Atlantic multidecadal variability in Arctic sea ice. *Geophys. Res. Lett.* 41: 463–469. doi: 10.1002/2013GL058084.

Molnár PK, Bitz CM, Holland MM, Kay JE, Penk SR, Amstrup SC. 2020. Fasting season length sets temporal limits for global polar bear persistence. *Nat. Clim. Change* 10: 732–738.

Murray RJ, Simmonds I, 1995. Responses of climate and cyclones to reductions in Arctic winter sea ice. *J. Geophys. Res.* 100: 4791–4806.

Nihashi S, Ohshima KI, Saitoh S-I, 2017. Sea-ice production in the northern Japan Sea, *Deep-Sea Res. Part I: Oceanogr. Res. Pap.* 127: 65–76.

Nishii K, Nakamura H, Orsolini YJ, 2011. Geographical dependence observed in blocking high influence on the stratospheric variability through enhancement and suppression of upward planetary-wave propagation. *J. Clim.* 24(24): 6408–6423. doi: 10.1175/JCLI-D-10-05021.1.

Notz D, Stroeve J, 2016. Arctic sea ice loss directly follows cumulative anthropogenic CO2 emissions. *Science* 354(6313): 747–750. doi: 10.1126/science.aag2345.

Ohshima KI, Nakanowatari T, Riser S, Wakatsuchi M, 2010. Seasonal variation in the in- and outflow of the Okhotsk Sea with the North Pacific. *Deep-Sea Res. II* 57: 1247–1256. doi: 10.1016/j.dsr2.2009.12.012.

Ohshima KI, NihashiS, Hashiya E, Watanabe T, 2006. Interannual variability of sea ice area in the Sea of Okhotsk: the importance of surface heat flux in fall. *J. Meteorol. Soc. Jpn.* 874: 907–919. doi: 10.2151/jmsj.84.907.

Olonscheck D, Mauritsen T, Notz D, 2019. Arctic sea-ice variability is primarily driven by atmospheric temperature fluctuations. *Nat. Geosci.* 12: 430–434. doi: 10.1038/s41561-019-0363-1.

Onarheim IH, Smedsrud LH, Ingvaldsen RB, Nilsen F, 2014. Loss of sea ice during winter north of Svalbard. *Tellus A* 66(1): 23933. doi: 10.3402/tellusa.v66.23933.

Onarheim IH, Eldevik T, Smedsrud LH, Stroeve JC, 2018. Seasonal and regional manifestation of arctic sea ice loss. *J. Clim.* 31: 4917–4932. doi: 10.1175/JCLI-D-17-0427.1.

Osborne JM, Screen JA, Collins M, 2017. Ocean–atmosphere state dependence of the atmospheric response to Arctic sea ice loss. *J. Clim.* 30(5): 1537–1552. doi: 10.1175/JCLI-D-16-0531.1

Park W, Latif M 2008. Multidecadal and multi-centennial variability of the meridional overturning circulation. *Geophys. Res. Lett.* 35: L22703. doi: 10.1029/2008GL035779.

Parkinson CL, 2019. A 40-y record reveals gradual Antarctic sea ice increases followed by decreases at rates far exceeding the rates seen in the Arctic. *PNAS* 116(29): 14414–14423.

Parkinson CL, Cavalieri DJ, 1989. Arctic sea ice 1973–1987: seasonal, regional, and interannual variability. *J. Geophys. Res.* 94(C10): 14499–14523. doi: 10.1029/JC094iC10p14499.

Parkinson CL, Cavalieri DJ, 2008. Arctic sea ice variability and trends, 1979–2006. *J. Geophys. Res.* 113: C07003. doi: 10.1029/2007JC004558.

Partington K, Flynn T, Lamb D, Bertoia C, Dedrick K, 2003. Late twentieth-century Northern Hemisphere sea-ice record from U.S. National Ice Center ice charts. *J. Geophys. Res.* 108(C11): 3343. doi: 10.1029/2002JC001623.

Peings Y, Magnusdottir G, 2014. Response of the wintertime Northern Hemisphere atmospheric circulation to current and projected Arctic sea ice decline: a numerical study with CAM5. *J. Clim.* 27(1): 244–264. doi: 10.1175/JCLI-D-13-00272.1.

Peterson IK, Pettipas R, 2013. Trends In air temperature and sea ice in the Atlantic Large Aquatic Basin and adjoining areas. *Can. Tech. Rep. Hydrogr. Ocean Sci.* 290: 3–8.

Pistone K, Eisenman I, Ramanathan V, 2014. Observational determination of albedo decrease caused by vanishing Arctic sea ice. *Proc. Natl. Acad. Sci.* 111: 3322–3326. doi: 10.1073/pnas.1318201111.

Post E, Bhatt US, Bitz CM, Brodie JF, Fulton TL, Hebblewhite M, Kerby J, Kutz SJ, Stirling I, Walker DA, 2013. Ecological consequences of sea-ice decline. *Science* 341(6145): 519–524.

Randall D, Curry J, Battisti D, Flato G, Grumbine R, Hakkinen S, Martinson D, Preller R, Walsh J, Weatherly J, 1998. Status of and outlook for large-scale modelling of atmosphere–ice-ocean interactions in the Arctic. *Bull. Am. Meteor. Soc.* 79: 197–219.

Rigor IG, Colony RL, Martin S, 2000. Variations in surface air temperature observations in the Arctic, 1979–97. *J. Clim.* 13(5): 896–914.

Rigor IG, Wallace JM, Colony RL, 2002. Response of sea ice to the Arctic Oscillation. *J. Clim.* 15: 2648–63.

Rigor IG, Wallace JM, 2004. Variations in the age of Arctic sea-ice and summer sea-ice extent. *Geophys. Res. Lett.* L09401. doi: 10.1029/2004GL019492.

Rösel A, Kaleschke L, Kern S, 2012. Influence of melt ponds on microwave sensors' sea ice concentration retrieval algorithms, Conference: Geoscience and Remote Sensing Symposium (IGARSS), *2012 IEEE International*, p. 4. doi: 10.1109/IGARSS.2012.6350608.

Schlichtholz P, 2011. Influence of oceanic heat variability on sea ice anomalies in the Nordic Seas. *Geophys. Res. Lett.* 38: L05705. doi: 10.1029/2010G.

Schweiger A, Lindsay R, Zhang J, Steele M, Stern H, Kwok R, 2011. Uncertainty in modelled Arctic sea ice volume. *J Geophys Res* 116: C00D06. doi: 10.1029/2011JC007084.

Screen J, Francis J, 2016. Contribution of sea-ice loss to Arctic amplification is regulated by Pacific Ocean decadal variability. *Nature Clim Change* 6: 856–860. doi: 10.1038/nclimate3011.

Screen JA, Simmonds I, 2013. Exploring links between Arctic amplification and mid-latitude weather. *Geophys Res Lett* 40. doi: 10.1002/grl.50174.

Screen JA, Simmonds I, 2010. The central role of diminishing sea ice in recent Arctic temperature amplification. *Nature* 464:1334–1337.

Serreze MC, Barry RG 2011 Processes and impacts of Arctic amplification: a research synthesis. *Glob. Planet Change* 77: 85–96.

Simmonds I, Keay K 2009. Extraordinary September Arctic sea ice reductions and their relationships with storm behaviour over 1979–2008. *Geophys. Res. Lett.* 36: L19715, doi: 10.1029/2009GL039810.

Sorteberg A, Kvingedal B, 2006. Atmospheric forcing on the Barents Sea winter ice extent. *J. Clim.* 19: 4772–4784.

Steffen K, Schweiger A, 1991. NASA team algorithm for sea ice concentration retrieval from Defense Meteorological Satellite Program special sensor microwave imager: comparison with Landsat satellite imagery. *J. Geophys. Res.* 96(C12): 21971–21987.

Stroeve J, Notz D, 2018. Changing state of Arctic sea ice across all seasons. *Environ. Res. Lett.* 13 103001.

Stroeve JC, Markus T, Boisvert L, Miller J, Barrett A, 2014. Changes in Arctic melt season and implications for sea ice loss. *Geophys. Res. Lett.* 41: 1216–1225. doi: 10.1002/2013GL058951.

Stroeve JC, Serreze MC, Kay JE, Holland MM, Meier WN, BarrettAP2012The Arctic's rapidly shrinking sea ice cover: a research synthesis. *Clim. Change* 110: 1005–1027. doi: 10.1007/s10584-011-0101-1.

Stroeve J, Holland MM, Meier W, Scambos T, Serreze M, 2007. Arctic sea ice decline: faster than forecast. *Geophys. Res. Lett.* 34(9): L09501. doi:10.1029/2007GL029703.

Timmermans ML, 2015. The impact of stored solar heat on Arctic sea ice growth. *Geophys. Res. Lett.* 42: 6399–6406. doi: 10.1002/2015GL064541.

Thompson DW, Wallace JM 2000 Annular modes in extratropical circulation, Part II: Trends, *J. Clim.* 13: 1018–1036.

Vihma T, 2014. Effects of Arctic sea ice decline on weather and climate: a review. *Surv. Geophys.* 35: 1175–1214.

Vinje T, 2001. Anomalies and trends of sea-ice extent and atmospheric circulation in the Nordic Seas during the period 1864–1998. *J. Clim.* 14: 255–267.

Walsh JE, 2013. Melting ice: What is happening to Arctic sea ice, and what does it mean for us? *Oceanography* 26: 171–181. doi: 10.5670/oceanog.2013.19.

Zhang R, 2015. Mechanisms for low-frequency variability of summer Arctic sea ice extent. *Proc. Natl. Acad. Sci. USA* 112(15): 4570–4575. doi: 10.1073/pnas.1422296112.

Zhang X, Sorteberg A, Zhang J, Gerdes R, Comiso JC, 2008. Recent radical shifts of atmospheric circulations and rapid changes in Arctic climate system. *Geophys. Res. Lett.* 35: L22701. doi: 10.1029/2008GL035607.

Zhang J, Rothrock DA 2003 Modeling global sea ice with a thickness and enthalpy distribution model in generalized curvilinear coordinates. *Mon. Weather Rev.* 131: 845–861.

Zhang Y, Maslowski W, Semtner AJ 1999 Impact of mesoscale ocean currents on sea ice in high-resolution Arctic ice and ocean simulations. *J. Geophys. Res.* 104: 18409–18429.

Zhang Y, Wallace JM, Battisti DS, 1997. ENSO-like interdecadal variability: 1900–93. *J. Clim.* 10:1004–1020.

Zwally HJ, Schutz B, Abdalati W, Abshire J, Bentley C, Brenner A, Bufton J, Dezio J, Hancock D, Harding D, Herring T, Minster B, Quinn K, Palm S, Spinhirne J, Thomas R, 2002. ICESat's laser measurements of polar ice, atmosphere, ocean, and land. *J. Geodyn.* 34: 405–445.

Glossary

Aerosol optical depth (AOD): It measures the extinction of the solar beam by dust and haze. In other words, particles in the atmosphere (dust, smoke, pollution) can block sunlight by absorbing or scattering light. AOD tells us how much direct sunlight is prevented from reaching the ground by these aerosol particles.

Aerosol radiative forcing (ARF): The effect of anthropogenic aerosols on the radiative fluxes at the top of the atmosphere and at the surface, and on the absorption of radiation within the atmosphere. Further, the aerosol radiative forcing values at the top of the atmosphere were low primarily (-1.3 Wm^{-2}) over Hanle and Merak.

Aethalometer: It is an instrument for measuring the concentration of optically absorbing ('black') suspended particulates in a gas colloid stream, commonly visualised as smoke or haze, often seen in ambient air under polluted conditions. The word aethalometer is derived from the Classical Greek verb 'aethaloun', meaning 'to blacken with soot'.

Aleutian low: It is a semipermanent low-pressure system located near the Aleutian Islands in the Bering Sea during the Northern Hemisphere winter. It is one of the largest atmospheric circulation patterns in the Northern Hemisphere and represents one of the 'main centres of action in atmospheric circulation'.

Antarctic Treaty System: The Antarctic Treaty System is the whole complex arrangements made to regulate relations among states in the Antarctic. At its heart is the Antarctic Treaty itself. The Treaty was signed in Washington on December 1, 1959, and entered into force on June 23, 1961.

Anthropogenic: Human-caused emissions include mercury released from fuels or raw materials or uses in products or industrial processes. Land, water and other surfaces can repeatedly re-emit mercury into the atmosphere after its initial release into the environment.

Arctic amplification: Over the past 30 years, the Arctic has warmed at roughly twice the rate as the entire globe, a phenomenon known as Arctic amplification. This means that global warming and climate change impact the Arctic more than the rest of the world. The Arctic, however, was about 2°C warmer. As sea ice declines, it becomes younger and thinner and, therefore, more vulnerable to further melting.

Arctic Circle: It is parallel, or line of latitude around the Earth, at approximately 66°30' N. Because of the Earth's inclination of about $23^{1}/_{2}°$ to the vertical, it marks the southern limit of the area within which, for one day or more each year, the Sun does not set (about June 21) or rise (about December 21).

Arctic haze: It is the phenomenon of a visible reddish-brown springtime haze in the atmosphere at high latitudes in the Arctic due to anthropogenic air pollution.

Arctic oscillation (AO): It refers to an atmospheric circulation pattern over the mid-to-high latitudes of the Northern Hemisphere. The AO's positive phase is characterised by lower-than-average air pressure over the Arctic paired with higher-than-average pressure over the northern Pacific and Atlantic Oceans. The most obvious reflection of the phase of this oscillation is the north-to-south location of the storm-steering, mid-latitude jet stream.

Arctic tundra: It is located in the Northern Hemisphere, encircling the North Pole and extending south to the coniferous forests of the taiga. The Arctic is known for its cold, desert-like conditions. The growing season ranges from 50 to 60 days. The average winter temperature is −34°C (−30°F), but the average summer temperature is 3°C–12°C (37°F–54°F) which enables this biome to sustain life.

Atmospheric circulation: The large-scale movement of air and ocean circulation is how thermal energy is redistributed on the surface of the Earth. The Earth's weather is a consequence of its illumination by the Sun and the laws of thermodynamics.

Atmospheric mercury depletion events (AMDEs): This phenomenon is termed as atmospheric mercury depletion events (AMDEs). Its discovery has revolutionised our understanding of the cycling of Hg in the Polar regions while stimulating a significant amount of research to understand its impact on this fragile ecosystem.

Biogenic silica (bSi): It is also referred to as opal, biogenic opal, or amorphous opaline silica, and forms one of the most widespread biogenic minerals. For example, microscopic particles of silica called phytoliths can be found in grasses and other plants.

Biogeochemical cycle: Any of the natural pathways by which essential elements of living matter are circulated. The term biogeochemical is a contraction that refers to considering the biological, geological and chemical aspects of each cycle.

Black carbon (BC) aerosol: It is often called soot and is the dominant form of light-absorbing particulate matter in the atmosphere. Incomplete combustion processes, both humans emit BC (e.g. diesel engines) and natural (e.g. wildfire).

Black carbon: It consists of pure carbon in several linked forms. It is formed through the incomplete combustion of fossil fuels, biofuel and biomass and is one of the main types of particle in both anthropogenic and naturally occurring soot. Black carbon causes human morbidity and premature mortality.

Boreal forests: The boreal forest (or "taiga") is the world's largest land biome. From a biological perspective, boreal forests are defined as forests growing in high-latitude environments where freezing temperatures occur for 6–8 months. Trees are capable of reaching a minimum height of 5 m and a canopy cover of 10%.

Carbon monoxide (CO): It is an odourless, colourless gas formed by the incomplete combustion of fuels. When people are exposed to CO gas, the CO molecules will displace the oxygen in their bodies and lead to poisoning.

Coarse-grained soils: These are defined as those soils whose individual grains are retained on a No. 200 (0.075 mm) sieve. Grains of this size can generally be seen with the naked eye, although a hand-held magnifying glass may occasionally be needed to see the smallest grains. Gravel and sand are coarse-grained soils.

Coastal cliff: The term coastal cliff, or sea cliffs, refers to a steeply sloping surface where elevated land meets the shoreline. Usually, tectonic sea cliffs result from differential erosion that removes soft rocks overlapping hard rocks along the fault plane.

Coniferous forest: The vegetation comprises cone-bearing needle-leaved or scale-leaved evergreen trees, found in areas with long winters and moderate to high annual precipitation. Pines, spruces, firs and larches are the dominant trees in coniferous forests.

Continental shelf: It is a portion of a continent submerged under an area of relatively shallow water known as a shelf sea. Much of these shelves were exposed by drops in sea level during glacial periods. The shelf surrounding an island is known as an insular shelf.

Diatoms: These are algae that live in houses made of glass. They are the only organism on the planet with cell walls composed of transparent, opaline silica. Intricate and striking patterns of silica ornament diatom cell walls.

Differential optical absorption spectroscopy (DOAS): It allows the quantitative determination of atmospheric trace gas concentrations by recording and evaluating the characteristic absorption structures (lines or bands) of the trace gas molecules along an absorption path of known length in the open atmosphere.

Dissolved organic carbon (DOC): The fraction of organic carbon operationally defined as that which can pass through a filter with a pore size typically between 0.22 and 0.7 μm. The fraction remaining on the filter is called particulate organic carbon (POC).

Dissolved silicate (DSi): It is a crucial nutrient in coastal water used by planktonic diatoms for cell division and growth.

Ecological parameters: A variable, measurable property whose value is a determinant of the characteristics of an ecosystem.

Ecosystems: An ecosystem is a geographic area where plants, animals and other organisms, as well as weather and landscape, work together to form a bubble of life. Ecosystems contain biotic or living parts, as well as abiotic factors or non-living parts. Ecosystems can be very large or very small.

Effective radiative forcing (ERF): The term is now used to quantify the impact of some forcing agents that involve rapid adjustments of the atmosphere and surface components assumed constant in the RF concept.

El Niño: It can affect our weather significantly. El Niño causes the Pacific jet stream to move south and spread further east. During winter, this leads to wetter conditions than usual in the southern U.S. and warmer and drier conditions in the North. El Niño also has a strong effect on marine life off the Pacific coast.

Fjords: These are glacially over-deepened semi-enclosed marine basins, typically with entrance sills separating their deep waters from the adjacent coastal waters, which restrict water circulation and thus oxygen renewal.

Glacier runoff: It is defined as all melt and rainwater that runs off the glacierised area without re-freezing.

Global biogeochemical cycles: The elements of carbon, hydrogen and oxygen, and the elemental nutrient elements nitrogen, phosphorus and sulphur are essential for life on Earth. The term 'global biogeochemical cycles' describes the transport and transformation of these substances in the global environment.

Global ocean conveyor belt: It is a constantly moving system of deep-ocean circulation driven by temperature and salinity. This motion is caused by a combination of thermohaline currents (Thermo = temperature; haline = salinity) in the deep ocean and wind-driven currents on the surface.

Greenhouse gas: It is a gas that absorbs and emits radiant energy within the thermal infrared range, causing the greenhouse effect. The primary greenhouse gases in the Earth's atmosphere are water vapour, carbon dioxide, methane, nitrous oxide and ozone.

Heat flux (Φ): It can be defined as the rate of heat energy transfer through a given surface (W), and heat flux density (φ) is the heat flux per unit area (Wm²).

Humic acids: These are the organic substances that coagulate when the strong-base extract is acidified. In contrast, fulvic acids are the organic substances that remain soluble when the strong-base extract is acidified.

Hybrid Single-Particle Lagrangian Integrated Trajectory Model (HYSPLIT): It is a computer model used to compute air parcel trajectories to determine how far and in what direction a parcel of air, and subsequently air pollutants, will travel.

Hydrography: It is the science that measures and describes the physical features of the navigable portion of the Earth's surface and adjoining coastal areas. Hydrographic surveyors study these bodies of water to see what the "floor" looks like. That data are used to update nautical charts and develop hydrographic models.

Hydroxyl radical: The hydroxyl radical, ·OH, is the neutral form of the hydroxide ion (OH⁻). Hydroxyl radicals are highly reactive (quickly becoming hydroxy groups) and consequently short-lived. They form an essential part of radical chemistry.

Indian summer monsoon: The Indian summer monsoon typically lasts from June to September, with large areas of western and central India receiving more than 90% of their total annual precipitation during the period, and southern and northwestern India receiving 50%–75% of their total annual rainfall.

Interplanetary magnetic field (IMF): It is now more commonly referred to as the heliospheric magnetic field (HMF), which is the component of the solar magnetic field that is dragged out from the solar corona by the solar wind flow to fill the solar system.

Lacustrine deposits: These are sedimentary rock formations that formed at the bottom of ancient lakes. A common characteristic of lacustrine deposits is that a river or stream channel has carried sediment into the basin.

Last Glacial Maximum (LGM): The Last Glacial Maximum (LGM), also referred to as the Late Glacial Maximum, was the most recent time during the Last Glacial Period that ice sheets were at their greatest extent.

In Britain, the LGM is the Dimlington Stadial, dated between 31,000 and 16,000 years.

Little ice age: It was a period of regional cooling that occurred after the Medieval Warm Period. It was not an actual ice age of global extent. The term was introduced into scientific literature by François E. Matthes in 1939.

Low-noise amplifier (LNA): An electronic amplifier amplifies a very low-power signal without significantly degrading its signal-to-noise ratio. A typical LNA may supply a power gain of 100 (20 decibels (dB)) while decreasing the signal-to-noise ratio by less than a factor of two (a 3 dB noise figure (NF)).

Low-pass filter: It is a filter that passes signals with a frequency lower than a selected cut-off frequency and attenuates signals with frequencies higher than the cut-off frequency. The exact frequency response of the filter depends on the filter design. The filter is sometimes called a high-cut filter or treble-cut filter in audio applications. A low-pass filter is the complement of a high-pass filter.

Medieval Warm Period (MWP): The Medieval Warm Period (MWP), also known as the Medieval Climate Optimum, or Medieval Climatic Anomaly, was a warm climate in the North Atlantic region lasting from c. 950 to c. 1250. It was likely related to warming elsewhere while some other areas, such as the tropical Pacific.

Methane clathrate: $(CH_4 \cdot 5.75H_2O)$ or $(4CH_4 \cdot 23H_2O)$, also called methane hydrate, hydromethane, methane ice, fire ice, natural gas hydrate, or gas hydrate, is a solid clathrate compound (more specifically, a clathrate hydrate) in which a large amount of methane is trapped within a crystal structure of water, forming a solid similar to ice.

Methylmercury: It is (sometimes methyl mercury (MHg)) an extremely toxic organometallic cation with the formula $[CH_3Hg]^+$. Its derivatives are the primary source of organic mercury for humans. It is a bio-accumulative environmental toxicant.

Microbial communities: These are groups of microorganisms that share a shared living space. The microbial populations that form the community can interact in different ways, such as predators and prey or symbionts.

Midnight Sun: It is also known as polar day – refers to an extended period of daytime that lasts 24 hours or more. The phenomenon occurs in or near the Arctic and Antarctic regions during summer.

Multidimensional gas chromatography (MDGC): A chromatography method used to separate complex samples, especially those with components that have similar retention factors, by running the eluent through two or more columns instead of the customary single column. MDGC equipment can be found in chemistry, environmental science and industrial labs, and is mainly known for its applications in fragrance and flavour sciences.

Multi-model mean (MMM): It is a simple way to reduce biases in individual model outputs[2], and thus, it is widely used for climate change projections. The usefulness of MMM may vary from one region to the other based on the regional climate and the diagnostic variables of interest.

Orthosilicic acid (H_4SiO_4): It is a chemical compound with the formula Si_4. It has been synthesised using non-aqueous solutions. It is assumed to be present when silicon dioxide SiO_4 dissolves in water at a millimolar concentration level.

Palaeoclimatology: It is the study of past climates. Since it is impossible to go back in time to see what climates were like, scientists use imprints created during past climate, known as proxies, to interpret paleoclimate. Past climate can be reconstructed using a combination of different types of proxy records.

Permafrost: The Arctic is so cold that the ground beneath the tundra surface remains frozen all year. This permanently frozen ground is called permafrost. The soil in the permafrost area remains harder than 32°F (0°C). If the Earth never warmed up, there would be no plants growing in the Arctic.

Persistent Organic Pollutants (POPs): These are toxic substances composed of organic (carbon-based) chemical compounds and mixtures. They include industrial chemicals like PCBs and pesticides like DDT. The existence of POPs is relatively recent, dating to the boom in industrial production after World War II.

Polar lows: These are defined as small but intense maritime meso-scale cyclones that form in cold Polar or Arctic air advected over relatively warmer water. This is why such systems are also known as Arctic hurricanes. Mean wind speed varies around 15 m/s, which is only Bft 7 or near gale.

Polar region: The regions of the Earth designated as polar are those areas located between the North or South Pole and the Arctic or Antarctic Circles, respectively. The northern polar region, called the Arctic, encompasses the Arctic Ocean and a portion of some surrounding landmasses.

Sea ice concentration: It is a valuable variable for climate scientists and nautical navigators. It is defined as the sea ice area relative to the total at a given point in the ocean.

Secondary Organic Aerosol (SOA): It is formed by definition when low volatility oxidation products of volatile organic compounds (VOCs) deposit onto existing particles or form new particles. SOA accounts for a significant fraction of the global atmospheric aerosol burden.

Seismic surveys: These are an advantageous geophysical method for studying the ground conditions to a significant depth and over a large area. Seismic is utilised in many applications for subsurface investigations, mineral exploration being one of the diagrams illustrating the principles of seismic exploration.

Semipermanent highs: The two semipermanent highs located in the Northern Hemisphere are the North Pacific High and North Atlantic (Bermuda-Azores) High. Subpolar Lows occur where the polar easterlies and prevailing westerlies converge (near 60°N/S). They are found in cold, stormy regions of rising air.

Spectrometer: It is an instrument used to probe a property of light as a function of its portion of the electromagnetic spectrum, typically its wavelength, frequency or energy. The measured property is usually the intensity of light, but other variables like polarisation can also be measured.

Sporadic: It means 10%–50% of the surface has permafrost under it. Certain permafrost, cast from a mountain's shadow or covered in thick vegetation, stay all year. Alpine permafrost is a discontinuous permafrost that exists on the tops of mountains where the ground remains very cold.

Substorm: A substorm, sometimes referred to as a magnetospheric substorm or an auroral substorm, is a brief disturbance in the Earth's magnetosphere that causes energy to be released from the "tail" of the magnetosphere and injected into the high-latitude ionosphere.

Supraglacial stream: A meltwater river that flows over the surface of the ice in a glacier or ice cap.

The International Polar Year: An International Polar Year or IPY is a year during which many nations coordinate their Polar expeditions, observations and analyses.

The polar night: It is a phenomenon where the night-time lasts for more than 24 hours in the northernmost and southernmost regions of Earth. This occurs only inside the polar circles.

Total electron content (TEC): It is an essential descriptive quantity for the ionosphere of the Earth. TEC is the total number of electrons integrated between two points along one metre required cross-section, i.e. the electron columnar number density. It is often reported in multiples of the so-called *TEC unit*, defined as $TECU = 10^{16}$ el/m^2= 1.66×10^{-8} mol/m^2.

Total organic carbon (TOC): It is the amount of carbon found in an organic compound. It is often used as a non-specific indicator of water quality or cleanliness of pharmaceutical manufacturing equipment.

Transformed Atlantic Water (TWA): The Transformed Atlantic Water (TWA) from the west Spitsbergen current and the glacier-melt freshwater at the inner fjord creates intense temperature and salinity gradients along the length of the fjord. The meltwater during summer not only stratifies the upper water column but significantly alters the turbidity.

Zooplankton: These are the animal component of the planktonic community ("zoo" comes from the Greek word for animal). They are heterotrophic (other-feeding), meaning they cannot produce their food and consume other plants or animals instead of food. In particular, this means they eat phytoplankton.

Index

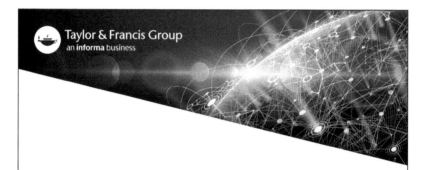